# Springer Series in Materials Science

Volume 261

The Springer Series in Materials Science covers the complete spectrum of materials physics, including fundamental principles, physical properties, materials theory and design. Recognizing the increasing importance of materials science in future device technologies, the book titles in this series reflect the state-of-the-art in understanding and controlling the structure and properties of all important classes of materials.

More information about this series at http://www.springer.com/series/856

Adrian Crisan
Editor

# Vortices and Nanostructured Superconductors

Springer

*Editor*
Adrian Crisan
National Institute of Materials Physics
Magurele
Romania

ISSN 0933-033X ISSN 2196-2812 (electronic)
Springer Series in Materials Science
ISBN 978-3-319-59353-1 ISBN 978-3-319-59355-5 (eBook)
DOI 10.1007/978-3-319-59355-5

Library of Congress Control Number: 2017941475



Printed on acid-free paper

This Springer imprint is published by Springer Nature
The registered company is Springer International Publishing AG
The registered company address is: Gewerbestrasse 11, 6330 Cham, Switzerland

# Preface

Starting with the success in liquefying He and the surprising discovery of the superconductivity of Hg by the group in Leiden led by H. Kammerling-Onnes, the last century provided us with the conceptual framework to understand and handle macroscopic quantum coherence: "superconductivity" in the case of existence of charge carriers, or "superfluidity" in the absence of charge carriers. No less than nine Nobel Prizes have been awarded for breakthrough discoveries and theories relating to superconductivity and superfluidity, from Kammerling-Onnes's initial 1913 prize, to the 2016 prize awarded to Thouless, Haldane and Kosterlitz for their theoretical discoveries of topological phases of matter and topological phase transitions, including topologically driven superconductivity.

Superconductivity is now an established (although, in some cases, not yet fully understood) form of macroscopic quantum coherence that regularly impacts the scientific community with new surprises, such as the discovery of high-temperature superconductivity in cuprates (Bednorz and Müller, IBM Zürich, Nobel Prize in 1987) followed rapidly by the discovery of superconductivity above the boiling point of nitrogen in $YBa_2Cu_3O_{7-x}$ by the Houston group of P. Chu; the discovery of superconductivity at 40 K in $MgB_2$; and later on, superconductivity in chalcogenides and pnictides, the last examples also being multi-component (multi-gap) superconductors. Superconductivity is a solid-state phenomenon that possesses the advantages that made electronic and semiconductor technology so successful: scalability and large-scale production. At the same time, radically different functionalities emerge due to macroscopic quantum coherence, from large-scale, energy-efficient applications due to electric current transportation without losses, to quantum computing and rapid single flux quantum electronics.

In this context, magnetic flux quanta in superconductors, also named "vortices" due to the circulating supercurrent supporting them, play crucial roles in both cases: (i) in superconducting electronics, they are the "carriers" of information instead of electrons; and (ii) in large-scale power applications, their mobility due to the Lorentz force leads to unwanted energy dissipation. So, for significant market penetration of such large-scale applications of superconductors, "pinning" of vortices by nanoscale engineered defects is crucial.

This book provides expert coverage of some modern and novel aspects of the study of vortex matter, dynamics, and pinning in nanostructured and multi-component superconductors. Vortex matter in superconducting materials is a field of enormous beauty and intellectual challenge, which began with the theoretical prediction of vortices by A. Abrikosov (Nobel Laureate for Physics in 2003, sadly deceased on March 23, 2017, age 88). Vortices, vortex dynamics, and pinning are key features in many of today's human endeavors: from the huge superconducting accelerating magnets and detectors at the Large Hadron Collider at CERN Geneva, which opened new windows of knowledge on the universe, to the Tokamak fusion magnets that will be utilized at the International Thermonuclear Energy Reactor in Cadarache, to the tiny superconducting transceivers using rapid single flux quanta, which have opened a revolutionary means of communication.

In this book, leading researchers survey the most exciting and important recent developments. The book offers something for almost everybody interested in the field: experimental techniques to visualize vortices and study their dynamics, nanotechnologies for improving current-carrying capabilities in high applied magnetic fields, current anisotropy, second magnetization peak, and intuitive and theoretical aspects concerning the novelty and beauty of multi-component superconductivity.

Chapter 1 presents unpublished data regarding the detailed evolution of a vortex approaching pinning centers. Using scanning Hall probe microscopy, the authors have directly visualized, at a microscopic level, the interaction of a single quantum vortex with pinning centers. When few adjacent pinning centers are present, the vortex can be trapped by one of them, while the interaction of the vortex with the adjacent pinning centers can be tuned by varying superconducting characteristic lengths with temperature. It was found that when the vortex size is comparable to the distance between two pinning centers, the vortex deforms along the line connecting the pinning centers and the magnetic flux spreads by embracing both pinning centers, thus generating a magnetic dipole. In contrast, a vortex located on an isolated pinning center preserves its round shape up to temperatures close to the critical temperature. The experimental data are in a good agreement with theoretical simulations based on the time-dependent Ginzburg–Landau approach.

Thirty years after the discovery of high critical temperature superconductors, the large-scale application of second-generation superconducting coated conductors based on rare-earth cuprates ($RE\mathrm{Ba_2Cu_3O_{7-x}}$) is still waiting for a lower price in terms of \$/kA m. This can be achieved by increasing the critical current. For high-field applications, this is even more crucial because of very large Lorentz forces. Fortunately, nano-engineered artificial pinning centers can provide large pinning forces to counteract Lorentz forces. Chapter 2 is a comprehensive review of various materials and architectures used for nanostructured $RE\mathrm{Ba_2Cu_3O_{7-x}}$ films with artificial pinning centers (APCs). Several categories of APCs are described in terms of their dimensionality (nanorods, nanoparticles, nanolayers, segmented nanorods, and more), and experimental results are discussed under the framework of Ginzburg–Landau theory. Various results obtained by many groups worldwide are presented and analyzed, with emphasis on outstanding pinning performance in a wide range of temperatures (4.2–77 K) and magnetic fields (1–30 T).

Energy resources are now a major issue in the global economy. It is obvious that fossil fuels will not last forever. In addition, changes to the climate due to increasing $CO_2$ emissions are now obvious. Fortunately, we are witnessing the first, tentative steps toward a clean, sustainable "hydrogen economy," so in the not-so-distant future we would expect that liquid hydrogen with its boiling point of 20.3 K will be a common, abundant, and cheap liquid. In these circumstances, energy-efficient cryo-magnetic devices and equipment (transformers, magnets, motors, induction furnaces, superconducting-magnet energy storage, fault-current limiters, etc.) based on $MgB_2$ (a cheap and abundant superconductor with critical temperature of 40 K) cooled by liquid H will have a performance/price ratio much better than similar equipment based on copper wires, will require much less energy to operate, and will be more environmentally friendly. Tuning electronic structure and nanostructure is the critical issue for engineering an $MgB_2$ superconductor toward applications. In Chap. 3, a variety of chemical and microstructural control techniques that have been developed to artificially enhance flux pinning strength in the material are presented. The influence of chemical additives and oxygen doping on the formation of nanoprecipitates and superconducting properties, carbon doping effects, and methods of introducing carbon using different sources and microstructural control via ball-milling and mechanical alloying techniques are reviewed and summarized.

As shown in Chaps. 2 and 3, with creative materials nano-engineering, researchers have been successful in improving critical currents over the field and temperature ranges relevant to applications. Intrinsic thermodynamic properties and the interaction between the magnetic vortices and the pinning landscape combine to determine how much current is supported. Due to both factors, the resultant critical currents generally vary with the angle of the imposed external field. Chapter 4 explores what is known about this current anisotropy, very important in superconducting devices in which various parts are exposed to magnetic fields of various orientations. The difficulty of this problem stems partially from having two sources of anisotropy: the anisotropy in the vortex cross section, arising from the intrinsic mass anisotropy of the carriers of the vortex current; and the anisotropy of the pinning centers themselves. The effects of these two factors are not easily separated. It is never possible to have a fully isotropic pinning landscape in an anisotropic superconductor. If the critical current ($I_c$) varies strongly with angle, this presents challenges for coil designers who must keep the coil within safe operating margins of $I/I_c$ everywhere within the coil. An unpredictable or only weakly predictable $I_c$ with field, field orientation, and temperature means exploiting the full capacity of these materials is compromised. The chapter summarizes the common methods to analyze the critical current anisotropy. Firstly, there are scaling methods, such as the Blatter scaling and other scaling approaches which are a modification of this approach. Secondly, there are more direct methods of calculating the expected response from defects, which examine the pinning forces on vortices from defects under certain assumptions, and finally, the author presents the vortex path model or maximum entropy method, which is an information theory or statistical approach for extracting information from the field orientation dependence of critical current.

The strong nonlinear relation between the current density and the electric field, along with nonlocal interactions, determines the complex behavior of vortex matter, which also includes "catastrophic" phenomena like vortex avalanches. Chapter 5 presents a brief review of the experimental studies devoted to these unstable vortex patterns, starting from historical findings up to recent works, and focusing on the analysis of vortex avalanches by means of magneto-optical imaging (MOI). The MOI technique, based on either Kerr (in reflection) or Faraday (in transmission) effects, is a powerful tool for the visualization of magnetic field distribution and enables the observation of magnetic field distribution in real time, on the whole surface of the superconductor, with microscopic resolution. The dynamics and morphology of the vortex avalanches are reported in detail, along with theoretical efforts for understanding and modeling this complex phenomenon. It is shown that vortex avalanches are ubiquitously occurring in superconductors if certain conditions are satisfied; in particular, temperature, applied magnetic field, and applied field rate are of paramount importance for observing/avoiding these phenomena. Together with the interest in the fundamental behavior of vortex matter, the understanding of vortex avalanches is instrumental for assessing the limits of superconductor usage in power applications.

Chapter 6 is devoted to the change of magnetic hysteresis curves and of vortex dynamics in "self-nanostructured" $La_{2-x}Sr_xCuO_4$ single crystals (with charge and spin stripes), offering useful information about the nature of the second magnetization peak occurring for single crystal specimens with random pinning. By decreasing $x$, with the external magnetic field oriented perpendicular to the stripe plane, the second magnetization peak completely disappears in the doping domain of a well-developed static stripe structure ($x \sim 1/8$) and reappears for $x \leq 0.10$. This behavior follows the instability of the quasi-ordered vortex solid (the Bragg vortex glass, stable against dislocation formation) in the presence of static stripe order (as revealed using small-angle neutron scattering experiments), which is confirmed by the determined temperature variation of the normalized vortex-creep activation energy. The results support the scenario in which the second magnetization peak is generated by the pinning-induced disordering of the Bragg vortex glass in the dynamic conditions of magnetic measurements.

The last two chapters are devoted to multi-component superconductivity, which is a rather new and very "fashionable" quantum phenomenon in various superconducting materials: multi-band superconductors in which different superconducting gaps open in different Fermi surfaces, films engineered at the atomic scale to enter the quantum confined regime, multilayer, two-dimensional electron gases at oxide interfaces, and complex materials in which different electronic orbitals or different carriers participate in the formation of the superconducting condensate. The increased number of degrees of freedom of the multi-component superconducting wave function allows for quantum effects that are otherwise unattainable in single-component superconductors. Chapter 7 focuses on inter-band phase difference soliton, the fractionalization of the unit magnetic flux quantum, and frustration between quantum phases of multiple components. Taking into account the fluctuations in the inter-band phase difference provides a bridge between superconducting multiband condensates and other multicomponent macroscopic quantum systems such as Bose–Einstein condensates with multiple components and particle physics systems governed by a non-Abelian gauge field. Chapter 8 is theory-focused, investigating various excitation modes in multi-gap superconductors. In particular, the Nambu–Goldstone mode, Leggett mode, and Higgs mode are important and play a key role in multi-gap superconductors. The multiple-phase invariance in a multi-gap system is partially or totally spontaneously broken in a superconductor. The dispersion relation and the mass formulas of these modes are evaluated by using the functional integral method. The broken multiple-phase invariance leads to a new quantum phase such as time-reversal symmetry breaking, the emergence of massless modes, and fractionally quantized-flux vortices.

Finally, apart from the colleagues, friends, and collaborators who contributed to this book, I am grateful to my many friends from the Vortex community (VORTEX 10th Jubilee Conference is held in September 2017 in Rhodes together with the 65th anniversary of its initiator, Prof. V.V. Moshchalkov). The birth and title of this book are related to the International Conference on Superconductivity and Magnetism organized by Prof. Ali Gencer (Ankara University) every even year since 2008. Together with my friend, Prof. M. Milosevic (Antwerpen University), I chaired and co-organized a session entitled "Vortices and Nanostructured Superconductors," which provided excellent, world-class contributions and vivid discussions among the very best people in the field. And last, but not least, my special thanks to Sara Kate Heukerott from Springer who proposed this endeavor to me.

Financial support from the Romanian Ministry of Research and Innovation through POC (European Regional Development Fund, Operational Fund Competitiveness) Project P-37_697 number 28/01.09.2016 is gratefully acknowledged. Various funding agencies supporting the research activities which led to the results reported are mentioned in each chapter.

Bucharest, Magurele, Romania Adrian Crisan

# Contents

# Contributors

**Malik I. Adam** Department of Mechanical Engineering, University Tenaga Nasional, Kajang, Malaysia

**Soo Kien Chen** Department of Physics, Faculty of Science, Universiti Putra Malaysia, Serdang, Selangor, Malaysia

**Adrian Crisan** National Institute of Materials Physics, Bucharest, Magurele, Romania

**Jozef T. Devreese** Universiteit Antwerpen, Antwerpen, Belgium

**Shi Xue Dou** Institute for Superconducting & Electronic Materials, Australian Institute of Innovative Materials, University of Wollongong, Wollongong, NSW, Australia

**Jun-Yi Ge** KU Leuven, Leuven, Belgium

**Valadimir N. Gladilin** KU Leuven, Leuven, Belgium

**Joffre Gutierrez** KU Leuven, Leuven, Belgium

**Alina M. Ionescu** National Institute of Materials Physics, Bucharest, Magurele, Romania; Faculty of Physics, University of Bucharest, Bucharest, Romania

**Ion Ivan** National Institute of Materials Physics, Bucharest, Magurele, Romania

**Francesco Laviano** Department of Applied Science and Technology, Politecnico di Torino, Torino, Italy

**Nick J. Long** Robinson Research Institute, Victoria University of Wellington, Wellington, New Zealand

**Minoru Maeda** Department of Physics, College of Science and Technology, Nihon University, Tokyo, Japan

**Paolo Mele** Research Center for Environmentally Friendly Materials Engineering, Muroran Institute of Technology, Muroran, Hokkaido, Japan

**Dana Miu** National Institute of Laser, Plasma, and Radiation Physics, Bucharest, Magurele, Romania

**Lucica Miu** National Institute of Materials Physics, Bucharest, Magurele, Romania

**Victor V. Moshchalkov** KU Leuven, Leuven, Belgium

**Yasumoto Tanaka** National Institute of Advanced Industrial Science and Technology, Tsukuba, Ibaraki, Japan

**Jacques Tempere** Universiteit Antwerpen, Antwerpen, Belgium

**Akiyasu Yamamoto** Department of Applied Physics, Tokyo University of Agriculture and Technology, Koganei, Tokyo, Japan

**Takashi Yanagisawa** Electronics and Photonics Research Institute, National Institute of Advanced Industrial Science and Technology, Tsukuba, Ibaraki, Japan

# Chapter 1
# Vortex Deformation Close to a Pinning Center

**Jun-Yi Ge, Joffre Gutierrez, Valadimir N. Gladilin, Jacques Tempere, Jozef T. Devreese and Victor V. Moshchalkov**

## 1.1 Introduction

In type-II superconductors, above the lower critical field $H_{c1}$, magnetic field penetrates into the superconductors in the form of quantized vortices with flux $\Phi_0 = h/2e$ ($h$, Plank constant; $e$, electron charge), forming the Abrikosov vortex lattice. When a current is applied, the vortices experience a Lorentz force perpendicular to the applied current. Vortex motion under this force leads to energy dissipation that limits technological applications. One common way to solve the problem is by introducing pinning centers to the superconductors, such as lithographically formed well-controlled pinning sites [1–4], ion-irradiated point defects [5], grain boundaries [6], and nanostructured self-assembled non-superconducting phases [7–9]. In all the above cases, the local superconductivity at pinning centers is suppressed, thus making them energetically favorable for vortices to be located on. Therefore, the critical current, above which the dissipation appears due to the vortex motion, can be strongly enhanced.

The vortex–pinning effect has been widely studied by many methods, for example, transport measurements [10–12], ac susceptibility measurements [13–15], and dc magnetization measurements [4, 16, 17]. In the transport measurements, vortices, going through a serious of different dynamic regimes [18], are depinned at large enough current density. Theoretical calculations predict that the vortex core becomes elongated along the direction of motion [19]. In the ac susceptibility measurements, under an ac driving force, vortices oscillate around the pinning centers. The interaction between pinning and vortices is manifested a sudden drop of the temperature dependence of in-phase ac susceptibility and a dissipation peak of the

J.-Y. Ge · J. Gutierrez · V.N. Gladilin · V.V. Moshchalkov (✉)
KU Leuven, Celestijnenlaan 200D, 3001 Leuven, Belgium
e-mail: Victor.Moshchalkov@fys.kuleuven.be

J. Tempere · J.T. Devreese
Universiteit Antwerpen, Universiteitsplein 1, 2610 Antwerpen, Belgium



A. Crisan (ed.), *Vortices and Nanostructured Superconductors*, Springer Series in Materials Science 261, DOI 10.1007/978-3-319-59355-5_1

out-of-phase ac susceptibility. Various phenomena have been revealed by using ac susceptibility measurement, such as the order–disorder transition and the memory effect of vortex lattice [13]. Isothermal magnetization measurement has also been used to study the vortex–pinning interaction. From the MH curves, the critical current density can be deduced which is directly related to the pinning effect of the superconductor.

However, most of these effects, such as vortex creep [20] and thermally activated flux motion [21], are based on the macroscopic response of the whole vortex lattice to the external drive. At the same time, little attention has been paid to the study of the behavior of an individual vortex when interacting with a pinning potential. A close investigation of this behavior can help us to manipulate vortices at a microscopic level which is important for potential applications such as quantum computing [22] and vortex generators [23]. Using the Ginzburg-Landau (GL) theory, Priour et al. [24] calculated the vortex behavior when approaching a defect. They predicted that a string will develop from the vortex core and extend to the vicinity of the defect boundary, while simultaneously the supercurrents and associated magnetic flux spread out and engulf the pinning center. However, this process happens fast and the vortex will be quickly attracted to the pinning center location, making it extremely hard to access directly with the local probe techniques so far.

Here, we propose a way to solve this problem by using a pair of pinning potentials where one of them traps a vortex. By changing the characteristic lengths, $\xi$ (coherence length) and $\lambda$ (penetration depth), simply through varying temperature, we are able to modify the effective interaction between the pinned vortex and the other pinning center nearby. This allows us to directly image the interaction process between a vortex and a pinning center with scanning Hall probe microscopy (SHPM). We have found that, with increasing temperature, the vortex shape becomes elongated with magnetic flux spreading over the adjacent pinning potential, thus providing strong evidence to the "string effect." The results of theoretical simulations based on the time-dependent Ginzburg-Landau (TDGL) approach are in line with our experimental findings.

## 1.2 Experimental

Our experiment is carried out on a 200-nm-thick superconducting Pb film prepared by ultra-high-vacuum (UHV) e-beam evaporation on a $Si/SiO_2$ substrate . During the deposition, the substrate is cooled down to liquid nitrogen temperature to avoid the Pb clustering. On top of Pb, 5-nm-thick Ge layer is deposited to protect the sample surface from oxidation. The sample surface is checked by atom-force microscopy (AFM), and a roughness of 0.2 nm is found. The critical temperature $T_c = 7.3$ K is determined by the local ac susceptibility measurements with a superconducting transition width of 0.05 K, indicating high quality of the sample. The pinning centers in the sample are created during sample preparation, and they appear randomly distributed. The local magnetic-field distribution was mapped using a low-temperature

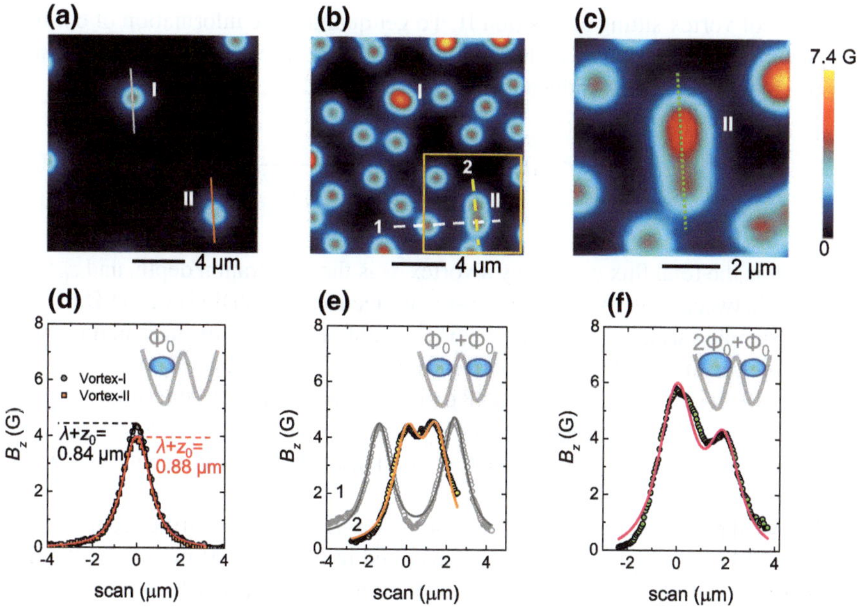


**Fig. 1.1** Distribution of vortices after field cooling, revealing locations of pinning centers. **a–b**: FC images at $H = 0.3$ Oe (**a**), 2.4 Oe **b** and 3.6 Oe (**c**). **a** and **b** were taken at the same area, while **c** is taken at the area indicated by the square in (**b**). At *low* field (**a**), only single quantum vortices, sitting on the pinning centers, are observed. At relatively high fields (**b**) and (**c**), when all the pinning centers are occupied, pinned giant vortices with vorticity $L = 2$ and interstitial single quantum vortices appear. **d** Field profile along the cross section for the pinned vortices indicated by the *solid lines* in (**a**). **e** Field profiles along the *dashed lines* as shown in (**b**). **f** Vortex profile (symbols) along the dotted line in (**c**). The *solid lines* in panels **d–f** represent fitting of the data with the monopole model. The insets in **d–f** schematically show the pinned vortices at position II, with respect to the pinning landscape

SHPM from Nanomagnetics Instruments with a temperature stability better than 1 mK and magnetic-field resolution of 0.1 Oe. All the images are recorded in the lift-off mode by moving the Hall cross above the sample surface at a fixed height of ~0.7 μm. In all the measurements, the magnetic field is applied perpendicular to the sample surface.

### *1.2.1 Distribution of Pinning Centers*

Figure 1.1 displays the vortex distributions at $T = 4.2$ K after field cooling (FC). It is well known that, in the presence of pinning, vortices will nucleate first at the pinning centers as shown in Fig. 1.1a. Two vortices, sitting at positions I and II, are observed in the scanned area. The field profiles for both vortices are displayed in Fig. 1.1d. It is shown that the field strength at the center of the vortex located at position I is higher than that of vortex sitting at position II. To get quantitative information of confined magnetic flux and penetration depth, the monopole is used [25] with the following expression for the vortex-induced magnetic field:

$$B_z(r, z_0) = \frac{\Phi}{2\pi} \frac{z_0 + \lambda}{\left[r^2 + \left(z_0 + \lambda\right)^2\right]^{3/2}} \quad (1.1)$$

where $\Phi$ is the total flux carried by a vortex, $\lambda$ is the penetration depth, and $z_0$ is the distance between effective two-dimensional electron gas (2DEG) of the Hall cross and the sample surface. The distance $z_0$ is constant in the used lift-off scanning mode. The fitting results are shown by solid lines in Fig. 1.1d, yielding $\Phi_{\mathrm{I}} = \Phi_{\mathrm{II}} = 1.1\Phi_0$, $\lambda_{\mathrm{I}} + z_0 = 0.84\,\mu\mathrm{m}$ and $\lambda_{\mathrm{II}} + z_0 = 0.88\,\mu\mathrm{m}$. Since $\lambda$ is related to the superfluid density, $\rho_s \propto \lambda^{-2}$ [26], the larger value of $\lambda_{\mathrm{II}} + z_0$, as compared to $\lambda_{\mathrm{I}} + z_0$, suggests a stronger suppression of superconducting condensate at the pinning site where vortex II is located.

The weakened superconductivity at the vortex-II position is further confirmed by the data of FC at a higher field as demonstrated in Fig. 1.1b. At position I, the field pattern in Fig. 1.1b apparently corresponds to a giant vortex with vorticity $L = 2$. On the other hand, around position II, two adjacent vortices are observed. The field profiles along the dashed lines in Fig. 1.1b are displayed in Fig. 1.1e. Both of them can be well fitted by two $\Phi_0$-vortices located the distance of 3.7 μm (profile (1)) and 1.27 μm (profile (2)) from each other. The fitting results are shown by the solid lines. We also notice that in Fig. 1.1b, the distance between the two vortices located at position II is much smaller than that between interstitial vortices (indicated by the circles in Fig. 1.1b). All these observations suggest that there exist two adjacent pinning sites at the positions II. The pinning potentials of these sites force the two vortices to overcome the vortex–vortex repulsive interaction and stay close to each other. At even higher fields (Fig. 1.1c), one of the two adjacent pinning sites attracts two flux quanta as demonstrated by the field profiles in Fig. 1.1f. This may indicate that its pinning strength is stronger as compared to the bottom located pinning site.

#### 1.2.1.1 Vortex Deformation with Temperature

Based on the above estimate of the relative pinning strength, now we focus on the effect of adjacent pinning potentials on a single vortex. To do this, we perform FC at 0.6 Oe to trap one $\Phi_0$-vortex (vortex located at position II in Fig. 1.2a). When warming up the sample, both $\xi(T)$ and $\lambda(T)$ increase as compared to the temperature-independent distance between the two adjacent pinning sites. In other words, a vortex pinned to one of these sites effectively "approaches" the neighboring site. Each step of this process is imaged directly by SHPM as displayed in Fig. 1.2a–f. For comparison, the field of the vortex located at position I, trapped by an isolated pinning center, is also recorded and analyzed.

At low temperature, both vortices exhibit a circular shape, suggesting the supercurrents are localized about the vortex core. With increasing temperature, the vor-

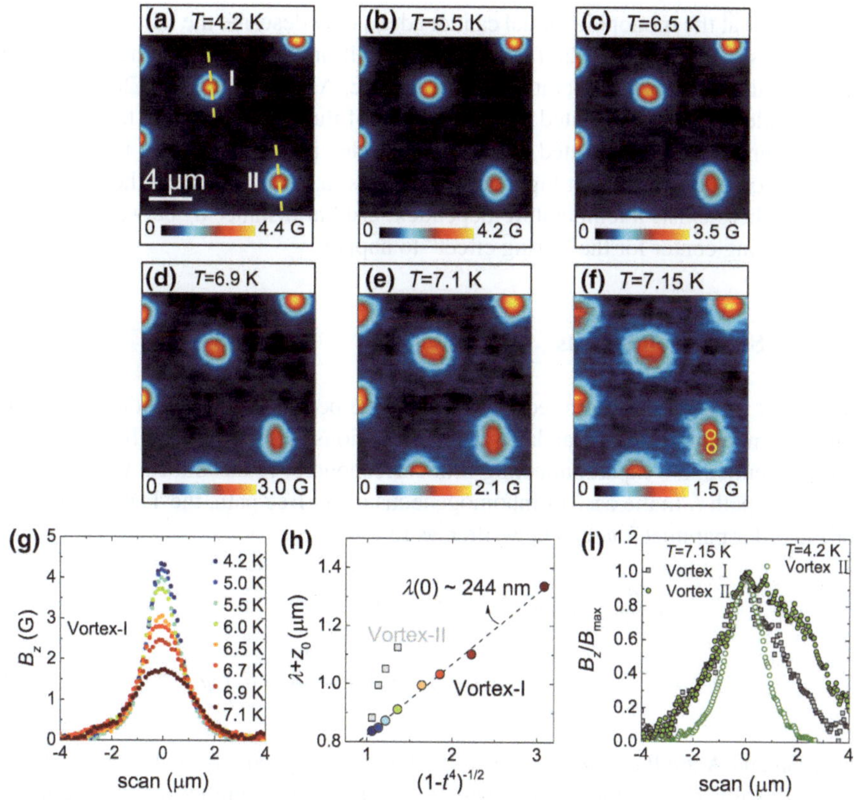


**Fig. 1.2** (Color online) Vortex deformation observed at the pinning centers. **a**–**f** Scanning Hall probe microscopy images measured after first performing field cooling down to 4.2 K (**a**) and then increasing temperature to 5.5 K **b**, 6.5 K **c**, 6.9 K (**d**), 7.1 K **e** and 7.15 K (**f**). The circles in **f** indicate the locations where the two $\Phi_0$-vortices are pinned in the FC image of Fig. 1.1b. **g** Field profiles of vortexI at different temperatures along the dashed lines indicated in (**a**). **h** $\lambda + z_0$ as a function of reduced temperature for vortex I (*squares*) and II (*circles*). The *dashed line* is a linear fit to the data. (I) Normalized field profiles for vortex II at 4.2 K (open *circles*) and 7.15 K (*filled circles*). For comparison, the field profile of vortex-I at 7.15 K (*squares*) is also shown

tex at position I preserves its circular shape up to 7.15 K, while the field strength decreases due to the magnetic profile broadening (Fig. 1.2g), correlated with an increase in the penetration depth. By fitting the profiles with the monopole of (1.1), the value of $\lambda(T) + z_0$ is extracted for different temperatures. According to the two-fluid model [27], in a superconductor with an isotropic energy gap, the temperature dependence of the penetration depth follows the expression $\lambda(T) = \lambda(0)/\sqrt{1 - t^4}$, where $\lambda(0)$ is the zero temperature penetration depth and the reduced temperature is $t = T/T_c$. As shown by the circles in Fig. 1.2h, for vortex at position I the linear dependence of $\lambda$ on $(1 - t^4)^{-1/2}$ is obeyed, yielding $\lambda(0) \approx 244$ nm from the slope. However, the corresponding plot for the vortex located at position II (squares in Fig. 1.2h) fails to follow a linear dependence. This is mainly due to a vortex deformation so that the monopole model cannot adequately describe the field distribution. As demonstrated in Figs. 1.2a–f, for vortex at position II, the broadening of the field profile with increasing temperature is asymmetric. As shown in Fig. 1.2i, the vortex field profile becomes elongated in the direction of the line connecting the two adjacent pinning centers (indicated by the open circles in Fig. 1.2f). The fact that this deformation only happens at high temperatures is in agreement with the prediction of [24]: There exists a temperature-dependent critical distance $d_c$ between a vortex and a pinning center for the "string effect" to happen.

#### 1.2.1.2 Simulation Results

To further analyze the observed phenomena, we performed simulations using the time-dependent Ginzburg-Landau (TDGL) equations (see, e.g., [26]). In our model, pinning centers in a superconducting film correspond to reduced local values of the mean free path $l$. In the case of inhomogeneous mean free path, the TDGL equation for the order parameter $\psi$ can be written as follows:

$$\left(\frac{\partial}{\partial\tau} + i\varphi\right)\psi = \frac{l}{l_m}(\nabla - i\mathbf{A})^2\psi + 2(1-t)\left(1 - \frac{l_m}{l}|\psi|^2\right)\psi, \tag{1.2}$$

where $\varphi$ and $\mathbf{A}$ are the scalar and vector potentials, respectively, and $l_m$ is the mean free path value outside the pinning centers. The relevant quantities are made dimensionless by expressing lengths in units of $\sqrt{2}\xi(0)$, time in units of $\pi\hbar/(4k_B T_c) \approx 11.6\tau_{GL}(0)$, magnetic field in units of $\Phi_0/(4\pi\xi^2(0)) = H_{c2}(0)/2$, scalar potential in units of $2k_B T_c/(\pi e)$, and $\psi$ in units of $|\psi_\infty|$, the order parameter magnitude far away from pinning centers at $H = 0$ . Here, $\mu_0$ is the vacuum permeability, $\tau_{GL}$ is the Ginzburg-Landau time, and $\xi(0)$ is the coherence length outside the pinning centers at zero temperature.

The vector potential $\mathbf{A}$, for which we choose the gauge $\nabla \cdot \mathbf{A} = 0$, can be represented as $\mathbf{A} = \mathbf{A}_e + \mathbf{A}_s$. Here, $\mathbf{A}_e$ denotes the vector potential corresponding to the externally applied magnetic field $\mathbf{H}$, while $\mathbf{A}_s$ describes the magnetic fields induced by the currents $\mathbf{j}$, which flow in the superconductor:

$$\mathbf{A}_s(\mathbf{r}) = \frac{1}{2\pi\kappa^2}\int d^3r' \frac{\mathbf{j}(\mathbf{r}')}{|\mathbf{r} - \mathbf{r}'|}, \tag{1.3}$$

where $\kappa = \lambda/\xi$ is the Ginzburg-Landau parameter and the current density is expressed in units of $\Phi_0/[2\sqrt{2}\pi\mu_0\lambda(0)^2\xi(0)] = 3\sqrt{3}/(2\sqrt{2})j_c(0)$ with $j_c$, the critical (depairing) current density of a thin wire or film [26]. Integration in (1.3) is performed over the volume of the superconductor.

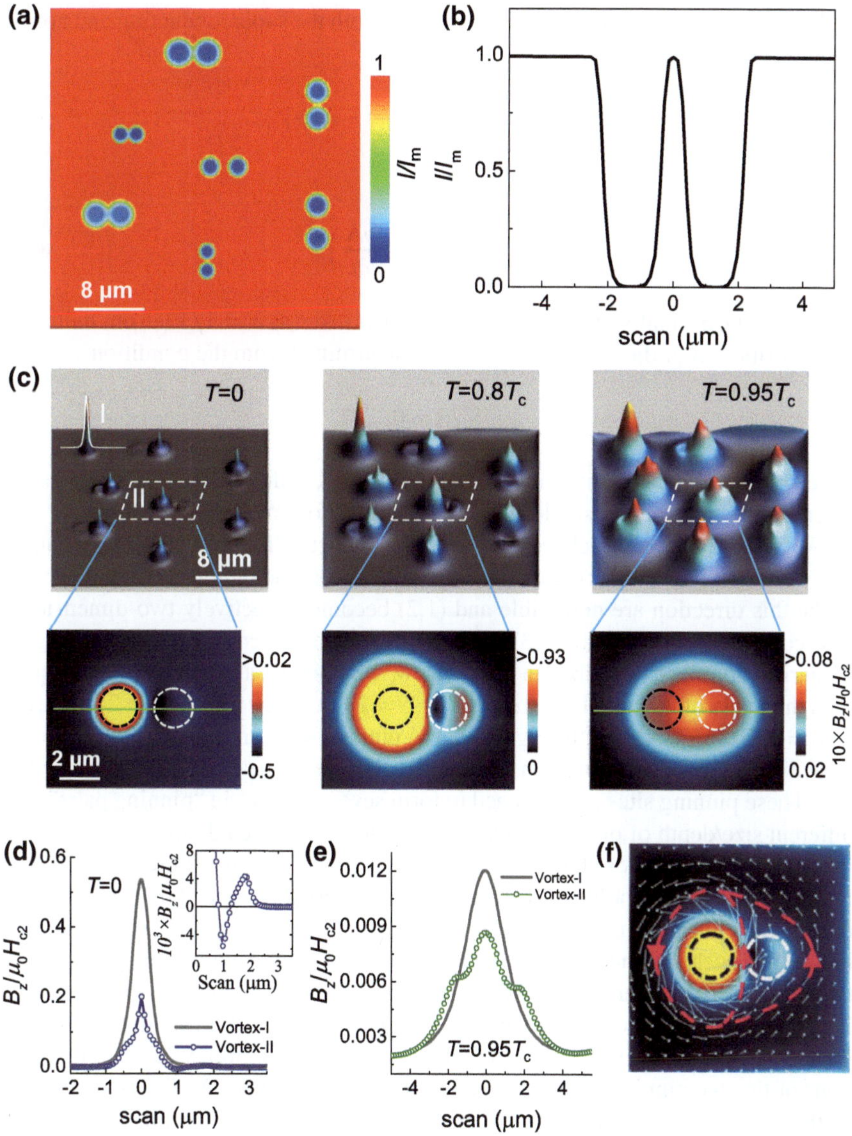


**Fig. 1.3** (Color online) Simulation results. **a** Variations of the normalized mean free path in a $28 \times 28\,\mu\text{m}^2$ superconductor containing seven symmetric pairs of pinning centers with different sizes and strength. **b** Line profile of one pinning pair showing the variation of the normalized mean free path. **c** 3D view of magnetic-field patterns, corresponding to different values of progressively increasing temperature after field cooling to $T = 0$ at $H = 0.00375H_{c2}$. Seven pinned $\Phi_0$ vortices and one interstitial vortex **i** are observed. The lower panels display an enlarged "top view" of the area indicated by the rectangles in each image. The dashed circles schematically show the pinning centers. The vortex field profiles along the direction of solid lines in **c** at $T = 0$ and $0.95T_c$ are given in (**d**) and (**e**), respectively. The inset in panel **d** provides a close view of the field profile for vortex II in the region between the two pinning centers. **f** Current density vector distribution for a vortex sitting on one pinning center of a pinning pair is indicated by the circles. The red dashed lines and the arrows show schematically the flow of vortex current around the pinning pair

In general, the total current density contains both the superconducting and normal components: $\mathbf{j} = \mathbf{j}_s + \mathbf{j}_n$ with

$$\mathbf{j}_s = (1-t)\left[\mathrm{Im}\,(\psi^*\nabla\psi) - \mathbf{A}|\psi|^2\right], \tag{1.4}$$

$$\mathbf{j}_n = -\frac{\sigma}{2}\left(\nabla\varphi + \frac{\partial\mathbf{A}}{\partial t}\right), \tag{1.5}$$

where $\sigma$ is the normal-state conductivity, which is taken as $\sigma = 1/12$ in our units [28]. The distribution of the scalar potential $\varphi$ is determined from the condition

$$\nabla\cdot\mathbf{j} = 0, \tag{1.6}$$

which reflects the continuity of currents in the superconductor. Both $j_\mathrm{n}$ and $\varphi$ vanish when approaching (meta)stable states, which are of our main interest here.

We assume that the thickness of the superconductor film is sufficiently small, so that variations of the order parameter magnitude across the sample as well as currents in the this direction are negligible and (1.2) becomes effectively two-dimensional. This equation, together with (1.3) and (1.6), is solved self-consistently following the numerical approach described in [29]. Below, we consider a superconducting square with thickness of 200 nm, lateral sizes $X \times Y = 20\,\mu\mathrm{m} \times 20\,\mu\mathrm{m}$, $\xi(0) = 210\,\mathrm{nm}$, and $\lambda(0) = 160\,\mathrm{nm}$. As shown in Fig. 1.3A, the pinning sites are introduced as circular areas, where the mean free path $l$ is smaller than its value $l_\mathrm{m}$ outside pinning centers. These pinning sites are arranged to form seven symmetric "pinning pairs" with different size/depth of pinning centers and distances between them.

The calculated vortex shape evolution with increasing temperature is displayed in Fig. 1.3c. After field cooling down to $T = 0$, a $\Phi_0$ vortex is trapped by one of the pinning centers of each pinning pair. Remarkably, the distribution of the magnetic field, induced by such a pinned vortex, exhibits a tail toward the neighboring pinning center. In contrast, the interstitial vortex, marked as I, remains circularly symmetric up to $T_\mathrm{c}$. The field distribution in the vicinity of pinned vortex-II at different temperatures is plotted as a color map below each image. The dashed circles indicate the positions of the two pinning centers, which form the pinning pair. At $T = 0$, while most of the flux is accumulated around one pinning center (black circle), a "magnetic-field dipole" appears at the position of the other pinning center (white circle). This can also be seen from the magnetic profiles (Fig. 1.3d), which are taken along the solid lines drawn in Fig. 1.3c (for comparison, the vortex profile for vortex-I is also shown). From the inset to Fig. 1.3d, a field dip between the two pinning centers is clearly seen.

The aforementioned magnetic-field dipole is reminiscent of what we have reported for the Meissner state of a superconductor [30]. When the Meissner currents flows through an area containing a pinning center, they generate in its vicinity two opposite sense current half-loops. The magnetic-field pattern, induced by this current configuration, has a dipole-like form (see Fig. 1.4a), which can be considered as a bound

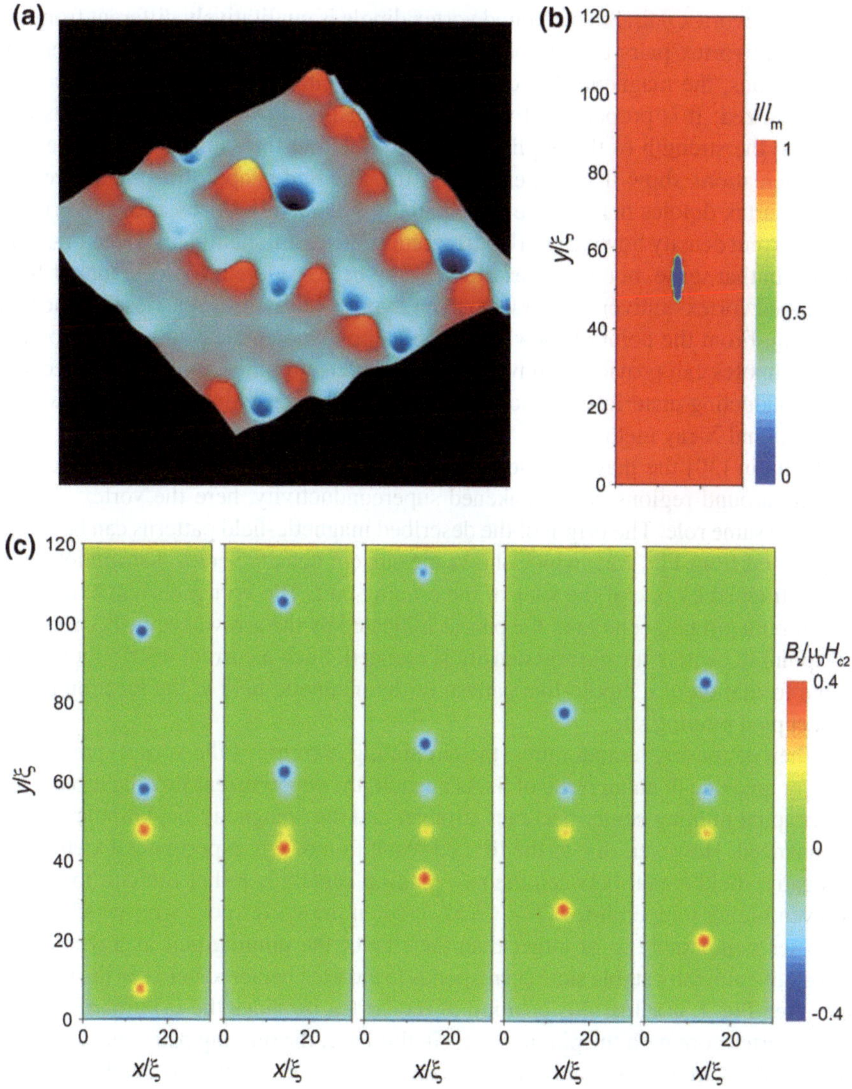


**Fig. 1.4** (Color online) **a** Vortex dipoles observed in the Meissner state of a superconducting Pb film with naturally formed pinning centers. The *bright red* (*dark blue*) color indicates high (*low*) magnetic field. **b** Variations of the mean free path $l$ assumed for simulation in a superconductor with one pinning center of elongated shape. (**c**) Simulated magnetic-field distributions for the pinning configuration in B, taken with the time step $\Delta tau = 5$. Single quantum vortices and antivortices are generated from the pinning center at large enough external current ($j_e = 0.55 j_c(T)$), applied in the $x$ direction. The simulations are performed for the penetration depth $\lambda = \xi$ and the thickness $d = \xi$, where $\xi$ is the coherence length. No external magnetic field is applied

vortex–antivortex pair. Such a bound vortex dipole is qualitatively different from the vortex–antivortex pairs usually observed in the mixed state of a type-II superconductor. Thus, the magnetic flux corresponding to either pole of a vortex dipole is not quantized: It is proportional to the local density of the Meissner currents and hence to the strength of the applied magnetic field. At the same time, our theoretical calculations show that, depending on the size and shape of a pinning center, bound vortex dipoles may evolve into spatially separated vortices and antivortices, if the current density flowing nearby the pinning site is sufficiently high (see Fig. 1.4b and c). In that sense, bound vortex dipoles may be considered as precursors of fully developed vortex–antivortex pairs, in which a vortex (antivortex) carries just one flux quantum. From the point of view of possible applications, measurements of bound vortex dipoles can provide a convenient way to detect hidden submicron defects in superconducting materials in a noninvasive way, which is rather difficult by using a conventional X-ray method.

While in [30] the magnetic-field dipoles are generated by the Meissner current flowing around regions with weakened superconductivity, here the vortex current plays the same role. The origin of the described magnetic-field patterns can be easily understood from Fig. 1.3f, which shows the current density vector distribution for a vortex trapped by one of the sites of the pinning pair. The vortex current extends to engulf both pinning centers of the pair as indicated by the arrows, while between the two pinning centers the current density is reduced. Such a current distribution leads to the formation of a dipole-like pattern of the magnetic field at the location of the unoccupied pinning site.

When increasing temperatures, the circulating currents of the vortex spread over a larger area due to an increase of $\lambda$. As a result, the supercurrent flowing around the unoccupied pinning center becomes stronger and the "magnetic-field dipole" signal is enhanced. However, due to the limited resolution of our experimental technique, a negative field signal between the two pinning centers is rather difficult to detect. At even higher temperatures ($T \geq 0.95T_c$), when the vortex core size, which is $\sim\xi$, becomes comparable to or larger than the size of the pinning pair as a whole, the thermodynamically stable state corresponds to a vortex centered between the pinning sites (see Fig. 1.3e): Just this position of the vortex provides the maximum overlap of the vortex core with the pinning sites. In this case, the pinning pair acts as a single pinning site with strongly anisotropic shape, which leads to an elongated magnetic-field pattern for the pinned vortex. Similar pinning geometry conversion has also been suggested in Refs. [31, 32]. We have also simulated the vortex deformations in the case of asymmetric pinning pairs, which lead to qualitatively similar results (not shown here).

The vortex distributions at relatively high magnetic fields are shown in Fig. 1.5. At those fields, a pinning center can accommodate more than one vortex. This is consistent with what we have observed in Fig. 1.1b. However, instead of giant vortices, which may be presumed from the experimental data, vortex clusters are formed at pinning sites considered in our theoretical model. This resembles the effect observed in superconducting film with blind antidots, where, due to the large size of blind holes, the trapped $n\Phi_0$ multiquanta vortex at the antidot dissociates into $n$ individual

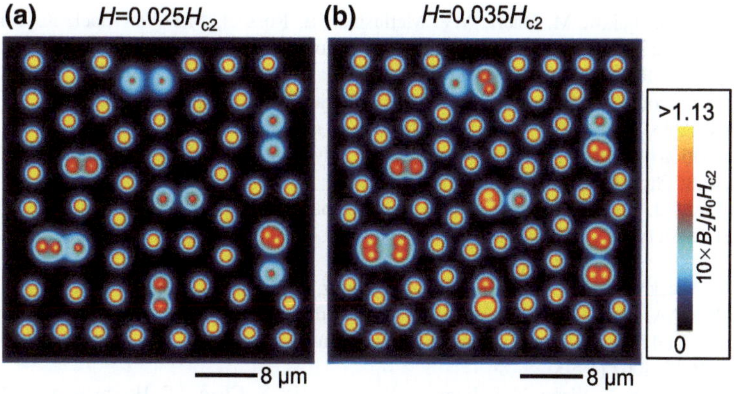


**Fig. 1.5** (Color online) Vortex patterns at relatively high magnetic fields. **a** Vortex patterns, corresponding to FC, for the pinning landscapes of Fig. 1.3a at $H = 0.025H_c$ **a** and $0.035H_{c2}$ (**b**)

$\Phi_0$ vortices in the bottom superconducting layer [33]. In this connection, it seems worth mentioning that at present, the resolution of our experimental technique may be not sufficient to confidently distinguish between giant vortices and compact vortex clusters.

To summarize, we have performed both experimental and theoretical studies of vortex states in the vicinity of a pinning center. Our experimental work provides clear evidence of vortex deformation by a nearby pinning center, where the vortex supercurrents and the induced magnetic-field profile expand and engulf the pinning site. The results of our TDGL theoretical modeling, which are fully consistent with the experimental observations, reveal an additional fine structure ("magnetic-field dipoles") in the magnetic field patterns, induced by those deformed vortices. By simply varying the temperature, the vortex geometry can be well controlled. A detailed understanding of vortex deformation by adjacent pinning centers paves the way to manipulate the vortex current distributions at a microscopic level. This provides new possibilities to control single flux quanta in superconducting devices.

**Acknowledgements** We acknowledge the support from FWO and the Methusalem funding by the Flemish government. This work is also supported by the MP1201 COST action. This research was also supported by the Flemish Research Foundation (FWO-Vl), project nrs. G.0115.12N, G.0119.12N, G.0122.12N, G.0429.15N and by the Research Fund of the University of Antwerpen.

## References

1. M. Baert, V.V. Metlushko, R. Jonckheere, V.V. Moshchalkov, Y. Bruynseraede, Composite flux-line lattices stabilized in superconducting films by a regular array of artificial defects. Phys. Rev. Lett. **74**, 3269 (1995)

2. V.V. Moshchalkov, M. Baert, V.V. Metlushko, E. Rosseel, M.J. Van Bael, K. Temst, Y. Bruynseraede, R. Jonckheere, Pinning by an antidot lattice The problem of the optimum antidot size. Phys. Rev. B **57**, 3615 (1998)
3. N. Haberkorn, B. Maiorov, I.O. Usov, M. Weigand, W. Hirata, S. Miyasaka, S. Tajima, N. Chikumoto, K. Tanabe, L. Civale, Influence of random point defects introduced by proton irradiation on critical current density and vortex dynamics of Ba(Fe0.925Co0.075)2As2 single crystals. Phys. Rev. B **85**, 014522 (2012)
4. D. Ray, C.J. Olson Reichhardt, B. Jank, C. Reichhardt, Strongly enhanced pinning of magnetic vortices in type-II superconductors by conformal crystal arrays. Phys. Rev. Lett. **110**, 267001 (2013)
5. B.M. Vlcek, H.K. Viswanathan, M.C. Frischherz, S. Fleshler, K. Vandervoort, J. Downey, U. Welp, M.A. Kirk, G.W. Crabtree, Role of point defects and their clusters for flux pinning as determined from irradiation and annealing experiments in YBa2Cu3O7-$\delta$ single crystals. Phys. Rev. B **48**, 4067 (1993)
6. C.-L. Song, Y.-L. Wang, Y.-P. Jiang, L. Wang, K. He, X. Chen, J.E. Hokman, X.-C. Ma, Q.-K. Xue, Suppression of Superconductivity by Twin Boundaries in FeSe. Phys. Rev. Lett. **109**, 137004 (2012)
7. J.L. MacManus-Driscoll, S.R. Foltyn, Q.X. Jia, H. Wang, A. Serquis, L. Civale, B. Maiorov, M.E. Hawley, M.P. Maley, D.E. Peterson, Strongly enhanced current densities in superconducting coated conductors of $YBa_2Cu_3O_{7-x}$ + $BaZrO_3$. Nat. Mater. **3**, 439–443 (2004)
8. J. Gutierrez, A. Llordes, J. Gazquez, M. Gibert, N. Roma, S. Ricart, A. Pomar, F. Sandiumenge, N. Mestres, T. Puig, X. Obradors, Strong isotropic flux pinning in solution-derived $YBa_2Cu_3O_{7-x}$ nanocomposite superconductor films. Nat. Mater. **6**, 367–373 (2007)
9. S.R. Foltyn, L. Civale, J.L. MacManus-Driscoll, Q.X. Jia, B. Maiorov, H. Wang, M. Maley, Materials science challenges for high-temperature superconducting wire. Nat. Mater. **6**, 631–642 (2007)
10. J. Ge, S. Cao, S. Shen, S. Yuan, B. Kang, J. Zhang, Superconducting properties of highly oriented Fe 1.03 Te 0.55 Se 0.45 with excess Fe. Solid State Commun. **150**, 1641 (2010)
11. D.B. Rosenstein, I. Shapiro, B.Y. Shapiro, HTransport current carrying superconducting film with periodic pinning array under strong magnetic fields. Phys. Rev. B **83**, 064512 (2011)
12. I. Sochnikov, A. Shaulov,Y. Yeshurun, G. Logvenov, I. Bozovic, Large oscillations of the magnetoresistance in nanopatterned high-temperature superconducting films. Nat. Nanotech. **5**, 516C519 (2010)
13. J. Ge, J. Gutierrez, J. Li, J. Yuan, H.-B. Wang, K. Yamaura, E. Takayama-Muromachi, V.V. Moshchalkov, Peak effect in optimally doped p-type single-crystal $Ba_{0.5}K_{0.5}Fe_2As_2$ studied by ac magnetization measurements. Phys. Rev. B **88**, 144505 (2013)
14. J. Ge, J. Gutierrez, M. Li, J. Zhang, V.V. Moshchalkov, Vortex phase transition and isotropic flux dynamics in $K_{0.8}Fe_2Se_2$ single crystal lightly doped with Mn. Appl. Phys. Lett. **103**, 052602 (2013)
15. J. Ge, J. Gutierrez, J. Li, J. Yuan, H.-B. Wang, K. Yamaura, E. Takayama-Muromachi, V.V. Moshchalkov, Dependence of the flux-creep activation energy on current density and magnetic field for a $Ca_{10}$ $(Pt_3As_8)[(Fe_{1-x}Pt_x)_2As_2]_5$ single crystal. Appl. Phys. Lett. **104**, 112603 (2014)
16. Y.L. Wang, X.L. Wu, C.-C. Chen, C.M. Lieber, Enhancement of the critical current density in single-crystal Bi2Sr2CaCu208 superconductors by chemically induced disorder. Proc. Natl. Acad. Sci. USA **87**, 7058–7060 (1990)
17. M. Motta, F. Colauto, W. Ortiz, J. Fritzsche, J. Cuppens, W. Gillijns, V. Moshchalkov, T. Johansen, A. Sanchez, A. Silhanek, Enhanced pinning in superconducting thin films with graded pinning landscapes. Appl. Phys. Lett. **102**, 212601 (2013)
18. J. Gutierrez, A.V. Silhanek, J. Van de Vondel, W. Gillijns, V.V. Moshchalkov, Transition from turbulent to nearly laminar vortex flow in superconductors with periodic pinning. Phys. Rev. B **80**, 104514 (2009)
19. D.Y. Vodolazov, F.M. Peeters, Rearrangement of the vortex lattice due to instabilities of vortex flow. Phys. Rev. B **76**, 014521 (2007)

20. T. Klein, H. Grasland, H. Cercellier, P. Toulemonde, C. Marcenat, Vortex creep down to 0.3 K in superconducting Fe(Te, Se) single crystals. Phys. Rev. B **89**, 014514 (2014)
21. L. Jiao, Y. Kohama, J.L. Zhang, H.D. Wang, B. Maiorov, F.F. Balakirev, Y. Chen, L.N. Wang, T. Shang, M.H. Fang, H.Q. Yuan, Upper critical field and thermally activated flux flow in single-crystalline $Tl_{0.58}Rb_{0.42}Fe_{1.72}Se_2$. Phys. Rev. B **85**, 064513 (2012)
22. O. Romero-Isart, C. Navau, A. Sanchez, P. Zoller, J.I. Cirac, Superconducting vortex lattices for ultracold atoms. Phys. Rev. Lett. **111**, 145304 (2013)
23. V.M. Fomin, R.O. Rezaev, O.G. Schmidt, Tunable generation of correlated vortices in open superconductor tubes. Nano Lett. **12**, 1282 (2012)
24. D. Priour Jr., H. Fertig, Deformation and depinning of superconducting vortices from artificial defects: a Ginzburg-Landau study. Phys. Rev. B **67**, 054504 (2003)
25. J. Ge, J. Gutierrez, J. Cuppens, V.V. Moshchalkov, Observation of single flux quantum vortices in the intermediate state of a type-I superconducting film. Phys. Rev. B **88**, 174503 (2013)
26. M. Tinkham, *Introduction to Superconductivity* (McGraw-Hill Inc, New York, 1996)
27. J. Bardeen, Two-fluid model of superconductivity. Phys. Rev. Lett. **1**, 399 (1958)
28. R. Kato, Y. Enomoto, S. Maekawa, Computer simulations of dynamics of flux lines in type-II superconductors. Phys. Rev. B **44**, 6916 (1991)
29. A.V. Silhanek, V.N. Gladilin, J. Van de Vondel, B. Raes, G.W. Ataklti, W. Gillijns, J. Tempere, J.T. Devreese, V.V. Moshchalkov, Local probing of the vortex-antivortex dynamics in superconductor/ferromagnet hybrid structures. Supercond. Sci. Technol. **24**, 024007 (2011)
30. J. Ge, J. Gutierrez, V.N. Gladilin, J.T. Deverese, V.V. Moshchalkov, Bound vortex dipoles generated at pinning centres by Meissner current. Nat. Commun. **6**, 6573 (2015)
31. J. Trastoy, M. Malnou, C. Ulysse, R. Bernard, N. Bergeal, G. Faini, J. Leseur, J. Briatico, J.E. Villegas, Freezing and thawing of artificial ice by thermal switching of geometric frustration in magentic flux lattices. Nat. Nanotech. **9**, 710–715 (2014)
32. M.L. Latimer, G.R. Berdiyorov, Z.L. Xiao, F.M. Peeters, W.K. Kwok, Realization of artificial Ice systems for magnetic vortices in a superconducting MoGe thin film with patterned nanostructures. Phys. Rev. Lett. **111**, 067001 (2013)
33. A. Bezryadin, Y.N. Ovchinnikov, B. Pannetier, Nucleation of vortices inside open and blind microholes. Phys. Rev. B **53**, 8553 (1996)

# Chapter 2
# Pinning-Engineered $YBa_2Cu_3O_x$ Thin Films

**Paolo Mele, Adrian Crisan and Malik I. Adam**

## 2.1 Introduction

In order to be used for practical applications such as lossless current transportation, winding of magnets, etc., superconducting materials should possess not only high enough critical temperature $T_c$, but also critical current density $J_c$, and upper critical field $B_{c2}$ as large as possible to cover a wide range of the currently available and future technologies of superconductivity.

Introduction of nanosized artificial pinning centers (APCs) was widely used to strongly enhance $J_c$ of high-temperature superconductor (HTSC) like $YBa_2Cu_3O_x$ (YBCO, $T_c = 92$ K) in magnetic field. Figure 2.1 shows the performance of state-of-the-art YBCO thin films with APCs in comparison with other superconducting materials [1]. It is worth noting the outstanding performance of APCs-added YBCO films at 77 K, surpassing the conventional superconducting metallic cables (NbTi, $Nb_3Sn$) at 4.2 K. The incorporation of one-dimensional defects such as nanorods in YBCO films prepared by pulsed laser deposition (PLD) could be exemplified as to how effective the APC technology evolves. After the pioneering work of J. McManus-Driscoll and coworkers on YBCO+BZO nanorods [2], in 2008 YBCO films added with 4 wt% BSO have been reported to

P. Mele (✉)
Research Center for Environmentally Friendly Materials Engineering,
Muroran Institute of Technology, Muroran, Hokkaido, Japan
e-mail: pmele@mmm.muroran-it.ac.jp

A. Crisan
National Institute of Materials Physics, 077125 Bucharest, Magurele, Romania
e-mail: adrian.crisan@infim.ro; acrisan652@gmail.com

M.I. Adam
Department of Mechanical Engineering, University Tenaga Nasional,
Kajang, Malaysia
e-mail: idriesm@uniten.edu.my; omavis1@gmail.com

 
A. Crisan (ed.), *Vortices and Nanostructured Superconductors*, Springer Series
in Materials Science 261, DOI 10.1007/978-3-319-59355-5_2

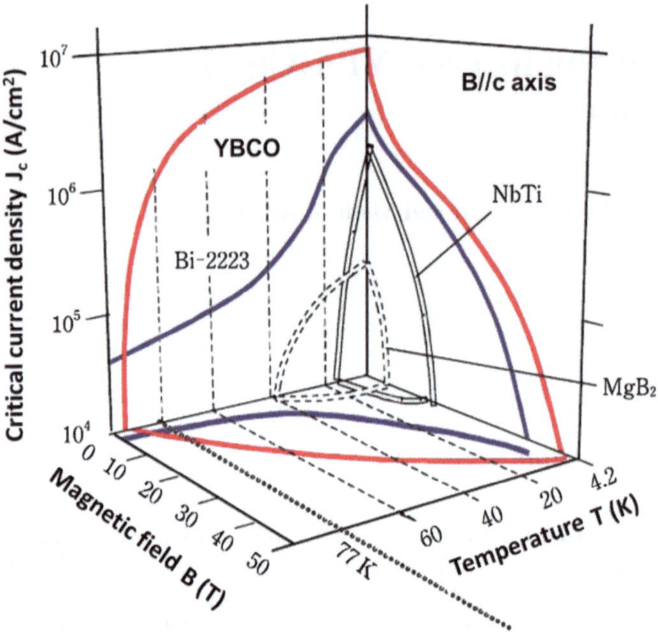


**Fig. 2.1** Dependence of critical current density ($J_c$) on external magnetic field ($B$) and temperature ($T$) for various superconducting wires: NbTi (white line), $MgB_2$ (*white dotted line*), Bi-2223 (*blue solid line*), YBCO (*red solid line*). In the YBCO case, the maximum of the critical current density can be obtained in a wide region. (Reproduced with permission from [1] with slight modification)

generate a $J_c$ = 0.3 MA/cm$^2$ and global pinning force $F_p = J_c \times B$ = 28.3 GN/m$^3$ at 77 K, 3 T ($B$ //c) [3]. Rapid progress in the maximum global pinning force in YBCO added with nanorods has been reported during the following years by decreasing the operating temperature: 120 GN/m$^3$ at 65 K with niobate addition [4]; 405 GN/m$^3$ at 40 K in YBCO added with BHO [5]; 700 GN/m$^3$ at 30 K, 1000 GN/m$^3$ at 20 K [6]; and even 1750 GN/m$^3$ at 4.2 K [7] in YBCO with 25% of Zr. Besides $F_p^{max}$, another parameter used to quantify the applicability range of YBCO films is the irreversibility field ($B_{irr}$). Remarkable improvements of $B_{irr}$ (77 K) = 11 T of YBCO+BSO have been published recently. $B_{irr}$ = 14.8 T (77 K) for MOCVD (Y,Gd)BCO doped with 15% vol Zr [7], $B_{irr}$ = 15 T for LTG-SmBCO-BHO [8], and $B_{irr}$ = 15.8 T for GdBCO-BHO on Hastelloy [9].

The impressive performances of YBCO added with nanorods are counterbalanced by the fact that $J_c$ is found to be intrinsically anisotropic in relation to the direction of the applied magnetic field with a broad maximum for *B*//*c*-axis direction. Current flow anisotropy was identified as one of the critical issues facing the practical applicability of these materials since the value of $J_c$ in certain applications might be needed to be constant in all directions of the applied magnetic field during machine operation. T. Haugan and coauthors opened the way to solve this issue with their seminal work on Y-211 randomly dispersed particles (3D-APCs) added to YBCO films [10]. After that, big progress has been achieved in YBCO added with isotropic APCs. Impressively, YBCO added with $Y_2O_3$ nanoislands showed extremely high global pinning force: $F_p > 1 TN/m^3$ at 4.2 K [11].

Besides the "basic" one-dimensional and three-dimensional APCs (two-dimensional APCs have been found to be much more difficult to be established), the pinning landscape became extremely rich, thanks to the nanoengineering approach of many groups worldwide in the past decade. Several combinations of APCs have been tried, such as adding randomly dispersed islands to the nanorods, cutting the nanorods when nanolayers are deposited parallel to the films' thickness, and several more diverse approaches. All of them will be covered in this extensive review.

This chapter is divided in three parts:

(i) Critical current and pinning force in pure YBCO superconductors and their limits
(ii) Artificial pinning center approach with broad overview of experimental results and theoretical models
(iii) Summary of the performance of APCs-added YBCO and general conclusion.

## 2.2 Critical Current and Pinning Force in Pure $YBa_2Cu_3O_x$ Thin Films

After the discovery of YBCO in 1987 [12], the first oxide superconductor with $T_c = 92$ K, a temperature that is larger than the boiling point of nitrogen, the great potential of this material for practical application has been immediately anticipated and many research groups have focused on the transport properties of sintered YBCO. Despite of considerably large irreversibility field ($B_{irr} \sim 7$ T at 77 K) in the bulk form of the sintered material, the critical current was found to have very low values in excess of $10^2$ A/cm$^2$ or so, not surpassing the typical performance of copper cables.

The reason of the low values of $J_c$ has been found into the low coherence length of YBCO, so that consequently $J_c$ does not pass easily through the grain boundaries [13]. Early studies of bulk YBCO and related REBCO (RE = lanthanide rare earth) materials have been demonstrated that $J_c$ can be enhanced thanks to the alignment of the grains by melt-texturing-growth (MTG) approach [14]. MTG allowed obtaining $J_c$ in the order of $10^4$ A/cm$^2$, two orders of magnitude larger than for sintered material. Indeed, the MTG REBCO materials are able to trap huge values of magnetic field [15, 16] and can be used for energy storage. However, as a matter of fact, MTG bulk materials are inherently brittle and then not feasible for development of long and flexible cables for the effective transportation of electrical current.

The same problem of low critical currents due to the misorientation of the grains has been found at first in the so-called first-generation wires and tapes (1G-WT), based on $(Bi,Pb)_2Sr_2Ca_2Cu_3O_x$ (Bi-2223) materials and produced by powder-in tube approach. Bi-2223 tapes can be used for power cables, but are not feasible for use at 77 K in any magnetic field of significant magnitude. The reason is the drastic suppression of $B_{irr}$ $(T)$ to the very low value of 0.2 T due to the large anisotropy [17].

From early 1990s, researchers focused on the development of "second-generation wires and tapes" (2G-WT). 2G-WT are called "coated conductors" (CC): The superconducting layer (usually YBCO or REBCO) is coated on a substrate in such a way as to obtain both in-plane and out-of-plane texturing of the superconducting grains in a biaxial alignment [18]. The biaxial alignment (i.e., the perfect epitaxy) can be easily obtained on a single crystal, which sizes are few cm, but is more difficult to achieve over long lengths (on the km scale) which are indispensable for practical applications.

Two kinds of approaches to prepare long-length CC have been proposed: ion beam-assisted deposition (IBAD) [19] and rolling-assisted biaxially textured substrates (RABiTS) [20]. IBAD technique involves the deposition of biaxially oriented buffer layer on polycrystalline metallic substrate, followed by deposition of superconducting film. In RABiTS, the metallic substrate (Ni) is biaxially textured as effect of cold rolling and recrystallization. The story of 2G-WT is the story of technological progress on IBAD and RABiTS, with continuous improvement in performance (critical current and pinning force) and length of the tapes. However, further discussion on 2G-WT is beyond the scope of this chapter. Nevertheless, the interested reader is kindly referred to the detailed analysis of the 2G-WT which can be found in the excellent textbook [21] and reviews [22, 23].

Parallel to the development of IBAD and RABiTS, a plethora of research has been done and duly published on the fabrication and characterization of YBCO films on single crystals, and the state-of-the-art advancement was the $J_c > 10^6$ A/cm$^2$ (self-field, 77 K, $B//c$) for YBCO films. It was concluded from this striking development that YBCO films can carry currents two orders of magnitude larger than the MTG YBCO materials.

It is quite impressive to see how, in about 28 years, four orders of magnitude improvement on the value of the critical current in YBCO materials has been achieved, thanks to the combined progress in material and flux pinning engineering.

The epitaxial YBCO films formed on single-crystal substrates via the non-equilibrium process such as pulsed laser deposition (PLD) [24] or chemical routes such as trifluoroacetate-metal oxide deposition (TFA-MOD) [25] and metal organic chemical vapor deposition (MOCVD) [26] have high $J_c$ of 1–5 MA/cm$^2$ at 77 K and zero field (see, e.g., [27] and references therein, [28]). This is because the various crystal defects which are naturally formed during the deposition of the epitaxial YBCO films such as antiphase boundaries, grain boundaries, stacking faults [29–33], voids, compositional impurities, fine precipitates, and dislocations [34] exist in a high density in the film, in addition to oxygen vacancies and twins

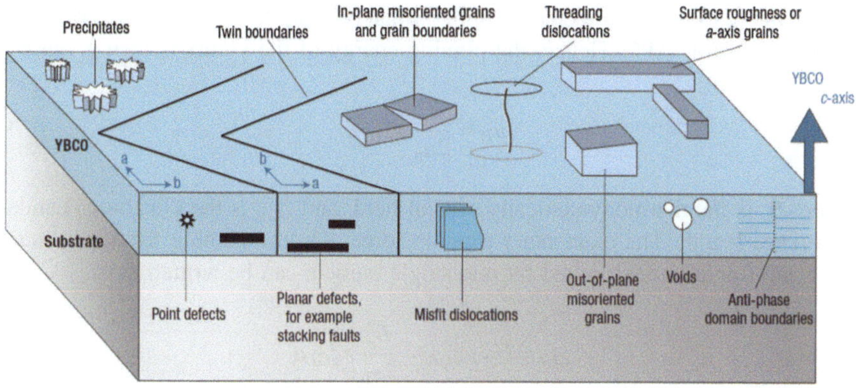


**Fig. 2.2** Schematic representation of natural thin-films defects proposed as flux pinning sites in YBCO. (Reproduced with permission from [38])

[35–37]. Schematic representation of natural defects in YBCO thin films is reported in Fig. 2.2 [38].

From early studies of PLD-prepared YBCO superconductor films, it has been proven that the three-dimensional Volmer–Weber growth mechanism creates steps on crystal substrate which acts as nucleation sites thereafter [27]. AFM images show growth spirals on PLD film, and this feature was accepted as clear evidence for the presence of screw dislocations in the structure of the grown films. Furthermore, cross-sectional TEM images have also presented evidence of the columnar microstructure of the PLD films. This was ascribed to the growth of spiral columns intertwined with the grain boundaries. The dislocations, both screw and edge dislocations which are found to extend along the *c*-axis of the films, have been suggested to act as strong pinning centers for the vortices in YBCO films. This was the case since the dislocation core is normal and its diameter is about 1–3 nm, which is comparable with the coherence length of the superconductor [30–40]. However, it is very difficult to control the density and distribution of the dislocations in order to improve $J_c$ under magnetic fields. This has been the problem for improving the in-field $J_c$ of the pure YBCO films for a long period of time.

## 2.3 Critical Current and Pinning Force in $YBa_2Cu_3O_x$ Thin Films with Artificial Pinning Centers (APCs)

Based on the Ginzburg–Landau model [41, 42], insertion of additional artificial pinning centers (APCs) on the nanoscale-prepared films is found to be useful as well as effective approach in enhancing the dynamic $J_c$ in the complex structure of these materials. The added defects were found to prevent the movement of vortices. According to global pinning theory, any non-superconducting region of a size *d* that is equivalent to few nanometers or approximately equal to the size of a vortex core radius can act as APC. Hence, the pinning energy of the vortex is then given by

$$u_P = \frac{B_c^2}{2\mu_0} \pi \xi_{GL}^2 d \tag{2.1}$$

where $B_c$ is the thermodynamically critical field, and $\xi_{GL}$ is the Ginzburg–Landau coherence length. The elementary pinning force, i.e., the pinning force associated with one single vortex pinned by one single particle can be written as

$$f_p = U_p \,/\, \xi_{GL} = \frac{B_c^2}{2\mu_0} \pi \xi_{GL} d \tag{2.2}$$

The global pinning force $F_P = J_c \times B$ therefore is directly proportional to the number of particles $N$ so that the critical current is expressed by

$$J_c = \frac{F_P}{B} = \frac{Nf_p}{B} = \frac{NB_c^2}{2\mu_0 B} \pi \xi_{GL} d \tag{2.3}$$

A classical representation of (2.2) and (2.3) is used for the analysis of an YBCO MTG bulk containing round-shaped $Y_2BaCuO_5$ (Y211) nanosized particles having a density $N_p$ and lattice spacing $a_f$, where $1/N_p = a_f^2\sqrt{3}/2$ and size $d$. In this case, the critical current density of the system is calculated as

$$J_c = \frac{\pi \xi_{GL} B_c^2}{2\mu_0 B\sqrt{\phi_0}} \frac{N_p d^2}{\sqrt{B}} \left(1 - \frac{B}{B_{c2}}\right) \tag{2.4}$$

It can be seen from the above equation that the $J_c$ of the system is directly proportional to the volume fraction of the nanoparticles, $N_p d^3$, divided by the size of the vortex core $d$ or, in other words, the number of YBCO/Y211 interfaces. By extension, this theoretical approach led the experimentalists to choose an approach allowing the reduction of the size $d$ of the Y211 inclusions in the films in order to increase $J_c$. The refinement of Y211 nanoparticle size and the consequent improvement of $J_c$ have been successfully accomplished by addition of secondary phases (e.g., Pt [14]).

The APC approach has been successfully extended to YBCO and REBCO thin films, and outstanding results, with global pinning force exceeding 30 GN/m$^3$ at 77 K, have been obtained by many groups all over the world in the past 10 years. Summary of pinning performance at 77 K is given in Fig. 2.3.

In the last 2 or 3 years, several groups focused their investigations on the lower temperatures (from 30 K down to 4.2 K) and high magnetic fields (from 10 T up to 30 T). The performance of APCs-added YBCO and REBCO overcame the "psychological threshold" of 1000 GN/m$^3$, with the present record value of 1750 GN/m$^3$ at 4.2 K, 30 T for YBCO+15%Zr prepared by MOCVD: 100-fold than the

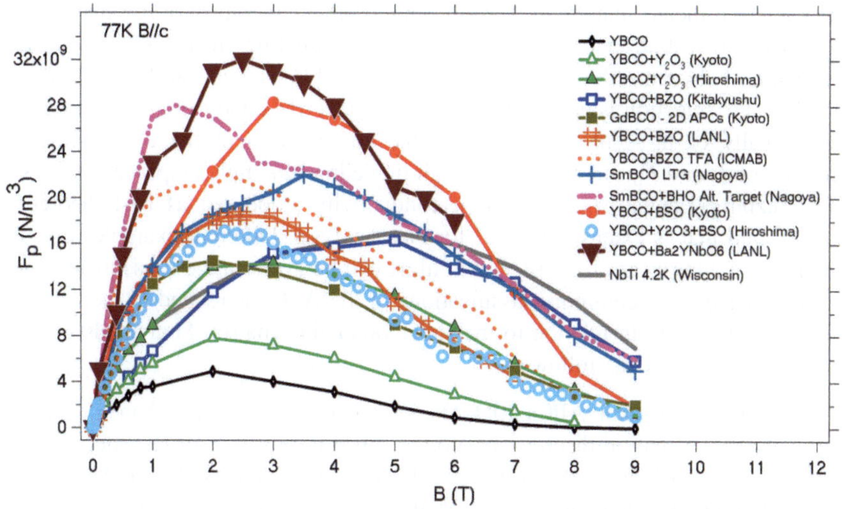


**Fig. 2.3** The global pinning force $F_p$ drawn against the applied magnetic field $B$ at (77 K, $B//c$) for YBCO films with several types of APCs. Majority of the samples are described in the main text. The $F_p$ performance of NbTi wire at 4.2 K was included as a milestone

**Fig. 2.4** The global pinning force $F_p$ drawn against the applied magnetic field $B$ at (4.2 K, $B//c$) for YBCO films and tapes with several types of APCs. The samples are described in the main text

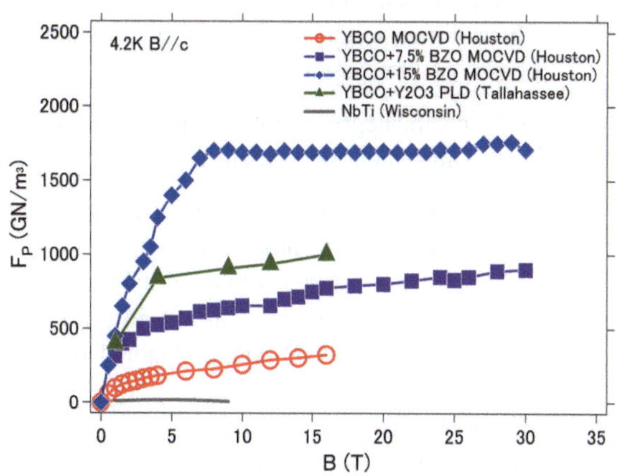


maximum pinning force of NbTi at the same temperature [7]. Summary of pinning performance at 4.2 K is given in Fig. 2.4.

The broad-spectrum strategies so far implemented for the production of superconducting thin films doped with different APCs can be classified into three main stream steps. The preliminary step is directed toward finding an element, elemental oxide, or elemental compound with feasible APC features. The second step is rather practical and is concentrated onto how to incorporate the viably defined APCs into YBCO or REBCO superconducting matrix so that the immediate objective here is to provide a potential pinning site. The third and final broad strategy was devoted to the ways and means that could possibly lead to practical control of the spacing and the spatial distribution of APCs in the superconducting matrix so that a clear correlation between APCs and the output values of $J_c$ and $F_p$ can be realized.

The APC selection is usually based on empirical trials and can also be predicted theoretically or in some cases through the combination of the two. While the analysis of $J_c$ and $F_p$ can be done through modeling of simple empirical equations, the Ginzburg–Landau theory of superconductivity stood out as best theoretical background that is extensively used for data analysis. Different examples of APC engineering approaches for YBCO thin films will be given in the subsequent pages.

One of the most common classifications of the APCs is related to their dimensionality and shape in relation to the superconducting matrix. Four kinds of APC can be distinguished as follows:

(i) Zero-dimensional APCs (0D-APCs): the defects smaller than $\xi$, such as vacancies, cation disorder, or atomic substitutions at Y or RE site
(ii) one-dimensional APCs (1D-APCs), i.e., linear defects: linear dislocations obtained by decoration of the substrate; columnar defects obtained by heavy ion irradiation; and columnar defects incorporated by the addition of secondary phases such as dislocations and columnar defects
(iii) two-dimensional APCs (2D-APCs), i.e., planar defects such as interfaces in multilayers, small-angle grain boundaries, antiphase boundaries, and surfaces of large precipitates
(iv) Three-dimensional APCs (3D-APCs) such as nanoparticles and secondary phases of the scale of $\xi$ or more.

Other four subcategories are generally obtained by the combination or modification of the aforementioned main types which include:

(v) Segmented 1D-APCs, obtained by alternating layers with pure YBCO and layers of YBCO+nanocolumns
(vi) Mixed 1D+3D APCs, obtained by simultaneous combination of columnar defects and nanoparticles in hybrid films
(vii) Multilayered 1D+3D-APCs obtained by alternating columnar defects and nanoparticles in multilayers
(viii) Combined multilayers and nanorods, obtained by simultaneous incorporation of APCs engineered parallel and perpendicular to the *c*-axis of the film.

Figure 2.5 represents the four basic cases and their most common combinations.

### 2.3.1 Zero-Dimensional APCs (0D-APCs)

0D-APCs (or dilute doping, or minute doping) of YBCO or REBCO thin films is referred to the substitution of very small amounts of foreign elements at Y or RE site. Several substitutions by element M (M = Tb, Ce, Pr, Nd, La, Co, and Eu) at

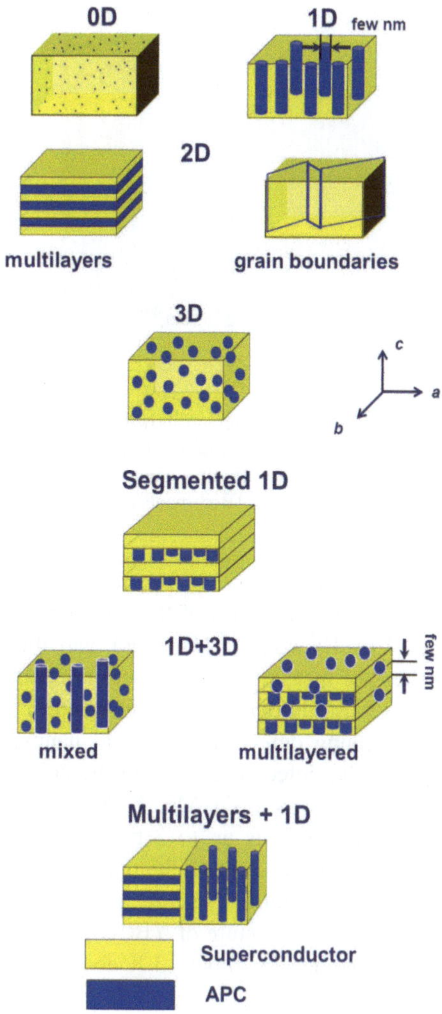


**Fig. 2.5** Schematic representation of nanometric artificial pinning centers (APCs) of different dimensionality incorporated in YBCO superconducting thin film. Details are given in the main text

the Y site of YBCO and RE site of REBCO (RE = Dy) were studied [43–48]. The doping enhances $J_c$ and $F_p$ over the range of the applied magnetic fields, and this is attributed to two intertwinned factors. Firstly, there is the enhanced concentration of nanoprecipitates of these substituents in YBCO film compared to the case of the pure YBCO one (see Fig. 2.6a). Consequently, the resulting stress field induces lattice misfit in the final structure of the hybrid film. In the case of YBCO added with Eu at Y site, the most consistent self-field transport $J_c$ values were achieved for Y/Eu = 0.7/0.3 films, with arguably the densest and smoothest microstructures (Fig. 2.6b).

Advanced doping with three different rare earth elements at RE site was tried in the middle of 1990s in melt-textured $(Nd_{0.33}Eu_{0.33}Gd_{0.33})Ba_2Cu_3O_x$ (NEG-123)

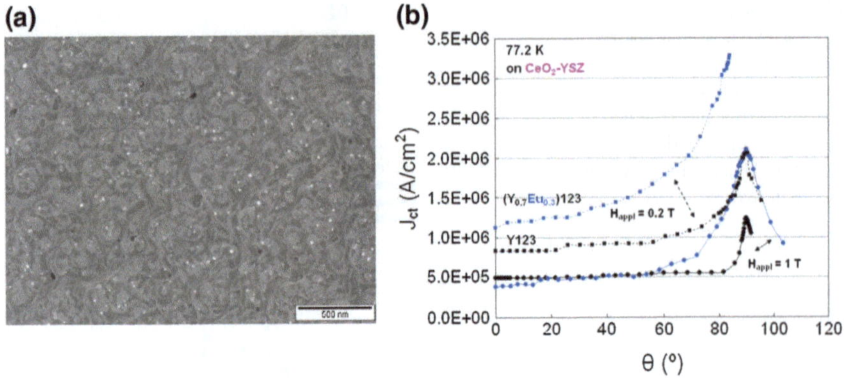


**Fig. 2.6** **a** SEM micrograph showing nanoprecipitates on the surface of $(Y_{0.7}Eu_{0.3})Ba_2Cu_3O_y$ film grown on $CeO_2$–YSZ substrate; **b** $J_c$ (77.2 K) of $(Y_{0.7}Eu_{0.3})Ba_2Cu_3O$ and YBCO films on $CeO_2$–YSZ substrates, for varying angle and $H_{appl} = 0.2$ and 1 T; $\theta = 0°$ corresponds to $H//c$, and $\theta = 90°$ is for $H//ab$. (Both panels reproduced with permission from [44])

and has shown marvelous performances in terms of $J_c$ and irreversibility field [49, 50]. In the bulk form of these materials, the self-field $J_c$ at 77 K has reached the level of $10^5$ A/cm$^2$ ($H//c$-axis), and the irreversibility field was recorded ~15 T when the measurement was performed at the applied field of 5 T. A very important as well as intuitive aspect deduced from the early results is the possibility to reduce the pinning defect size down to nanoscale level and to engineer a method that brings the defect size close to the coherence length. These nanoscale particles of secondary phases, Zn, Mo, Ti, Nb, etc., have eventually enhanced the flux pinning of these materials up to 3 times compared to that measured on pristine YBCO compounds. As a result, the pinning force density remained strong up to the liquid oxygen temperature (90.2 K), which enables magnetic levitation near to the vicinity of critical temperature of YBCO [51]. The applicability of levitation technology is thus extended by about 13 K above the usual limit of the boiling point of liquid nitrogen.

Along this line, studies on NEG-123 films were tried with emphasis on the great potential of this material for practical applications [52–54]. In comparison with GdBCO films, pinning performance and irreversibility field were enhanced. It appeared, however, that there is not much difference between the $J_c$ and $F_p$ values of the NEG-123 films and the bulk form of the same material. This fact may be related to issues regarding the structural optimization.

### 2.3.2 *One-Dimensional APCs (1D-APCs)*

In the attempt to improve $J_c$ over a wide range of magnetic fields and field orientations, the one-dimensional artificial pinning centers (1D-APCs) in the form of columnar defects, screw and edge dislocations, and their consequential lattice strain constitute the most investigated type of pinning centers associated with films and coated conductors alike. This type of pinning centers is generally classified into three subcategories.

The first category comprises columnar defects that are introduced into the film by heavy ion and/or neutron irradiation.

The second subtype is the linear dislocations which can be obtained by decoration of substrate prior to the thin-film deposition.

Nanocolumns composed of secondary non-superconducting phases incorporated in superconducting film encompass the third subtype: the 1D nanorods. The ability of these 1D nanorods in restraining the vortex flux line mobility and, hence, the dramatic improvement of $J_c$ and the global flux pinning $F_p$ were clearly demonstrated in the optimally doped $BaMO_3$–YBCO samples with (M = Zr, and Sn) [3]. A comparison of the $F_p$ values of the two samples, however, showed that the global pinning is larger in the $BaSnO_3$-doped sample. Additions of other perovskite ($BaHfO_3$) or double perovskite nanorods were tried as well.

#### 2.3.2.1 Columnar Defects by Neutron and Ion Irradiation

In the first half of 1990s, significant efforts have been made to create non-superconducting columns by high-energy irradiation of YBCO single crystals or thin films with neutrons and heavy ions [55–59]. Both neutrons and heavy ions have been found as efficient tools for improving $J_c$, $F_p$, $B_{irr}$ of the thin films since the size of defects and vortex core is comparable. Furthermore, the size, density, and shape of defects are controllable though variation of energy and fluence of the ions or neutrons.

Due to the high cost of irradiation, this top-down approach was confined to small-size samples and laboratory studies for long time. Very recently [60], development of reel-to-reel neutron irradiation facility drastically lowered the cost to few USD/m so that this way appears feasible for large-scale applications. In the framework of this "revival," fast irradiation (exposure of just one second per 0.8 $cm^2$) by 3.5 meV oxygen ions significantly improves the in-field performance of YBCO tapes without modification of the coating conductor growth in the industrial process [61].

Develos-Bagarinao et al. [62] proposed a cost-effective way to create nanopores in MOD YBCO films: conventional Ar ion milling of the films through nanoporous anodic aluminum oxide (AAO) templates. After irradiation, etching of the YBCO film revealed columnar voids deeply penetrating into YBCO matrix (Fig. 2.7a). Despite to low pore density (~4.56 $\mu m^{-2}$), in-field $J_c$ (2 T) was enhanced by a factor of two with respect to the $J_c$ of the unirradiated films, with typical *ab*-pinning (Fig. 2.7b). With AAO membranes having much higher density of pores, *c*-axis-correlated pinning and much higher $J_c$ are expected.

Another original approach has been proposed by Sueyoshi et al. [63] by investigating the flux pinning property in a GdBCO-coated conductor splayed with

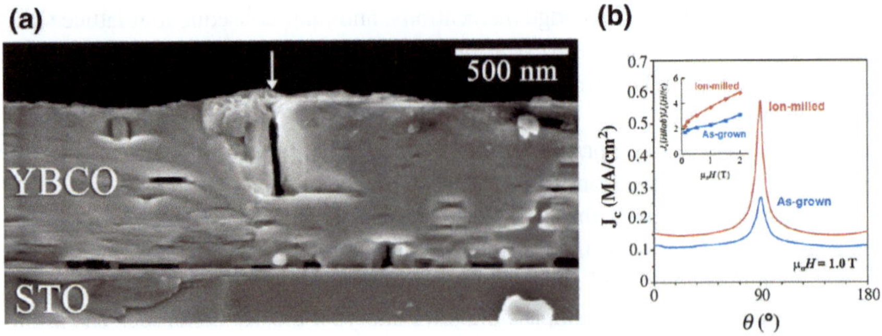


**Fig. 2.7** **a** Cross-sectional SEM image of an Ar ion-milled, Br-ethanol chemically etched YBCO film. The arrow indicates a typical columnar void created by the ion milling and subsequent etching (reproduced with permission). **b** In-field current-carrying performance and structural properties comparing as-grown to ion-milled YBCO films: Angle-dependent $J_c$ ($\theta$) at 77.3 K, 1 T of as-grown YBCO and ion-milled YBCO. The inset shows the ratio $J_c(B//ab)/J_c(B//c)$ as a function of applied field. (Both panels reproduced with permission and slightly modified from [62])

$Xe^+$ ions where the resulted columnar defects were oriented parallel to both *c*-axis and the *ab* plane. The angular dependence of $J_c(\theta)$ has been shown to produce multiple maxima in the field angular orientation dependence of the critical current.

#### 2.3.2.2 Linear APCs in YBCO Films Deposited on Decorated Substrates and on Functional Buffers

The first bottom-up approach to produce 1D-APCs was the deposition of nanodots on the substrates followed by deposition of the superconducting film. The immediate effect of the decoration of substrates by 1D-APCs is the formation of linear dislocations parallel to the *c*-axis of the superconducting film. This procedure is now usually named "substrate decoration method."

In the pioneering work of Crisan et al. [64–66], they deposited an amorphous phase of Tl-based thin-film superconductor on substrates decorated by sputtered Ag nanodots. This approach was successful, and the $J_c$ was enhanced to a value more than one order of magnitude (Fig. 2.8). The drawback of Tl-based materials lies in their unsuitability for applications because of the toxicity of Tl. Later on, the technology was applied to YBCO films, numerous studies on improvement of $J_c$ by linear defects induced by decoration were published so far, and several decoration techniques were subsequently employed. Examples are annealed $CeO_2$ buffer for YBCO films [67, 68], iridium-sputtered particles for YBCO films [69], Au, Ag, and Pd PLD-deposited nanodots [70, 71], spin-coated $BaTiO_3$ and $BaZrO_3$ nanoparticles [72], gas-phase-deposited $Y_2O_3$ particles [73], and PLD-deposited $Y_2O_3$ particles [74].

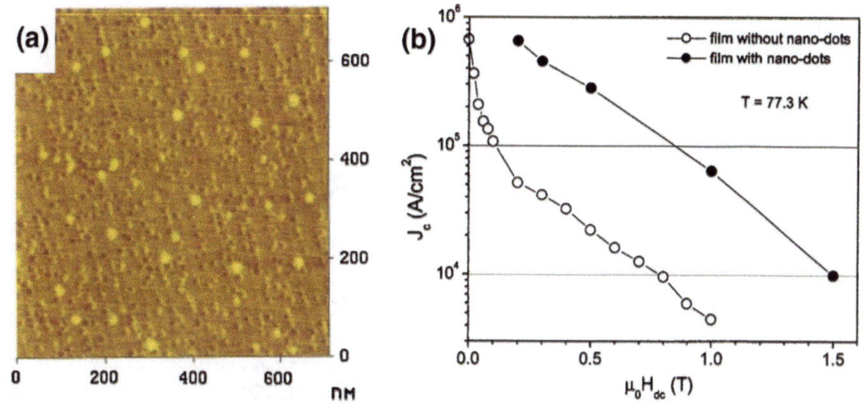


**Fig. 2.8** **a** Atomic force micrograph of Ag islands grown on annealed $SrTiO_3$ (100) by 5 s rf sputtering; **b** Critical current density of (Cu, Tl)$BaSrCa_2Cu_3O_y$ films grown on substrate without (*empty symbols*) and with (*filled symbols*) Ag nanodots at 77 K and various dc fields, estimated from Brandt's formula. (Both panels reproduced with permission from [64])

K. Matsumoto et al. proposed the formation of distributed nanosized $Y_2O_3$ islands on the substrate surface, followed by YBCO [75–80] or SmBCO [81] deposition. Depending on deposition conditions, the formation of yttria nanoislands of diameter 15–50 nm and height 4–10 nm was found to require a small number of laser shots, typically 5–10. After chemically etching deposited YBCO films, the density of the etch pits and the distance between closest neighbors are nearly identical to those observed for the nanoislands. Nanoislands induce the formation of extended linear defects such as dislocations, and nearly, all the islands can induce dislocations inside the film. In such a way, the density of dislocations was enhanced up to 200–220/μm$^2$—about three times larger than the natural dislocation density values. In the best conditions (substrate decorated with 10 laser pulses on $Y_2O_3$ target), both $J_c$ and the maximum global pinning force of the samples at 77 K were enhanced by up to 1.7 times than those of pure YBCO film deposited in the same conditions, and clear *c*-axis-correlated pinning was observed.

Although the fabrication of PLD film on the decorated substrate results in an enhancement of $J_c$ in magnetic field, the decoration technique is very laborious and delicate process. The growth window for deposition is very narrow where a small difference in the number of the laser shots can radically change the size and distribution of the nanoislands in the film.

More recently, Tsai and coworkers deposited YBCO films by PLD on STO substrates buffered with $(Fe_2O_3)_x(CeO_2)_{1-x}$ functional ferromagnetic composite [82]. The $Fe_2O_3$ vertically aligned nanopillars (VAN) embedded in $CeO_2$ matrix provide significant increase in the $J_c$ (Fig. 2.9) due to pronounced magnetic pinning effect. The amount of nanopillars can be tuned by varying the concentration $x$ of $Fe_2O_3$ in the matrix. Best result was obtained when $x = 0.3$: $F_p^{max}$ (65 K) ~ 20 GN/m$^3$, $F_p^{max}$ (40 K) ~ 180 GN/m$^3$, and $F_p^{max}$ (5 K) ~ 1100 GN/m$^3$.

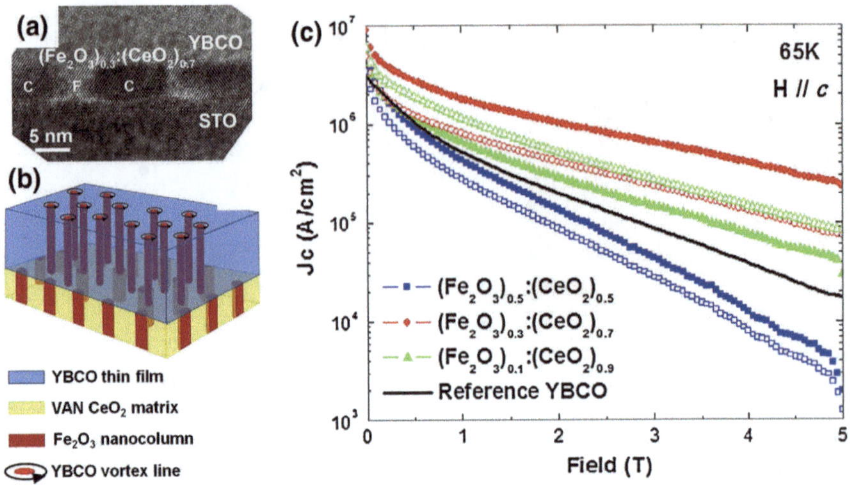


**Fig. 2.9** **a** Cross-sectional high-resolution TEM micrographs and **b** illustration of pinning landscape of $(Fe_2O_3)_{0.3}$:$(CeO_2)_{0.7}$ buffer containing vertically aligned nanopillars (VAN) below YBCO film. **c** In-field performance ($J_c(B//c)$) as a function of applied magnetic field plots of the VAN nanolayers doped samples compared with the reference YBCO sample measured at 65 K. (All panels reproduced with permission and slightly modified from [82])

#### 2.3.2.3 Oxide Nanocolumns Embedded in Superconducting Films

The majority of published reports on 1D-APCs is devoted to the formation and characterization of self-aligned nanorods of 5–15 nm diameter grown in the epitaxial YBCO films by ablation of YBCO target doped with a small wt% of perovskites $BaMO_3$ (BMO, M = Zr, Sn, Hf) by PLD or MOCVD methods [2–4, 83–128].

A key question related to columnar 1D-APCs is: Why perovskites (BMO) form nanorods in YBCO while other dopants ($Y_2O_3$ for example) do not? This question remained unanswered for several years. Wu et al. [129–134] recently proposed an explanation based on the interfacial strain (Fig. 2.10).

(i) the *local* YBCO/dopant interfacial strain, extended for few nm around the dopant, determines the morphology of the nanostructures (vertical alignment = nanorods) or horizontal alignment (particles or nanopatches) inside the films (Fig. 2.10a);
(ii) the *global* film (YBCO+incorporated dopants)/substrate interfacial strain, extended for hundreds of nm, tunes the 1D-APC landscape: Nanorod orientation evolves as *c*-aligned → splayed → *ab*-aligned (i.e., nanopatches or nanolayers) as the global strain increases. It is worth noting that if the global strain is negligible, nanorods remain parallel to the *c*-axis up to very high concentrations of the dopant (up to 45 vol.%) (Fig. 2.10b).

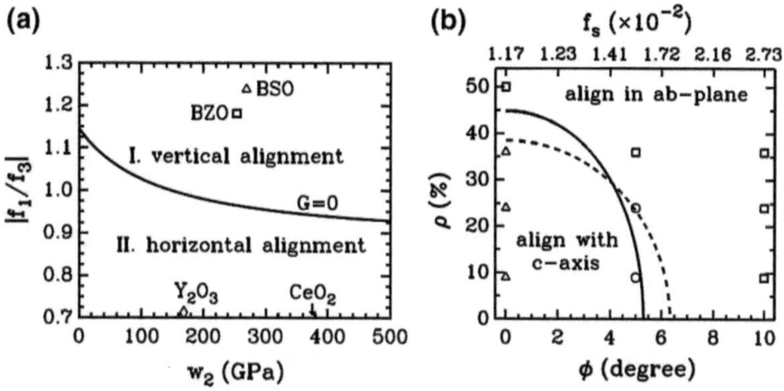


**Fig. 2.10** **a** Threshold of $f_1/f_3$ as a function of $w_2$ for nanodefects (BZO, BSO, $Y_2O_3$, $CeO_2$) orientation in the *c*-oriented YBCO film on lattice-matched substrates, where $f_1$ and $f_3$ are the lattice mismatch between the film and dopant in the [100] and [001] directions, respectively, and $w_2$ is a function of the elastic constants of the dopant. Below (above) the *solid curve*, the vertical (horizontal) alignment is unattainable (Reproduced with permission from [125]). **b** Phase diagram for BZO nanostructure in *c*-axis-oriented YBCO films, where $r$(%) is the volume density of BZO in the film, $f$ is the vicinal angle of STO substrate, and $f_s$ is the substrate/film lattice mismatch. The solid curve is the calculated phase boundary. Experimental points are indicated by *triangles* (*c*-axis-aligned nanorods), *squares* (*ab* plane-aligned nanopatches and circles (mixed nanorods and nanopatches). (Reproduced with permission from [87])

Additionally, in [132–134], Wu and coworkers applied an elastic strain model to understand the effect of the lattice strain on the diameter of the BZO nanorods embedded in YBCO films. They found that the radius $R$ of the nanorods was not dramatically affected by the concentration of BZO in the targets and is varying only slightly in the range $\chi = 0.2 \sim 0.7$, where $\chi = R/D$, the ratio between radius of the nanorods ($R$), and distance between two adjacent nanorods ($D$). $R$ can be predicted from c and $E_{\text{elastic}}$ of the nanorods according to the relation

$$R = \frac{\gamma_S \chi^2}{\gamma_0 + \gamma_1 \chi^2 - E_{elastic}} \tag{2.5}$$

where $\gamma_s = C_s\varepsilon_{\text{rod}}$, $\gamma_0 = C_0\varepsilon_{\text{rod}}$, and $\gamma_1 = C_0\varepsilon_{\text{rod}}$; $C_s$, $C_0$, and $C_1$ are constants, while $\varepsilon_{\text{rod}}$, the elastic energy of the nanorods, is calculated from the elastic constants $c_{ij}$ of BZO and YBCO and mismatch between YBCO and the nanorods.

This approach can be extended to YBCO-BSO and YBCO-BHO [134]. Figure 2.11 shows as individual points the experimental values of $R$ as reported from several groups and our predicted value of $R$ for the three systems YBCO-BZO, YBCO-BSO, and YBCO-BHO using different values of the constants. It appears that $R_{\text{BSO}} > R_{\text{BZO}} > R_{\text{BHO}}$. Our results are consistent with [134] even if in such work same value of $\gamma_0 = 180$ is used for the three systems. This suggests a universal behavior for $R$ versus $\chi$. It is quite remarkable that following this approach it

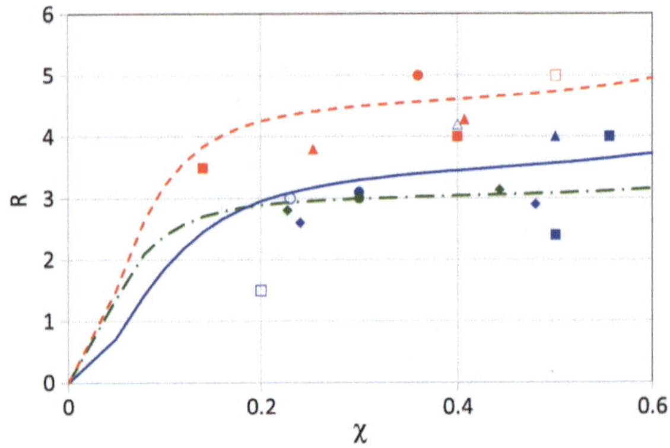


**Fig. 2.11** Nanorod radius calculated from (2.5) as a function of $\chi$ for BZO (*blue* continuous line), BSO (*red dashed line*), and BHO (*green dashed* and *dotted line*) nanorods in YBCO films, using $C_s = 0.9$, $C_1 = 85$, $C_0 = 275$ for BZO; $C_s = 0.5$, $C_1 = 75$, $C_0 = 316$ for BSO; and $C_s = 0.4$, $C_1 = 100$, $C_0 = 290$ for BHO (see main text for details). Individual points are the measured nanorod radius of BZO in YBCO from [88] (*blue filled squares*), [84] (*blue empty square*), [132] (*blue filled diamond*), [6] (*blue empty diamond*), [111] (*blue filled triangle*), [3] (*blue empty triangle*), and [83] (*blue filled circle*); BSO in YBCO from [117] (*red filled triangles*), [116] (*red filled squares*), [115] (*red empty square*), and [79] (*red filled circle*); BHO in YBCO from [83] (*green filled circle*) and an unpublished work (Mele et al. 2013) (*green filled diamonds*)

is possible to predict the diameter and density of the nanorods (or any other kind of defect) embedded in a film knowing the elastic constants and crystalline parameters of the materials.

Another systematic work comparing the effect of BMO (M = Zr, Sn, Hf) on the superconducting properties of YBCO was recently proposed by Horide et al. [83]. It was shown that only $B_\Phi$ (i.e., the density of the nanorods) and not BMO selection nor wt% content are affecting the $T_{irr}/T_c$ curves.

When the nanorods are added to a YBCO film, the maximum pinning force is given for $B//c$ direction, and the critical current can be calculated from the solution of the London and Ginzburg–Landau equations for the case of columnar pins [42]:

$$J_c^{columns} = \frac{\Phi_0}{32\pi^2 \xi_{ab} \lambda_{ab}^2} \ln\left(1 + \left(\frac{C_0}{\sqrt{2}\xi_{ab}}\right)^2\right) \quad (2.6)$$

In this expression, $C_0$ is the radius of the columnar pin, $\Phi_0$ is the flux quanta, $\lambda_{ab} = \lambda_0\ (1 - t^4)^{-0.5}$ is the London penetration depth in the $ab$ plane, $\xi_{ab} = \xi_0\ (1 - t)^{-0.5}$ is the coherence length in the $ab$ plane, and $t$ is the reduced temperature: $t = T/T_c$. Equation (2.5) is valid when the sizes of defects and the coherence length in the $ab$ plane are comparable. In the case of YBCO, $\lambda_0 = 150$ nm and $\xi_0 = 1.5$ nm [17].

## $BaZrO_3$ (BZO) Nanocolumns

In the early times (2005), $BaZrO_3$ (BZO) was very popular material to be used as dopant. The first report has been published by McManus-Driscoll et al. [2] and immediately followed by the publication of many more reports in which the likelihood of the new dopant as a good pinning catalyst is revealed [84–100]. The nanorods are found to accumulate longitudinal and parallel to the *c*-axis of the films, and they act as *c*-axis-correlated pinning centers (Fig. 2.12). By utilizing the BZO doping in YBCO films, the $J_c$ was pushed up and eventually reached 0.7 MA/cm$^2$ (77 K, 1 T, *B*//c) and 0.1 MA/cm$^2$ (77 K, 5 T, *B*//c). The achieved values correspond to several times $J_c$ of the epitaxial YBCO films. Consequently, the $F_p^{max}$ is increased to reach values of 12–16 GN/m$^3$ (77 K, *B*//c) which are quite close to those of NbTi (16 GN/m$^3$, 4.2 K). It was noticed that there is an optimal value in the amount of BZO doping, but the matching field $B_\phi$ value has also reached 8–10 T. As a result, the $B_{irr}$ curve is greatly pushed up to the high magnetic field side from 5–7 T to 8–11 T at 77 K and *B*//c. Introduction of BZO nanorods by means of mixed target or surface-modified targets was carried out, and similar results were reported for several other REBCO materials with higher $T_c$ than YBCO, such as NdBCO [101–104], ErBCO [105–108], EuBCO [109], and the LTG-SmBCO [110].

In thin films prepared by PLD, the performance of BZO-added YBCO was surpassed by BSO and BHO (see the following paragraphs). Nevertheless, more

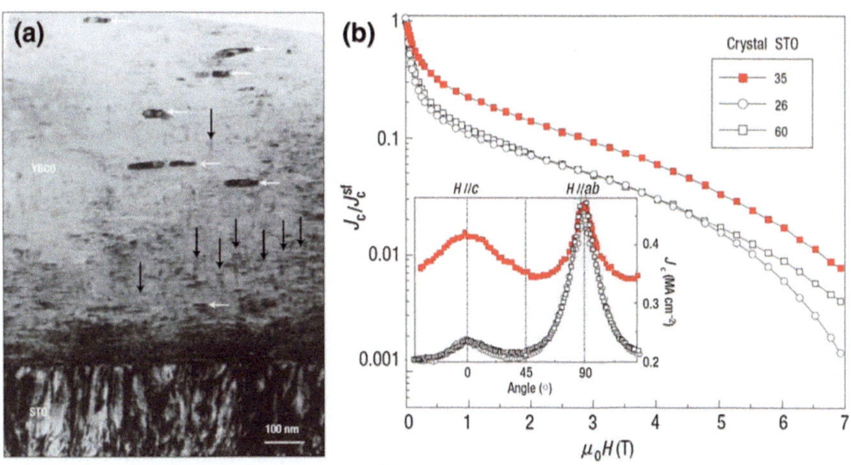


**Fig. 2.12** **a** Low-magnification TEM micrograph showing nanoparticles of $BaZrO_3$ in YBCO $+BaZrO_3$ films on STO-buffered MgO single crystals. Some of the nanoparticles have been labeled using white arrows and some of the columnar defects by black arrows; **b** normalized critical current density $J_c/J_c$ (sf) at 75.5 K versus magnetic field applied parallel to the *c*-axis for pure YBCO and YBCO+$BaZrO_3$ films deposited on single-crystal $SrTiO_3$ substrates; inset shows the angular dependence of $J_c$ at 1 T with sample 60 data multiplied by 0.75. Open points are for pure YBCO, and filled points are for YBCO+$BaZrO_3$. (Both panels reproduced with permission from [2])

recently, the group of V. Selvamanickam at Houston University [6, 111–113] reported outstanding results in (Y,Gd)BCO thin films (thickness about 0.9 μm) deposited by MOCVD method, incorporating heavy concentrations of Zr (7.5%, 15%, 25% in volume). In 15% BZO-added (Y, Gd)BCO, the maximum pinning force density $F_p^{max}$ reached 14 GN/m$^3$ at 5 T, 77 K, which is comparable with $F_p^{max}$ of NbTi at 4.2 K, and the impressive 1.7 TN/m$^3$ at 4.2 K [6]. Another significant finding is the extremely high $B_{irr}$ at (77 K) ~ 14.8 T, a value that surpasses the $B_{irr}$ ~ 11 T of the NbTi superconductor at 4.2 K [6].

Eventually, thickness of the films was increased up to 2.2 μm with incorporation of BZO nanorods fully extended throughout the whole thickness of the film (Fig. 2.13a), and addition of 20% Zr was recognized [6] as the most effective with critical current of 4206 A/12 mm at 30 K, 2.5 T, corresponding to $J_c$ of about 15 MA/cm$^2$. This performance exceeds a fourfold improvement in critical current at the operating condition of a 10 MW wind generator (3000 A/12 mm at 30 K, 2.5 T). Indeed, it is a factor of 5.6 higher than the critical current values of commercial coated conductors available today. Linear correlation can be established between $J_c$ (77 K, 3 T, *B*//c) and $J_c$ (*T*, *B*//c) for temperatures lower than 20 K and fields larger than 9 T [112]. Indeed, it has been demonstrated that in these MOCVD-YBCO films highly doped with Zr, the requirement to have $J_c$ of 15 MA/cm$^2$ at 30 K, 2.5 T (*B*//*c*) is to have a large self-field $J_c$ at 77 K: between 2.5 and 3.8 MA/cm$^2$ [113]. In the present status, the MOCVD-(Y,Gd)BCO +BZO tapes appear as the most promising material for practical application of the second-generation tapes. Summary of the performances of MOCVD-YBCO-Zr tapes is reported in Fig. 2.13b.

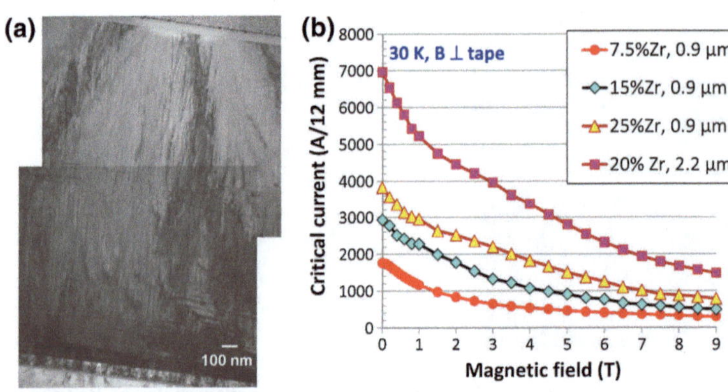


**Fig. 2.13** **a** Cross-sectional TEM image of a 2.2-mm-thick 20 mol% Zr-doped (Gd, Y)BaCuO thin film showing well-aligned $BaZrO_3$ nanocolumns throughout the film thickness; **b** magnetic field dependence of critical currents of a 2.2 mm GdBCO tape with 20% mol addition of Zr and the best 0.9-mm-thick GdBCO tapes with 7.5%, 15%, and 25% addition at 30 K in the orientation of filed parallel to the *c*-axis. (Both figures reproduced with permission from [6])

## $BaSnO_3$ (BSO) Nanocolumns

The $BaSnO_3$ (BSO) nanocolumns introduced into YBCO were the next try for finding a better nanocomposite superconducting film. Numerous works related to the introduction of the relatively new dopant into YBCO were published in the literature [3, 114–122]. The rather dense and straight in shape BSO nanorods are easily formed in YBCO films. Because the lattice mismatch of BSO compound with YBCO is better than that of the BZO, deterioration of $T_c$ of the films due to the presence of non-superconducting species in the superconducting matrix is moderate. As a result, the YBCO thin films with BSO nanorods are found to have excellent pinning properties.

Historically, first reports on the BSO-added YBCO were published by Varanasi et al. [114–118]. They used a pie-shaped YBCO target, removing a 30° YBCO sector and substituting it with a BSO sector. The periodical ablation in PLD chamber of BSO sector generated BSO nanorods parallel to the *c*-axis of the film (Fig. 2.14a). The angular dependence of $J_c$ shows wide peak with a maximum at *B*//c direction, which is the signature of *c*-axis-correlated pinning (Fig. 2.14b).

The YBCO-BSO system was extensively studied by Mele et al. [117] proposing systematic increment of the BSO content by means of mixed YBCO-BSO targets. Same as in the case of BZO, the perovskite phase is incorporated in the YBCO film with a shape of nanorods, independently from the way used to add the second phase to the target (pie-shaped or mixed).

Figure 2.15 shows the morphology and $F_p(B)$ characteristics of YBCO+BSO films with systematic increment of BSO content. It is clear that concentration and diameter of the nanorods increase with the BSO content in the target. The highest $F_p^{max}$ of the 4wt% BSO-doped YBCO film is 28.3 GN/m$^3$ (77 K, *B*//c), and it corresponds to about twice as much compared to the BZO-doped films of the same

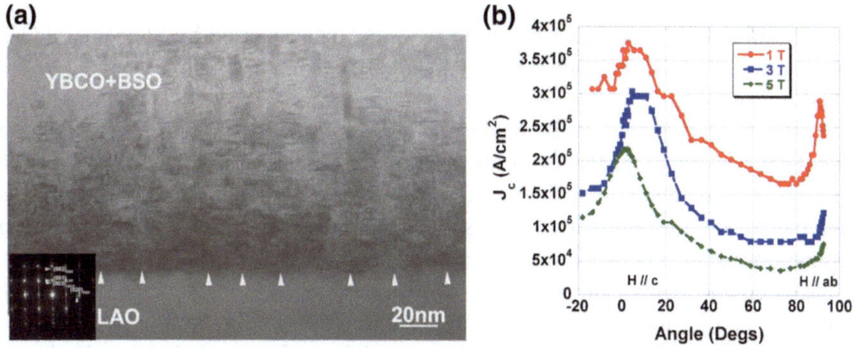


**Fig. 2.14** **a** Cross-sectional TEM image of an YBCO+BSO sample on a $LaAlO_3$ substrate. The particles were identified to be $BaSnO_3$ in the selected area diffraction pattern. Nanocolumns extending through the thickness are marked by arrows; **b** angular dependence of $J_c$ values of YBCO+BSO sample on a $LaAlO_3$ substrate at various applied fields of 1, 3, and 5 T at 77 K. (Both figures reproduced with permission from [115])

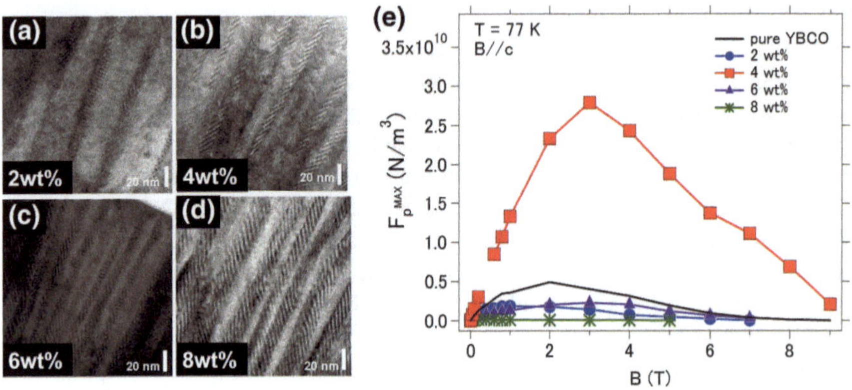


**Fig. 2.15** TEM cross-sectional images of YBCO films incorporated with increasing amount of BSO: **a** 2wt%, **b** 4wt%, **c** 6wt%, and **d** 8wt%; **e** global pinning force $F_p$ ($T = 77$ K, $B$//c) for YBCO films added with increasing content of BSO (2, 4, 6, 8 wt%). (Panels **a**~**d** are reproduced with permission from [117] with slight modifications)

**Table 2.1** Physical parameters, calculated $J_c$ and measured $J_c$ (0 T, 77 K) for YBCO+BSO films

| $BaSnO_3$ wt% | 2 | 4 | 6 | 8 |
|---|---|---|---|---|
| APCs ($m^{-2}$) ($\times 10^{-15}$) | 0.91 | 2.30 | 5.10 | 6.80 |
| $d$ (nm) | 30 | 21 | 14 | 12 |
| $T_c$ (K) | 88.68 | 88.57 | 86.46 | 83.08 |
| $C_0$ (nm) | 3.8 | 4.28 | 4.3 | 5.5 |
| $U_0$ (N) ($\times 10^{-12}$) | 1.01 | 1.4 | 0.84 | 0.59 |
| Calculated $J_c$ | 12 | 16.3 | 9 | 5 |
| Measured $J_c$ | 0.89 | 1.64 | 0.54 | 0.23 |
| $J_c$ calc/$J_c$ meas | 13.5 | 9.94 | 16.7 | 22 |
| Eff. Curr. blocking | 16% | 37% | 50% | 71% |

concentration of 4 wt% BZO [3]. Analysis of the data reported by Mele et al. [117] for a series of YBCO+BSO thin films and the calculations of $J_c$ using (2.5) are listed in Table 2.1.

From the data displayed in Table 2.1, it is obvious that the measured and the calculated values extracted from all the samples follow a similar trend with the concentration of BSO, with a maximum for 4wt% BSO-added sample. This means that increasing the amount of APCs, which lead to the reduction of the mean separation distance between the columns, does not increase $J_c$ indefinitely. Instead, it has been shown that the enclosure of too many defects in the superconducting matrix has a detrimental effect since the resulting current-blocking effect stifles or thwarts the movement of vortices. In this regard, it can be concluded that the 4 wt% BSO content represents the best compromise between the amount of APCs and the current flow hindrance that primarily stems from the effect of current blocking. To further justify this argument, qualitative as well as quantitative analysis of the samples results from this series is given below.

It is known that the pinning force per unit length of the pinned flux line is proportional to the work done to interchange the flux line from a point to another point where the flux pinning is smaller. In this case, the equation of proportionality is written as $F_p\ \alpha\ \Delta WL/x$, where $\Delta W$ being the work done to transfer the flux line to the new point, $L$ is the length of the flux line, and $x$ is the extent of pinning interface. The pinning force density $F_p$ is then customarily calculated from the temperature and/or field dependence $J_c(T, B)$ data according to the function $F_p = J_c \times B$. As shown in Fig. 2.8a, the behavior of the field-dependent $F_p(B)$ at $B//c$ direction is depicted for the 1D inclusion of BSO+YBCO samples with varying BSO concentration, as listed in the legend. Data from the undoped pristine YBCO sample and a standard (NbTi, measured at 4.2 K) sample are also shown for comparison. To elucidate additional information from this graph, Fig. 2.15e shows further split into two more figures as presented in Fig. 2.16a, b, respectively. Figure 2.16 depicts a comparative plot of the global pinning force density $F_p$ as a function of the reduced field $b$ for four different YBCO films engineered with four different 1D nanorod additives. As evident from the figure, the 4 wt% BSO-added YBCO film has clear superior pinning performance, both in the maximum pinning value of about 28.30 GN/m$^3$ and in the position of this maximum at about 0.4 reduced field. Since the effectiveness of the pinning force in all the four samples is dominated by the 1D columnar pins due to the presence of these four types of 1D ingredient, we can safely conclude that the 4 wt% BSO-added YBCO sample has a better control of the vortex pinning force associated with 1D pinning centers.

The scaling behavior of the 4 wt% BSO-YBCO sample is further fitted to the Dew–Hughes like equation $F_p = Ab^m(1\text{-}b)^n$ where $A$ is a scaling constant and $m = 0.5$ and $n = 2.5$ are fitting parameters. The original Dew–Hughes model roughly gives $m = 1$ and $n = 2$ and the maximum $F_p^{max}$ at $b \approx 0.25$ and 0.35 for either 1D or 3D point pinning. It should be noted here that the parameters $m$ and $n$ may be necessary for fitting purposes, but they might not be sufficient for the determination of the exact type of pinning associated with the samples. Important information about the dominant pinning mechanism, however, may be extracted

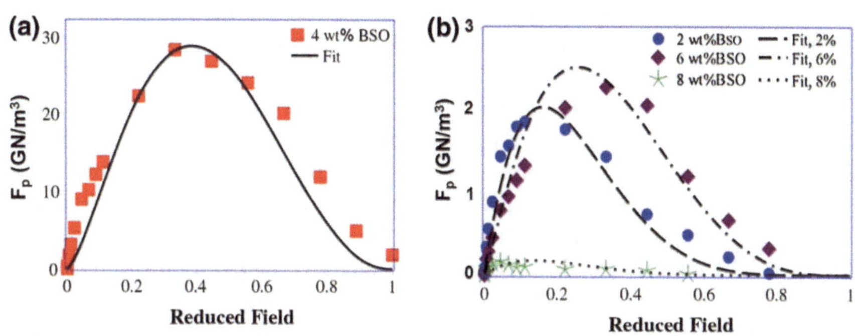


**Fig. 2.16** Global pinning force $F_p$ as a function of reduced field for **a** the 4 wt% BSO+YBCO sample and **b** the rest of the BSO-doped YBCO series of samples (2, 6, and 8 wt%). Ginzburg–Landau related fitting of the experimental data is described in the main text

from the pinning curve when vertically extrapolated down to the reduced field axis. As demonstrated in Fig. 2.16a, b, the BSO+YBCO set of samples provides different maxima reflecting the relative strength of pinning force related to their $F_p(B)$ curves and the amount of the 1D species as well. Had it been for a single pinning mechanism produced by the BSO doping with different percentage in the samples, the scaling behavior of the preceding equation should have coincided with the maximum pinning from the experimental data. But this does not seem to be the case here. Accordingly, there may at least be two different varieties of pinning force mechanisms competing for a dominant role in this set of samples. Maximum pinning for the 2 wt% BSO, the 6 wt% BSO, and the 8 wt% BSO samples corresponds to $b \approx 0.15$, $b \approx 0.31$, and $b \approx 0.035$, respectively. It can be deduced from these values that the 2, 6, and 8 wt% BSO concentration on the YBCO films acts only as uncorrelated defect point pinning with which each sample has an $I_c(B)$ value and an $F_p(B)$ value that is less compared to that of the virgin YBCO sample. This assertion is valid for the data shown in Fig. 2.16b. A reduced field $b \approx 0.4$ is, however, being verified for the 4 wt% BSO-added YBCO sample (Fig. 2.16a), a value closer to 0.7 where pinning mechanisms in a superconductor are dominated by the variation in the order parameter $\delta l$. We believe that the 4 wt% BSO contents have, to a certain degree, resulted in different reaction between its non-superconducting precipitates and the superconducting matrix. From Fig. 2.16a, b, it is conceivable to distinguish two types of pinning behaviors accompanying this sample: The highest $I_c$ recorded for the 4 wt% BSO+YBCO sample among this series may partially be due to well-arranged nanoscale defects resulted from the concomitant crystal growth during PLD deposition of the film using a mixed target. These correlated defects later contribute to $\delta l$ pinning mechanism which is responsible for the enhancement of $I_c(B)$ in the low-field regime. The second type of pinning mechanism that can be associated with the behavior of this sample is $\delta T_c$ pinning which is commonly known to be the necessary precursor that enriches the value of $I_c(B)$ at high-field regime.

In the mixed state of a type II superconductor, flux pinning has a substantial effect in resisting the entry and exit of vortex flux into the material. It is apparent that the 2, 6, and 8 wt% BSO concentrations into the YBCO have created disjointed defects of sizes not in tandem with the superconducting order parameter or particularly the penetration depth, $\lambda$.

In such a situation, the unfavorable effect of the impurity sites (i.e., weak pinning centers) pushes the vortex flux lines to assume a distortive energy state. For the 4 wt% BSO sample, where pinning centers are strong, although the flux vortex lattice is little distorted, it is expected that each vortex flux line is individually pinned. It was reported that the vastly improved $I_c(B)$ and $F_p(B)$ properties in the 4 wt% BSO sample were mainly due to the effective length of the BSO defects (nanorods) which contributed to a positive superconducting energy exchange between the flux pinning mechanism and the flux density from the externally applied magnetic field [7, 117].

Figure 2.17a shows the angular dependence of the transport critical current $I_c(\theta)$ at 77 K for the 4 wt% BSO+YBCO sample. The measurements were performed

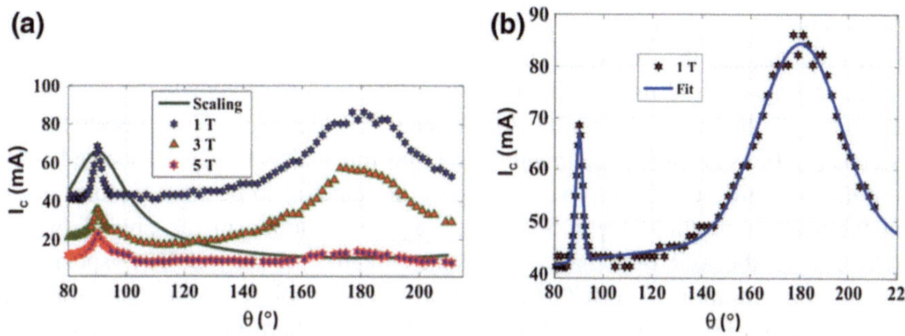


**Fig. 2.17** $I_c(\theta)$ dependence for 4 wt% BSO+YBCO sample. **a** Data at 77 K and 1, 3, 5 T, with a draw of a Ginzburg–Landau scale to indicate the estimate of isotropic pinning line; **b** replot of (77 K,1 T) data fitted by a Gaussian expression

under magnetic fields of 1, 3, and 5 T, respectively. The field orientation angle $\theta$ points to the direction of the applied magnetic field with respect to the *c*-axis of the film.

With the aim to better understand the cause of the flux pinning mechanisms which is the main cause of driving the vortex pinning force density of a high $T_c$ superconductor in a magnetic field, it is of crucial importance to discuss the angular dependence $I_c(\theta)$ data taking into consideration the response of the sample to a varying field values across a wide range of the measuring angles. We shall start the discussion from the viewpoint of the Blatter model [33, 34]. In a YBCO superconductor although generally regarded as less anisotropic, the $I_c(\theta)$ can be found to have strong dependence on the orientation of the applied magnetic field with respect to the crystallographic axis directions. Indeed, this was the case when the APCs are accommodated in the pristine YBCO films so as to improve the pinning potential in the material. As viewed from Fig. 2.17a, the anisotropy of the crystallographic axes *a* and *b* with their level located along the cuprate plane can show sharp but relatively small peak at $\theta = 90°$ when the applied field *B is parallel to ab* plane (*B//ab*). The physical meaning of the isotropic or intrinsic pinning of this orientation is that the vortices revolving parallel to the *ab* plane are strongly pinned by the presence of point defects such as 1D- and 3D-APCs, and the weakly superconducting layers between the $CuO_2$ planes. The physical properties of the isotropic contribution to the overall pinning landscape is therefore largely explained by the Blatter scaling approach for the vortex flux line in the anisotropic Ginzburg–Landau (GL) theory. The Blatter model has shown that the anisotropy of a superconductor can have an anisotropic parameter, ε:

$\varepsilon = \frac{1}{\gamma} = \frac{\lambda_{ab}}{\lambda_c} = \frac{\xi_c}{\xi_{ab}} = \sqrt{\frac{m_{ab}}{m_c}}$ where $m_{ab}$, $m_c$ and $\lambda_{ab}$, $\lambda_c$, $\xi_{ab}$, $\xi_c$ are the mass tensor and the anisotropic superconducting parameters in the *a*, *b*, and *c* crystallographic directions. The effective magnetic field in play in this case is given by $B(\theta) = \varepsilon_\theta B$. The Blatter scaling model can then be given for anisotropic superconductor as $\varepsilon_\theta = \sqrt{sin^2\theta + \varepsilon^2 cos^2\theta}$ with $\theta$ being the angle between the applied field $B$ and the *ab* crystal plane.

The scaling method used to produce Fig. 2.17a can hardly scale the seemingly isotropic contribution of the 4 wt% BSO-doped YBCO sample. The reasons for the discrepancy between the experimental data and the scaled fit can be viewed from the points that the scaling approach assumes the material to be strictly isotropic so that the scaling coefficient $\gamma \approx 5$–7. The reason mentioned in the preceding line could lead to the assertion that the pinning behavior of the 1D BSO species of existing concentration in YBCO film is also not isotropic enough apart from the narrow window in the vicinity of 85° $\leq \theta \leq$ 105° where pinning mechanism is due to random defects or uncorrelated disorder. However, the scaling approach has no answer to the huge and rather extended peak centered around $\theta = 180°$. Moreover, the angular-dependent $J_c(\theta)$ for YBCO films deposited on different substrates shows that it is extremely difficult to account for the combined effect of random defects and mass anisotropy when the $J_c(\theta)$ data are mapped across a wide angular range. It was concluded that the effect of correlated pinning structure in the samples with the maximum at both $\theta = 90°$ and $\theta = 180°$ cannot be accounted for by the anisotropic scaling [102].

Since the data presented in Fig. 2.17a are highly anisotropic across the advancing angular range, we use a Gaussian statistical expression to account for the enormous pinning effect of the 4 wt% BSO nanocolumns in the vortex flux pinning mechanism of this YBCO sample. Figure 2.17b shows the model fit to the 1 T $I_c(\theta)$ data.

The Gaussian statistical expression used to fit the data shown in Fig. 2.1 can be written as

$$I_c(\theta) = I_{(c(0))n} \exp \sum_{n=1}^{3} \left( - \left( \frac{\theta - \beta_n}{\delta_n} \right) \right)^2 \tag{2.7}$$

where the numerical coefficients for the quantities appearing in the (2.6) are shown in Table 2.2.

In the fit shown in Fig. 2.17b, the $\theta$ values across the angular range are normalized by means of 131.4° where the coefficients are 95% confidence bound and the goodness of fit was $R^2 = 0.99$. From the data shown in Table 2.2, it is clear that the data of the sample are divided into three regions, the isotropic peak, the plateau, and the enormously huge anisotropic peak at the far right of the curve. Unlike the original anisotropic GL model where the isotropic peak shows elliptical symmetry and a concave pinning plateau elsewhere, the statistical estimation using (2.7) has at

**Table 2.2** The angular quantities used in fitting the $I_c(\theta)$ for the 1 T field shown in Fig. 2.16

| Parameter | n = 1 | n = 2 | n = 3 |
|---|---|---|---|
| $I(0)_n$ | 38.4 | 24.33 | 46.03 |
| $\beta_n$ | 1.16 | −0.99 | 1.49 |
| $\delta_n$ | 0.56 | 0.05 | 8.40 |

least roughly accounted for the effect of the 1D pinning on the vortex configuration feature of the sample measured at 1 T.

Indeed, the addition of 4 wt% BSO into YBCO sample has contributed to better isotropic feature at a narrow angular range, as characterized by a sharp peak centered at $\theta = 90°$. The anisotropy of $I_c(\theta)$ in the intermediate region is rather short lived but shows a horizontal feature just before the $c$-axis anisotropy dominates.

### $BaHfO_3$ (BHO) Nanocolumns

Recently, $BaHfO_3$ (BHO) nanorods became trendy since BHO has been found to give strong pinning without significant depression of the $T_c$ of the superconducting matrix. Tobita et al. [119] first reported that a BHO-doped $GdBa_2Cu_3O_y$ (GdBCO) film is deposited by PLD on IBAD-MgO substrate. The most interesting feature of BHO nanocolumn addition was reported as $J_c$ is undepressed by increasing thickness of the film (Fig. 2.18).

The SmBCO/BHO system was extensively studied in Nagoya University by Y. Yoshida group [5, 8, 124–126]. Low-temperature growth (LTG) SmBCO/BHO films show striking performance with $F_p^{max} = 28$ GN/m$^3$ and $B_{irr} = 15$ T [8] as a result of the unique linearity and the narrow angle distribution of the BHO nanorods.

Very recently, it was found that BHO-added SmBCO films prepared by LTG method present large values of $F_p$ below 77 K. As shown in Fig. 2.19, maximum flux pinning forces ($F_p$) of 105 GN/m$^3$ in 9.0 T at 65 K and 405 GN/m$^3$ in 9.0 T at 40 K have been obtained thanks to the high density and small size of the nanorods embedded in the superconducting matrix [5]. Furthermore, the increase in self-field $J_c$ of GdBCO-BHO films deposited on single crystals with artificial grain

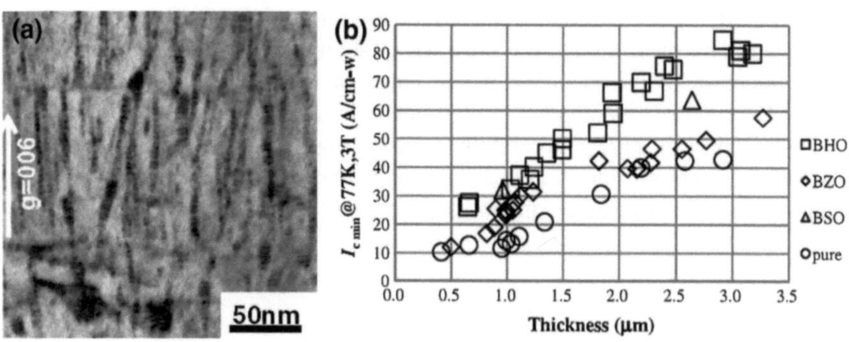


**Fig. 2.18** **a** Cross-sectional TEM images of GdBCO-coated conductors incorporating BHO nanorods. **b** Thickness dependence of $I_c$ at 77 K and 3 T for BHO-doped GdBCO CCs compared with BSO- and BZO-doped GdBCO CCs and pure GdBCO CCs. (Figures reproduced with permission from [123])

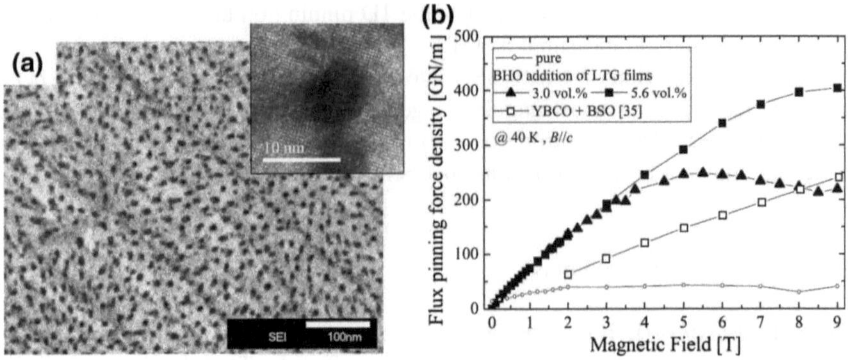


**Fig. 2.19** **a** Plan-view TEM images of SmBCO films with 4.1 vol.% BHO content, fabricated with alternating target technique by pulsing the BHO target 3 times per cycle. **b** Magnetic field dependence of flux pinning force density $F_p$ at 40 K for 0 (pure), 3.0, and 5.6 vol.% BHO-doped $SmBa_2Cu_3O_y$ LTG films, and a $BaSnO_3$-doped $YBa_2Cu_3O_y$ film reported by Mele et al. [3]. (Figures reproduced with permission from [9] with slight modifications)

boundaries by flux pinning of the Abrikosov–Josephson vortices by BHO nanorods [126] was demonstrated. All the results described previously were obtained on single crystals. The deposition of GdBCO+BHO has been tried on metallic substrates as well [9], obtaining very large $F_p^{max} = 23.5$ GN/m$^3$ and outstanding value of irreversibility field $B_{irr} = 15.8$ T.

Other Oxide Nanocolumns

Other than perovskite BMO (M = Zr, Sn, Hf) nanorods, Harrington et al. [127] reported excellent pinning performance at 77 K via the incorporation of very stable $RE_3TaO_7$ (RE = Er, Gd, Yb) pyrochlore nanorods in YBCO PLD films without significant depression of $T_c$. The presence of additional nanoparticles (Fig. 2.20a) yields increase of the $J_c$ in the whole range of angles, while the effect of the nanorods becomes predominant at high fields (Fig. 2.20b, c).

Feldmann et al. [128] successfully added a high density of double perovskite $Ba_2YNbO_6$ nanorods into PLD thin films, and substantially, enhanced pinning was achieved in their sample: maximum global pinning forces of 32.3 GN/m$^3$ (75.5 K) and 122 GN/m$^3$ (65 K).

### 2.3.3 Two-Dimensional APCs (2D-APCs)

Two subcategories of 2D-APCs can be mentioned: (i) nanolayers and (ii) grain boundaries

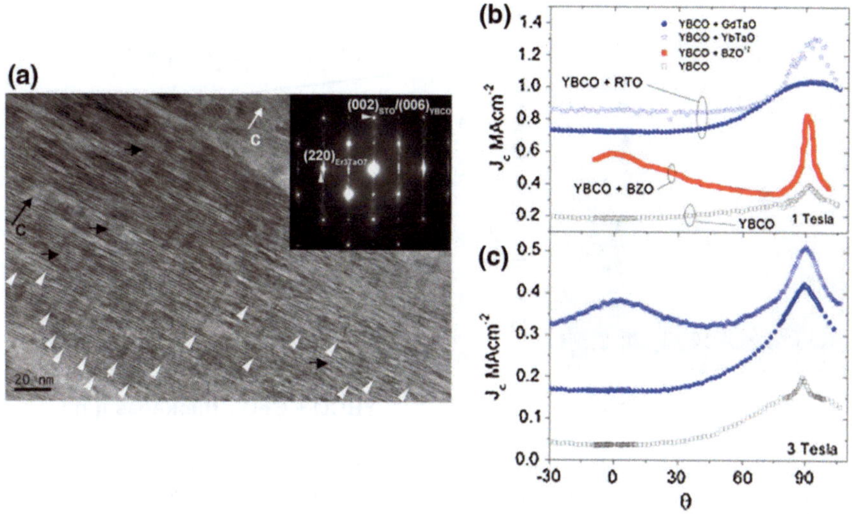


**Fig. 2.20** *Left* **a** Microstructure of a typical YBCO+RTO film. TEM cross-sectional image of YBCO+5 mol% ErTaO film grown at 765 °C and 10 Hz, inset, proves the structure of the rare earth tantalate phase is the pyrochlore $Er_3TaO_7$. Vertical arrows show positions of the self-assembled nanorods, and horizontal arrows show the positions of some of the random nanoparticles. *Right* Angular dependence of the critical current density at 77 K for standard YBCO, YBCO+BZO (~7 mol%) and YBCO+RTO films: **b** 1 T and **c** 3 T. (Figures reproduced with permission from [127] with slight modifications)

#### 2.3.3.1 Nanolayers as 2D-APCs

As understood in the early 1990s [135, 136], the critical current of thin films does not increase indefinitely by increasing thickness. The reason [137–142] is that the quality of thin film, and hence $J_c$, is best near the substrate, and as the film is grown thicker, current-blocking defects (misoriented crystallites, cracks or voids) become more prevalent, depressing the critical current.

In their remarkable works, Foltyn et al. [38, 143] proposed a revolutionary way to have thick YBCO films increased critical current: Six YBCO-$CeO_2$ layers were deposited on buffered metal substrates by PLD, resulting in a 3-μm-thick film where critical current exceeded 1000 $A/cm^2$. It was thus demonstrated [38, 143] that by periodically inserting 30-nm $CeO_2$ layers every 0.5 μm or so in a thick YBCO film (Fig. 2.21a), a high critical current density is maintained. As shown in Fig. 2.21b, single-layer coatings do not reach 600 A/cm width unless the thickness is more than 4 μm. However, six 0.55-mm-thick YBCO layers, separated by 40-nm-thick $CeO_2$ layers, have a $J_c$ of 4 $MA/cm^2$—nearly that of a single 0.55-mm layer. The key is in the multilayers' ability to preserve thin-film $J_c$ values, as indicated by the arrow (Fig. 2.21). This amazing result marked a significant advance in coated conductor technology.

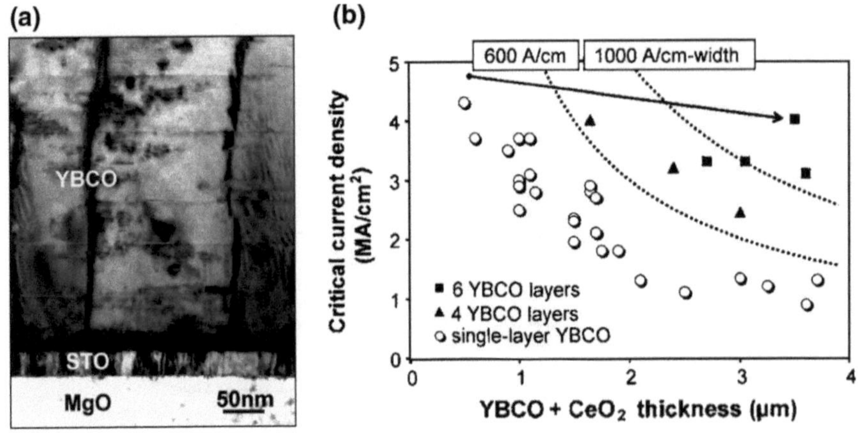


**Fig. 2.21** **a** Transmission electron microscope image of a pulsed laser deposition YBCO+$CeO_2$ multilayer on a single-crystal MgO substrate with a $SrTiO_3$ buffer (**b**). Comparison of single-layer and YBCO/$CeO_2$ multilayer films. (Reproduced with permission from [143])

After this breakout, research on multilayers flourished and a variety of multilayered and quasi-multilayered structures have been tried: $Y_2O_3$/YBCO [144, 145], YBCO/DyBCO [146, 147], YBCO/EuBCO [148], YBCO/NdBCO [149], $Y_2BaCuO_5$ (Y-211)/YBCO [150], YBCO/YSZ [151], $SrRuO_3$/YBCO [152], YBCO+$BaZrO_3$/$CeO_2$ [153], YBCO/$BaZrO_3$ [154], YBCO/$SrTiO_3$ [155], YBCO/$LaCaMnO_3$ [156] YBCO/ZnO [157], $Sm_{1.04}Ba_{1.96}Cu_3O_y$/$Sm_{1.08}Ba_{1.92}Cu_3O_y$ [158], and so on. The general common effect of multilayer approach is an enhancement of $J_c$ in magnetic field, sometimes combined with weak *c*-axis-correlated pinning.

All these structures involve *c*-axis-oriented YBCO layers stacked with different kind of oxides. Horii et al. [159] proposed a multilayered structure alternating *a*-axis-oriented YBCO [160] and semiconducting PrBCO interlayers. The PrBCO layers act as 2D-APCs. The consequence is a weakening of the intrinsic pinning along the *ab* plane and enhancement of the *c*-axis-correlated pinning [161, 162].

### 2.3.3.2 Grain Boundaries as 2D-APCs

The grain boundaries in the HTS show the Josephson-like behavior when the misorientation angle of adjacent grains becomes ten degrees or more. This causes a weak coupling due to which $J_c$ rapidly decreases at the grain boundaries. However, the small-angle grain boundary can be considered as a dislocation array, and there is a possibility to act as strong pinning center when the misorientation angle is small. This is one of the natural pinning previously mentioned. Matsumoto et al. [163] have reported the use of grain boundaries as effective pinning centers in GdBCO films. By varying the oxygen partial pressure during deposition, the size of the grains continuously changed from 196 to 92 nm. $J_c$ and $F_p^{max}$ increase with the reduction of the grain size, suggesting that the (001) tilt grain boundaries act as pinning centers in the film and that the improvement of $F_p$ appears because the density of pinning centers increases as the grains become smaller. The maximum $F_p$ for the sample of which grain size is 92 nm is about 15 GN/m$^3$ at 77 K, *B*//*c*, approaching the value of NbTi at 4.2 K.

Another type of grain boundary, in this case as 2D artificial pinning center, comes from the catalytic effect of silver during the growth of YBCO as was demonstrated in A. Crisan's group [164, 165]. It was shown that Ag nanodots (in substrate decoration and/or quasi-multilayer approach) have a twofold beneficial effect: (i) allow the growth of much thicker (up to 5 μm) YBCO films without a very strong decrease in the critical current density (thus increasing the critical current) and (ii) promoting a columnar growth of YBCO. The surface of the cylinders, which in the case of BZO-doping is also entangled with BZO nanorods, provides very strong pinning centers. Their 2D character was demonstrated by a Dew–Hughes-Like analysis of the reduced field dependence of the global pinning force [166].

### 2.3.4 Three-Dimensional APCs (3D-APCs)

For practical applications of YBCO superconductors, there is a very desirable feature which is the isotropic character of $J_c$ with high values under a wide range of applied magnetic field under various orientations. Therefore, considerable efforts have been made to prepare superconducting thin films incorporating 3D-APCs.

Pioneering work on 3D-APCs addition was published in 2004 by Haugan et al. [10] with incorporation of disk-shaped Y211 nanoparticles in YBCO film by PLD (Fig. 2.22a). The Y211 addition clearly shows improvement of $J_c$ in magnetic fields due to enhanced pinning (Fig. 2.22b).

Extensive work on the introduction of 3D-APCs in YBCO films has been made by several groups using TFA-MOD method. It was reported that the highly densified nanoparticles can be introduced into the YBCO films by mixing foreign elements of proper quantities with the starting solution including Y, Ba, and Cu. In the BZO case, nearly isotropic angular dependence of $J_c$ is reached [167] with $F_p^{max}$ of 22 GN/m$^3$ (77 K, *B*//*c*) [168]. At its time, this result from X. Obradors group (ICMAB Barcelona) was the best pinning performance reached with isotropic APCs (Fig. 2.23a, b). Other effective isotropic pinning centers added to YBCO by MOD process are $Y_2O_3$, $BaCeO_3$, and $Ba_2YTaO_6$ [169]. It is worth noting here that the round-shaped particles dispersed in the superconducting film by MOD are randomly oriented because there is no preferred crystallographic orientation between the superconducting matrix and the foreign phase.

PLD process was also extensively used to incorporate 3D-APCs in superconducting thin films. First of all, a Sm-rich SmBCO target, with composition $Sm_{1+-xBa2-x}BaCuO_y$, was used for ablation in the PLD process [170–174]. Low $T_c$ 3D

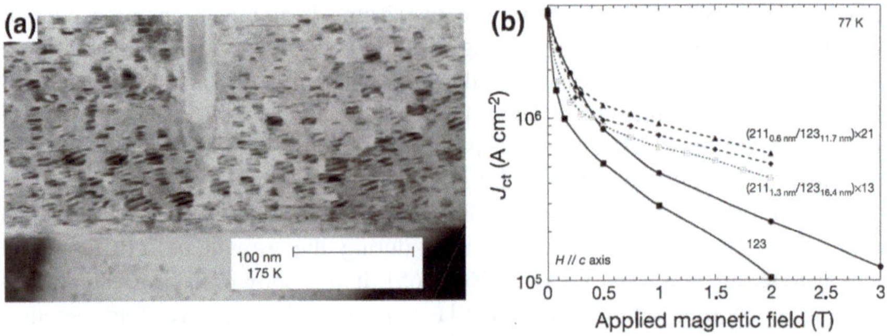


**Fig. 2.22** **a** Transmission electron micrographs of YBCO+Y211 films, showing repeated layering structure and nanoparticle formations; **b** critical current density as a function of applied magnetic field for YBCO+Y211 films compared to pure YBCO films (*solid lines*) prepared by PLD (*filled squares*) and MOD (*filled circles*). Multilayer films (*dashed lines*) were measured for different Y211 and YBCO parameters, and for samples from the same deposition run (*triangles and diamonds*). (Reproduced with permission from [10])

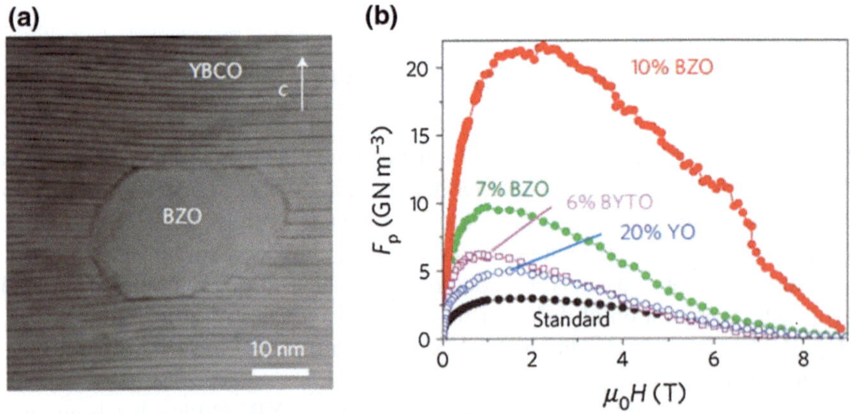


**Fig. 2.23** **a** High-magnification cross-sectional TEM micrograph of a BZO nanoparticle embedded in YBCO thin film grown on STO; **b** pinning force versus magnetic field at 77 K in a series of YBCO MOD films added with several kinds of nanoparticles. (Both panels reproduced with permission from [168])

nanoparticles are randomly dispersed in the superconducting matrix due to the local compositional fluctuation, and significant improvement in the pinning properties has been achieved. Very high values $J_c = 0.37$ MA/cm$^2$ (77 K, 5 T, $B//c$) and $F_p^{max} = 23$ GN/m$^3$ (77 K, $B//c$) were reported for the first time in 2006. Additionally, $B_{irr}$ of the SmBCO films reached very high values, up to 12 T (77 K, $B//c$).

Alternatively, the introduction of nanoparticles into YBCO film was made by switching the superconducting target and the dopant target in an alternating manner. Under selected experimental conditions, instead of multilayered structure with separation between the two phases, nanoparticle structure embedded into the film was obtained. The 3D-APCs incorporated were Y211 [175–177], $Y_2O_3$ [178–182], perovskites ($BaIrO_3$ [183], and $BaHfO_3$ [184]). Metallic elements such as Ag, Au [185, 186], and $Gd_2Ba_4CuWO_y$ compound [187] were consequently employed.

A third method that allows the creation of 3D-APCs into the superconducting films was developed by using the target surface modification method [188, 189]. In this method, a small piece of the sintered APC material such as $Y_2O_3$ is placed onto the YBCO target and together, and they are used for the usual PLD deposition process. This technique has the merit to introduce the nanostructure into epitaxial thin films through a very simple procedure.

$Y_2O_3$ nanoparticles are incorporated in YBCO matrix and show epitaxial relationship with YBCO: (001)YBCO//(001)$Y_2O_3$, and [100]YBCO//[110]$Y_2O_3$ crystallographic orientation. That is, the *a*-axis of the $Y_2O_3$ nanoparticles is oriented 45° with respect to the YBCO *a*-axis, while the same out-of-plane (*c*-axis) orientation is apparently preserved in the hybrid structure. In YBCO films where 3D-APCs are incorporated, additional defects such as stacking faults are also present, most probably playing the role of additional pinning centers [178, 189] (see Fig. 2.24a). The $Y_2O_3$ particles are homogenously dispersed into YBCO matrix, and consistently, the $J_c$ angular dependence does not show any peak at *B*//*c* orientation (Fig. 2.24b).

More recently, Xu et al. [11] proposed a heat treatment-based method via direct ablation of pure YBCO target at high temperature of 830°C instead of the usual 800 °C used by other workers. Without any addition of secondary phase onto the target or surface modification, dispersed $Y_2O_3$ particles are obtained due to different kinetic energy during the growth.

Mele et al. studied systematically the effect of increasing additions of $Y_2O_3$ in YBCO thin films [189]. Similar to BSO compound, the $Y_2O_3$ oxide addition increases the amount of APCs according to the size of the area of the sliced piece

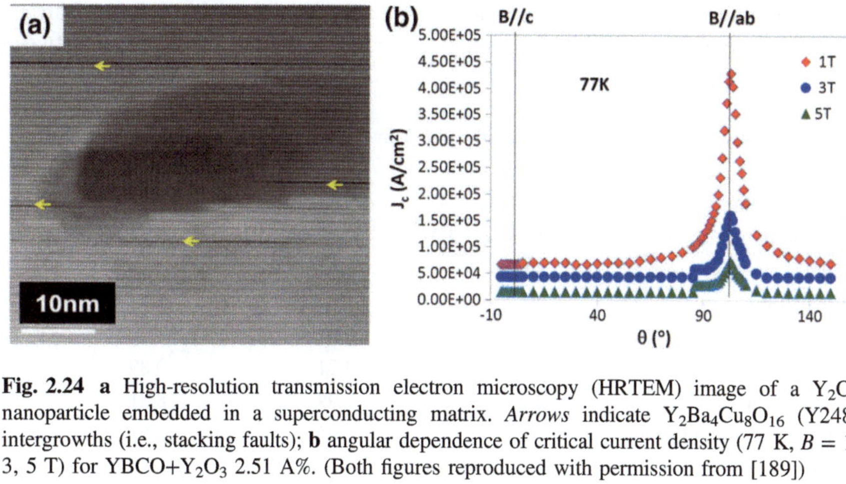


**Fig. 2.24** **a** High-resolution transmission electron microscopy (HRTEM) image of a $Y_2O_3$ nanoparticle embedded in a superconducting matrix. *Arrows* indicate $Y_2Ba_4Cu_8O_{16}$ (Y248) intergrowths (i.e., stacking faults); **b** angular dependence of critical current density (77 K, $B = 1$, 3, 5 T) for YBCO+$Y_2O_3$ 2.51 A%. (Both figures reproduced with permission from [189])

placed onto the superconducting target, but the resulting critical currents and pinning force density were not always proportional to $Y_2O_3$ contents, as deduced from the study of the whole series of the samples [189]. For example, the 9.22A% $Y_2O_3$ film presents a global pinning values lower than the standard YBCO film, while the 5.44 A% $Y_2O_3$ sample produced a global pinning force of 14.3 GN/m$^3$ at 77 K, twofolds the value obtained in the 2.51A% $Y_2O_3$ sample which is comparatively similar to the global pinning of NbTi tape at 4.2 K [190]. This value is also very close to the value obtained for the YBCO films with $BaZrO_3$ nanorods, but with a much less depression in the superconducting critical temperature.

Nanoparticle defects such as Y211 and $Y_2O_3$ are experimentally proven to increase the strength of pinning centers so that vortices surrounding them can bend to create discrete trapped segments [191–193]. In this context, it is foreseeable to estimate the maximum critical current $J_c$ from the depinning of curve segments which are fixed by two consecutive nanoparticles separated by a distance $d$. The analytical expression resulting from the above argument can be written as

$$J_{c,\max} = \frac{\Phi_0}{4\pi\mu_0\lambda_{ab}\lambda_c d}\ln\frac{d}{\xi_c} \tag{2.8}$$

where $\Phi_0$ is the flux quanta, $\mu_0$ is the magnetic permeability, $\lambda_{ab}$ and $\lambda_c$ are the London penetration depths in the $ab$ plane and along the $c$-axis, respectively, and $\xi_c$ is the coherence length along the $c$-axis. Analysis of the data reported by Mele et al. [189] for a series of YBCO+$Y_2O_3$ thin films and the calculations of $U_0$, $J_c$ by using (2.8) are listed in Table 2.3.

Parameters extracted for YBCO at 77 K are $\lambda_{ab} = 0.35$ μm, $\lambda_c = 2.4$ μm, $\xi_c = 0.6$ nm, and the mean separation between $Y_2O_3$ particles $d \sim 18$ nm, as approximated from TEM images [189] for $Y_2O_3$ 5.44A% sample, and the estimated $J_c$ (77 K, self-field) is 6.40 MA/cm$^2$. The relative consistency with the experimental value of 3.21 MA/cm$^2$ demonstrates that $Y_2O_3$ particles act as good pinning centers. Indeed, the effective current blocking is only 13% of $J_c$ and therefore no much of a hindrance for the supercurrent to percolate avoiding the pinning barriers (normal cores). For the 9.22A% sample, the mean separation distance between

**Table 2.3** Physical parameters, calculated $J_{c,}$ and measured $J_c$ (0 T, 77 K) for YBCO+$Y_2O_3$ films

| $Y_2O_3$ content | 2.51A% | 5.44A% | 9.22A% |
|---|---|---|---|
| Amount of APCs (m$^{-2}$) | $1.75 \times 10^{15}$ | $2.68 \times 10^{15}$ | $5.23 \times 10^{15}$ |
| $T_c$ (K) | 89.26 | 89.2 | 87.78 |
| $d$ (nm) [36] | 20 | 18 | 12 |
| Calculated $J_c$ | 5.93 | 6.4 | 7.4 |
| Measured $J_c$ | 2.62 | 3.21 | 0.89 |
| $J_c$ calc./$J_c$ meas. | 2.26 | 1.99 | 8.31 |
| Effective current blocking | 10% | 13% | 32% |

particles is estimated to be $d \sim 12$ nm. Hence, current calculation has given $J_c$ (77 K, self-field) = 7.40 MA/cm$^2$, while the experimental value is merely 0.89 MA/cm$^2$. The depression of $J_c$ in the 9.22A% sample may be due to the presence of a larger density of non-superconducting particles such as Y248 intergrowth compound which contribute to about 32% current blocking. Interestingly enough, the presence of Y248 intergrowths can play two opposite roles in the superconductivity of the sample. While large sizes of the Y248 intergrowths can result in a poor superconducting material, they may also act as good pinning centers provided that their sizes are comparable to the coherence length. However, in the case of this particular sample, Y248 is present in the form of disks situated perpendicular to *c*-axis of the film with lengths of several hundreds of nanometers which in turn impede the motion of vortices.

It is worth noting that the pinning performance of the nanorods at 77 K is superior with respect to the $Y_2O_3$ nanoparticle doping. However, in the case of eventual decrease in $T_c$ due to the spherical inclusion of foreign species, the pinning performance of nanorods considerably deteriorates. This behavior was first observed by Mele et al. [188] as it was later confirmed in recent years in the work of Xu et al. [11]. They reported detailed analysis of YBCO+$Y_2O_3$ films in a wide range of operating temperatures. Their findings showed that at 4.2 K and 16 T, the $Y_2O_3$-YBCO film attains $F_p$ = 1000 GN/m$^3$, about 30% higher than the 7.5% Zr-added REBCO prepared by MOCVD [11] where BZO nanorods are integrated into the film.

There are different techniques used for the approximation of the very high values of $J_c$ in superconductors with nanoscale pinning centers. Since the electric field vector within a conductor satisfies a vector wave $\nabla^2\widehat{\mathrm{B}} + \kappa^2\widehat{\mathrm{B}} = 0$, we consider a vortex flux line as a tiny cylinder of radius *a* lying along the *z*-axis where $J_c$ has only a *z* -axis component. The waveform due to the field generated by the interaction of the vortex flux line pinned by a nanoparticle in the presence of the externally applied field can be expressed in terms of the Bessel's equation of order zero such that

$$\frac{d^2\widehat{B}_z(\rho)}{d\rho^2} + \frac{1}{\rho}\frac{d\widehat{B}_z(\rho)}{d\rho} + \kappa^2\widehat{B}_z(\rho) = 0 \tag{2.9}$$

where $\rho(z)$ is a local coordinate for the vortex flux line, and $\kappa$ is a wave number. The solution to the above equation yields the field generated by the vortex interaction with a potential pinning site and can be given by

$$\widehat{B}_{z,\mathrm{v}}(\rho) = \widehat{B}_{(0),\mathrm{v}}\frac{I_0(\kappa\rho)}{I_0(\kappa a)} \tag{2.10}$$

where $\widehat{\mathrm{B}}_{(0),\mathrm{v}}$ is the field at the surface of the film. The $I_0$ is the zero-order Bessel's function of the first kind. Hence, the $J_c$ within the vortex flux line is the consequence of the film's conductivity and the interactive field generated.

$$\widehat{\mathrm{J}}_{z,\mathrm{v}}(\rho) = \sigma \widehat{\mathrm{B}}_{(0),\mathrm{v}} \frac{I_0(\kappa\rho)}{I_0(\kappa a)} \tag{2.11}$$

Therefore, at near zero field, $\widehat{\mathrm{B}}_{(0)}$, and $\widehat{\mathrm{J}}_{c,\mathrm{v}}$ distributions are nearly uniform. Note that the term $\sigma$ can "mathematically" be dropped from (2.10) since $\sigma \to \infty$ for a superconductor. In this case, (2.10) reduces to the form shown in (2.9) and the field vector essentially extends the current vector over a finite length of the vortex flux line. The asymptotic behavior of the high $J_c$ across the field is therefore estimated using the large argument of the Bessel's function of the first kind.

$$I_n(x) = \sum_{\kappa=0}^{\infty} \frac{(-1)^{\kappa}}{\kappa!\Gamma(\kappa+p+1)} \left(\frac{x}{2}\right)^{2\kappa+p} \tag{2.12}$$

where $I_n$ is the Bessel's function of integer order $n$ = 0, 1, 2, …, $x$ is any positive quantity, $p = 0.5$, $\kappa$ = 0, 1, 2, …, and $\Gamma(\kappa+p+1) = \frac{(2\kappa+1)}{2^{2\kappa+1}\kappa!}\sqrt{\pi}$. The asymptotic behavior of (2.11) yields the analytical expression shown below in (2.12).

$$I_n(x) = \sqrt{\frac{2}{\pi x}} \cos\left[\pi x - \left(\frac{n\pi}{2}\right) - \frac{\pi}{4}\right]; \quad \text{thus,} \quad J_c(B) = \sqrt{\frac{2}{\pi x}} \cos\phi \tag{2.13}$$

where $\phi = \left[x - \left(n\frac{\pi}{2}\right) - \frac{\pi}{4}\right] \approx 0.007°$ is the angle at which the pinned vortex can bend.

The maximum attainable $J_c$ is 40 MA/cm$^2$. This value corresponds to the maximum current density through the normal direction or the $z$-axis of a vortex core spinning around the vicinity of magnetic field $B$ = 0 T [193]. Figure 2.25a shows

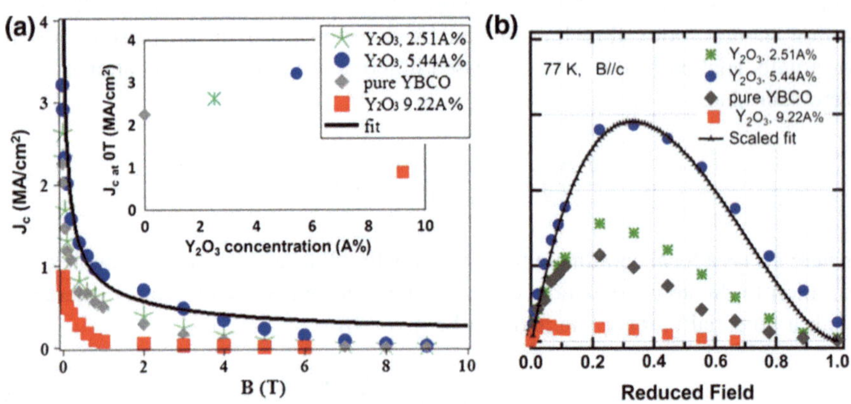


**Fig. 2.25** **a** Critical current density, $J_c$, versus magnetic field, $B$ (77 K, $B//c$), for a pure YBCO film and doped YBCO films with $Y_2O_3$ nanoparticles. Fit based on single vortex dynamics is plotted along with experimental data. Inset shows the behavior of $J_c(B = 0)$ in a function of $Y_2O_3$ concentration. **b** Global pinning force, $F_p$, versus the reduced field $b = B/B_{irr}$ at 77 K and $B//c$. Theoretical, $b = 0.35$ is within 2% of the experimental error. Note that $F_p$ = 1.3 GN/m$^3$ at $b$ = 1. (Reproduced with permission from [189])

the field dependence $J_c(B)$ for the standard YBCO together with YBCO samples added with 2.51A%, 5.44A%, and 9.22A% $Y_2O_3$. A fit based on (2.11) and (2.12) is also depicted in the figure. In comparison with the standard YBCO film, the 5.44A %, $Y_2O_3$ sample shows a good field-dependent $J_c$ up to 4 T. Beyond 4 T however, the $J_c$ of all samples decreases below the curve of the depinning current density. As the low-field regime is known to characterize the single vortex dynamics, the $J_c(B)$ of the 5.44A%, $Y_2O_3$ sample is seen to have a quite good consistency with the positive effect of $Y_2O_3$ nanoparticles.

For YBCO samples added with 2.51A%, 5.44A%, and 9.22A% $Y_2O_3$,, the minimum pinned vortex core length is estimated to be about 6.5%, 8.0%, and 2.2%, respectively [189].

### 2.3.5 Segmented 1D-APCs

For practical applications, the merits and drawbacks of continuous nanorods (1D-APCs, high $J_c$ but anisotropic pinning) and randomly distributed nanoparticles (3D-APCs, isotropic pinning but lower $J_c$) suggested investigating the intermediate anisotropy range of vortex pinning between the two limits.

This kind of approach has been studied at first by Horide et al. [194]. By means of PLD, they prepared multilayered samples comprised of pure YBCO layers alternating with YBCO layers containing nanorods. In this way, segmented BZO nanorods are produced because the nanorods nucleate and grow exactly above the upper portion of nanorods underneath the pure YBCO layer. The thickness of the layers containing nanorods can be easily varied by changing the number of pulses in PLD ablation, so that from "short nanocolumns" (still 1D-APCs) it is possible to gradually pass to "nanoparticles" (3D-APCs, but regularly distributed).

As a result, the *c*-axis-correlated peak in the angular current plot becomes gradually small. Also, the critical exponents at the vortex liquid–solid transition change from the Bose glass to the vortex glass.

The effect of segmented BSO nanorods on $B_{irr}$, $T_c$, $J_c$ characteristics has been systematically investigated, and flux pinning mechanisms have been discussed in recent work of Matsumoto et al. [195, 196]. The thicknesses of superconducting layers of YBCO ($l_s$) and pinning layers of YBCO+BSO ($l_p$) have been controlled, and several samples have been prepared combining $l_p$ = 90, 60, 30 nm and $l_s$ = 45, 30, 15 nm (Fig. 2.26a).

By applying a "harmonic oscillator" approach based on the Bose glass state, the $J_c$ of samples can be calculated as

$$J_c^{segmentedB//c} = \frac{\Phi_0}{16\pi\mu_0\lambda_{ab}^2\xi_{ab}}\left[\frac{l_p}{l_p + l_s}\right] \tag{2.14}$$

In the *B*//*c* case.

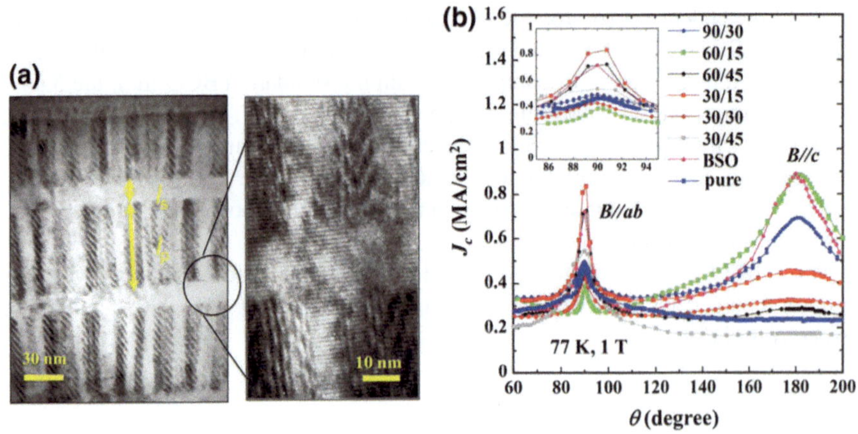


**Fig. 2.26** **a** TEM cross section of PLD-YBCO film with length controlled nanorod segments (60/15 configuration). **b** Field angular dependences of $J_c(\theta)$ at 77 K, 1 T for YBCO-BSO film, YBCO-BSO/YBCO multilayered films with 90/30, 60/15, 64/45, 30/15, 30/30, 30/45 configurations. Inset shows a magnification of *B*//*ab* peak. (Both figures reproduced with permission from [195])

This expression must be modified in the *B*//*ab* case:

$$J_c^{segmentedB//ab} = \frac{B_c^2 \pi \xi_{ab} a_0^2}{2\mu_0 \Phi_0 d^2}\left[\frac{2C_0}{l_p + l_s}\right] \quad (2.15)$$

where $a_0$ is the vortex line spacing, $a_0 = (\Phi_0/B)^{1/2}$, $C_0$ is the radius of columns, and $d$ is the mean distance between them.

The tendencies of $J_c$ (*B*//c) > $J_c$ (*B*//*ab*) were observed for the films with large $l_p$, but the opposite ones of $J_c$ (*B*//*c*) < $J_c$ (*B*//*ab*) for the films with small $l_p$ (Fig. 2.19b). This behavior suggests that a vortex of length $l_p$+$l_s$ is strongly pinned over a portion of length $l_p$ and the pinning effect is degraded at *B*//*c* when $l_p$ becomes small, however, recovered at *B*//*ab* by additional pins involved in the films.

Significant variation was considered by Horide et al. [197]: Instead of superconducting YBCO layers, semiconducting perovskite PrBCO layers are alternated with YBCO+BSO layers. Main result is that the *c*-axis-correlated pinning was weakened by the PrBCO layers in the multilayers.

### *2.3.6 Combined One-Dimensional and Three-Dimensional (1D+3D) APCs*

Another nanoengineering approach to control the angular dependence of $J_c$ in the intermediate range is the combination of APCs of different dimensionality.

Two ways are possible to obtain combined one-dimensional and three-dimensional (1D+3D) APCs in YBCO films: (i) simultaneous combination of columnar defects (1D) and nanoparticles (3D) in hybrid films by ablation of a single target and (ii) separating columnar defects and nanoparticles in multilayers by alternate ablation of two targets.

#### 2.3.6.1 Hybrid YBCO Films with Mixed 1D+3D-APCs

One of the earliest reports on hybrid films with 1D+3D-APCs incorporation was made by Mele et al. [198] using a YBCO+4wt% BZO-mixed target whose surface was modified by sticking a 2.5A% $Y_2O_3$ sector on it. They formed YBCO films incorporating 1D-APCs (BZO nanorods) and 3D-APCs ($Y_2O_3$ islands) at the same time. In the double-doped sample, the angular dependence of $J_c$ showed an intermediate behavior, with broad peak at *B*//*c* due to the flux pinning by *c*-axis-correlated defects, combined with a plateau, due to the flux pinning by isotropic pinning centers. Both BZO nanorods parallel to the *c*-axis and $Y_2O_3$ nanoparticles randomly dispersed inside the YBCO matrix were consistently observed in the cross-sectional TEM images. Similar results in 1-μm-thick films were obtained by Zhou et al. using a YBCO target added with 5 mol% BZO and 5 mol% $Y_2O_3$ [199].

More recently, Jha et al. [200, 201] studied systematically the influence of double APCs on the pinning properties of YBCO films grown by PLD on $SrTiO_3$. They used three YBCO-BSO-mixed targets (2 wt% 3wt%, and 4 wt%), fabricating films with each one added with two different areal concentrations of $Y_2O_3$ (2.2A% and 3A%), and he compared in total ten samples, including YBCO reference and the three samples with BSO only. Both 1D- and 3D-APCs were incorporated in the film according to TEM images (Fig. 2.27a) and are active pinning centers. The

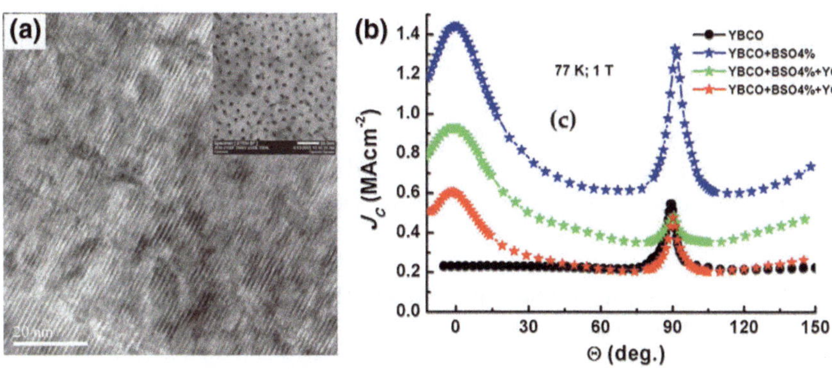


**Fig. 2.27** **a** High-resolution TEM image (cross-sectional view) of (**a**) YBCO+BSO 2wt%+$Y_2O_3$ 3A% nanocomposite films. The *inset* shows the planar view of the TEM images; **b** comparison of angular dependence of $J_c$ at 77 K, 1 T for YBCO+BSO 4wt series. (Both figures reproduced with permission from [201])

addition of $Y_2O_3$ nanoparticles to BSO nanorods improved the $F_p^{max}$, with shifting of $F_p^{max}$ at higher values of $B$, due to the increase of pinning center density. Best hybrid sample (YBCO+BSO 3wt%+$Y_2O_3$ 3A%) has $F_p^{max}$ = 11GN/m$^3$ (77 K, $B//c$). Since this value is lower than in the case of YBCO+BSO 4wt% without $Y_2O_3$ addition (about 20 GN/m$^3$), it seems that an excessive concentration of APCs hinders the path of supercurrents.

On the other hand, the addition of $Y_2O_3$ has the benefit to reduce the anisotropy of $J_c$ in the intermediate region of $\theta$, being $\theta$ the angle between the $c$-axis of the film and the applied magnetic field (Fig. 2.27b).

Horide et al. [202] obtained similar results on IBAD technical substrate with seven YBCO films added with $Y_2O_3$ and BSO (all the possible combinations of BSO 2wt%, 4wt%, and $Y_2O_3$ 0.69A%, 2.2A%).

In these hybrid YBCO films with nanorods and nanoparticles, when external magnetic field is parallel to $c$-axis of the film (i.e., parallel to the nanorods), vertical segments of the vortices are strongly pinned by the BSO nanorods and connected by kinks. The contribution from the pinning by continuous nanorods is given by (2.5). This contribution must be combined with the additional contribution from the nanoparticles or from the oxygen vacancies for the pinning of the vortex kink.

The nanoparticles give a contribution which can be expressed by the following equation

$$J_c^{3D-kink} = \frac{\Phi_0}{16\pi\mu_0\lambda_{ab}^2\xi_{ab}}\left[\frac{\xi_c}{l_s}\right] \tag{2.16}$$

where $l_s$ is the length of the kink which corresponds to the spacing between nanorods. The additional oxygen vacancy contribution can be given by

$$J_c^{vac-kink} = \frac{D^2\Phi_0}{64\pi d\mu_0\lambda_{ab}^2\xi_{ab}\xi_c}\sqrt{n\xi_{ab}\xi_c d} \tag{2.17}$$

where $D \approx 20$ nm is the diameter, and $n \approx 10^{26}$ m$^{-3}$ is the concentration of the oxygen vacancies. At $B = 1$ T, $T = 77$ K, the two contributions of the nanodimensional species have been calculated, $J_c^{3D\text{-kink}} = 0.46$ MA/cm$^2$ and $J_c^{vac\text{-kink}} = 0.28$ MA/cm$^2$. This difference in $J_c$ values indicates that the vortex pinning for the kinks by the $Y_2O_3$ nanoparticles is more effective than the one by random point defects provided by oxygen vacancies.

#### 2.3.6.2 Multilayers with Alternate YBCO+1D-APCs and YBCO +3D-APCs Layers

Very recently, Mele et al. [203] tried to add 1D-APCs and 3D-APCs in a separate fashion fabricating 1D+3D multilayers. To this purpose, YBCO+4wt% BSO-mixed target and YBCO+2A% $Y_2O_3$ surface-modified target were switched in the PLD chamber, and different configurations of YBCO+BSO layer (nm)/YBCO+$Y_2O_3$ layer (nm) varied as 90/30, 60/10, and 10/10 were studied.

Overall, it was found that the contribution of 1D pinning is quite superior with respect to that of the 3D ones. Indeed, same as shown in previous sections of the chapter for 1D and 3D cases, the calculated $J_c$ is larger than the measured $J_c$ because of the effect of current blocking by APCs. The general trend $J_c(90/30) > J_c(60/10) > J_c(10/10)$ can be explained in terms of a structural compromise between the APC pinning efficiency and current-blocking effect when the pinning center size is less compatible with other superconducting parameters.

### *2.3.7 Combination of APCs Parallel (Nanorods) and Perpendicular (Nanolayers) to C-axis of YBCO Film*

The pinning landscape is enriched by the combination of nanorods parallel to the *c*-axis of the film and nanolayered structures, perpendicular to the *c*-axis. Two cases of hybrid parallel/perpendicular APCs are reported in the literature.

Wee et al. [204] fabricated a hybrid NdBCO film with one half of the film having BZO nanorods aligned along the *c*-axis and the other half of the film having BZO nanoscale defect structures aligned along the *ab* plane. This unique hybrid nanoscale structure resulted in less anisotropy in the angular dependence of $J_c$ compared to pure NdBCO films and to films with nanoscale defects aligned along only one direction.

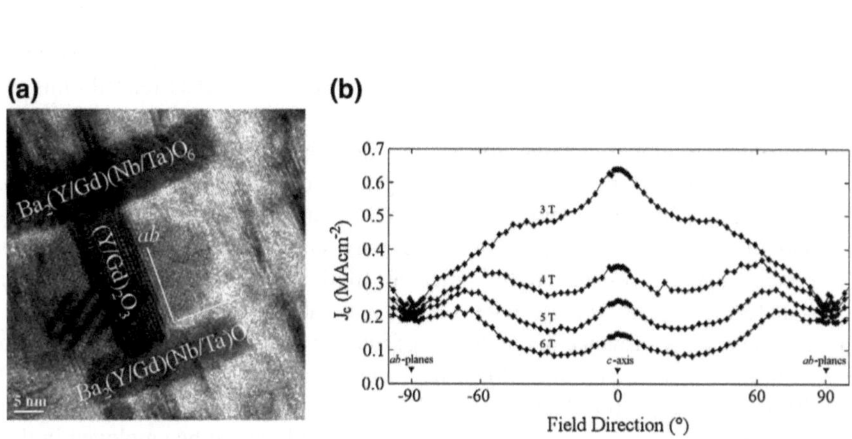


**Fig. 2.28** **a** TEM image of the $YBa_2Cu_3O_7$+2.5 mol% $Gd_3TaO_7$+2.5 mol% $Ba_2YNbO_6$ sample showing a platelike nanoparticle nucleated between two rod segments (reproduced with permission from [201]); **b** high-field angular dependence of $J_c$ measured at 77 K for a YBCO sample with (Nb–Ta) doping, applied fields ranging from 3 to 6 T. (Reproduced with permission from [206])

In the early work of Ercolano et al. [205], the addition of double perovskite $YBa_2NbO_6$ generated two kinds of defects: nanorods parallel to the *c*-axis and dispersed nanoparticles, with consequent improvement of pinning in the direction of both *c*-axis (*c*-axis-correlated pinning) and *ab* plane. In an improvement of their previous work [206], Ercolano et al. added $Gd_3TaO_7$ to $Ba_2YNbO_6$ in YBCO target and found two kinds of defects in YBCO film after the deposition: $Ba_2(Y/Gd)(Nb/Ta)O_6$ segmented rods and $(Y/Gd)_2O_3$ platelike nanoparticles, parallel and perpendicular to the *c*-axis of the film, respectively (Fig. 2.28a). The two kinds of defects are interconnected, forming a jungle-gym network, and the angular dependence of $J_c$ shows peculiar behavior: Increasing the external magnetic field, isotropic pinning becomes predominant over *c*-axis-correlated pinning (Fig. 2.28b).

## 2.4 Conclusion

This chapter presents an overview of the last 10-year research on artificial pinning centers (APCs) made by several groups in the world with the purpose of enhancing the performance (critical current and global pinning force) of $YBa_2Cu_3O_y$ (YBCO) thin films and coated conductors.

Impressive results have been achieved and can be summarized as follows

(1) Control of dimensionality and spacing of the APCs have been achieved, and advanced combinations such as segmented nanorods, mixed pinning centers, multilayers, plus nanorods have been incorporated in the YBCO and REBCO films grown on bare, decorated, or buffered crystalline substrates.
(2) The irreversibility field at 77 K was impressively enhanced to over 15 T in optimized SmBCO and GdBCO films and coated conductors added with BHO nanorods without significant depression of the superconducting critical temperature. This enhancement testifies the robustness of REBCO-related films for practical applications.
(3) Outstanding global pinning forces of ~30 $GN/m^3$ (77 K *B*//*c*), drastically surpassing 15 $GN/m^3$, the global pinning force of conventional NbTi (4.2 K), have been achieved in short pieces of YBCO added with nanorods ($BaSnO_3$ and niobate) so that the use of liquid nitrogen instead of liquid helium as coolant can be considered for practical applications on the large scale for transformers, fault current limiter (FCL), and cables.
(4) Exemplary performance in conditions of interest for practical applications such as motors, generators, and SMES has been obtained: $J_c$ over 15 $MA/cm^2$ at 30 K, 3 T in 2.2-μm-thick heavily doped $(Gd, Y)Ba_2Cu_3O_x$ superconductor tapes. Cryocoolers, instead of liquid coolants, are likely to be employed in this *T*/*B* region.
(5) Huge pinning forces have been achieved at low temperatures and ultra-high magnetic fields in YBCO-coated conductors heavily doped with BZO

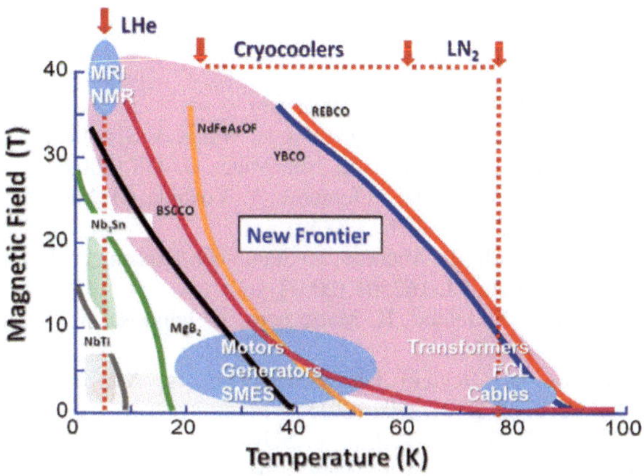


**Fig. 2.29** Temperature dependence versus magnetic field diagrams for different superconducting materials including YBCO, REBCO, NdFeAsOF, BSCCO, $MgB_2$, NbTi, and $Nb_3Sn$. The new frontier region (40 K–20 T) for the development of applications is indicated together with the three regions where the different power applications and magnets are more likely to occur soon by using coated conductor technology and APCs technology. The temperature regions where the different cryogenic approaches are needed are also indicated. (Reproduced with permission from [43] with slight modification)

nanorods: $F_p^{max} > 1700$ GN/m$^3$ at 4.2 K, 30 T. Applications such as MRI and NMR are likely to be implemented soon by using HTSC tapes in liquid helium.

The *T*/*B* region described in points (3), (4), and (5) is shown in Fig. 2.29, and new frontier (40 K/20 T) is likely to be approached soon. The incorporation of APCs in YBCO thin films allowed to achieve marvelous performances in a wide range of temperatures (4.2–77 K) and magnetic fields (1–30 T) so that the applicability limit of YBCO+APC thin films and coated conductors was considerably expanded in the past ten years. Remarkable progress in the quality of materials, wide-scale characterization, and large-scale production allows concluding that the future of coated conductors added with APCs for applications on the large scale is bright.

**Acknowledgements** P.M. recognizes insightful discussions on thin-film deposition, flux pinning, vortex behavior, and APC engineering with many colleagues over the past ten years: K. Matsumoto, Y. Yoshida, S. Awaji, S. Horii, R. Kita, M. Mukaida, A. Ichinose, T. Horide, D. C. Larbalestier, B. Maiorov, V. Selvamanickam, and X. Obradors. P.M. also gratefully thanks H. Abe, H. Honda, S. Saini, A. K. Jha, and S. J. Singh for their invaluable support to the experiments.

P.M. acknowledges partial financial support by Kakenhi S, grant number 23226014.

A.C. acknowledges financial support from Romanian Ministry of Research and Innovation, POC project 28/01.09.2016.

M.I.A. acknowledges partial support from MOHE, Grant Scheme number FRGS/2/2014/SG06/UNITEN/02/3.

## References


1. T. Ijima, Oyo Buturi **75**, 26 (2006) (in Japanese)
2. J.L. MacManus Driscoll, S.R. Foltyn, Q.X. Jia, H. Wang, A. Serquis, L. Civale, B. Maiorov, M.E. Hawley, M. Maley, D. E. Peterson, Nat. Mater. **3**, 439 (2004)
3. P. Mele, K. Matsumoto, T. Horide, A. Ichinose, Y. Yoshida, M. Mukaida, S. Horii, R. Kita, Supercond. Sci. Technol. **21**, 032002 (2008)
4. D.M. Feldmann, O. Ugurlu, B. Maiorov, L. Stan, T.G. Holesinger, L. Civale, S.R. Foltyn, Q.X. Jia, Appl. Phys. Lett. **91**, 162501 (2007)
5. S. Miura, Y. Yoshida, Y. Ichino, K. Matsumoto, A. Ichinose, S. Awaji, Supercond. Sci. Technol. **28**, 065013 (2015)
6. V. Selvamanickam, M. Heydari Gharahcheshmeh, A. Xu, Y. Zhang, E. Galstyan, Supercond. Sci. Technol. **28,** 072002 (2015)
7. A. Xu, L. Delgado, N. Khatri, Y. Liu, V. Selvamanickam, D. Abraimov, J. Jarozynski, F. Kametani, D.C. Larbalestier, APL Mater **2**, 046111 (2014)
8. S. Miura, Y. Yoshida, Y. Ichino, A. Tsuruta, K. Matsumoto, A. Ichinose, S. Awaji, Jpn. J. Appl. Phys. **53,** 090304 (2014)
9. S. Awaji, Y. Yoshida, T. Suzuki, K. Watanabe, K. Hikawa, Y. Ichino, T. Izumi, Appl. Phys. Expr. **8**, 023101 (2015)
10. T. Haugan, P.N. Barnes, R. Wheeler, F. Meisenkothen, M. Sumption, Nature **430**, 867 (2004)
11. A. Xu, J. Jarozynski, F. Kametani, D. Larbalestier, Appl. Phys. Lett. **106**, 052603 (2015)
12. M.K. Wu, J.R. Ashburn, C.J. Torng, P.H. Hor, R.L. Meng, Z.J. Huang, Y.Q. Wang, C.W. Chu, Phys. Rev. Lett. **58**, 908 (1987)
13. D. Dimos, P. Chaudhari, J. Mannhart, Phys. Rev. B **41**, 4038 (1990)
14. M. Murakami (ed.), *Melt-processed High-Temperature Superconductors* (World Scientific, Singapore, 1992)
15. M. Tomita, M. Murakami, Nature **421**, 517 (2003)
16. J. Durrell, A. Dennis, J. Jaroszynski, M. Ainslie, K. Plamer, Y. Shi, A. Campbell, J. Hull, M. Strasik, E. Hellstrom, D. Cardwell, Supercond. Sci. Technol. **27**, 082001 (2014)
17. D. Larbalestier, A. Gurevich, D.M. Feldmann, A. Polyanskii, Nature **414**, 368 (2001)
18. M. Parans Paranthaman, T. Izumi (eds.), High-performance YBCO-coated superconductor wires. MRS Bull. **29,** 533–589 (2004)
19. Y. Ijima, K. Onabe, N. Futaki, N. Tanabe, N. Sadakata, O. Kohno, Y. Ikeno, J. Appl. Phys. **74**, 1905 (1993)
20. A. Goyal, D.P. Norton, J.D. Budai, M. Paranthaman, E.D. Specht, D.M. Kroeger, D.K. Christen, Q. He, B. Saffian, F.A. List, D.F. Lee, P.M. Martin, C.E. Klabunde, E. Hartfileld, V.K. Sikka, Appl. Phys. Lett. **69**, 1795 (1996)
21. A. Goyal (ed.), Second-generation HTS conductors (Kluwer Academic Publishers, Dordrecht, 2005)
22. M. Parans Paranthaman, YBa2Cu3O7 coated conductors, *Chapter 4 in "High Temperature Superconductors*", ed. by R. Battacharya, M.P. Paranthaman (Wiley VCH, Weinheim, 2010), pp. 93–101
23. V. Selvamanickam, High temperature superconductor (HTS) wires and tapes, *Chapter 2 in "High Temperature Superconductors (HTS) for Energy*", ed. by Z. Melhem (Woodhead Publishing, Cambridge, 2012), pp. 34–62
24. B. Schey, Pulsed laser deposition of high-temperature superconducting thin films and their applications, *Chapter 14 in "Pulsed Laser Deposition of Thin Films"*, ed. by R. Eason (Wiley, New Jersey, 2007), pp. 313–331
25. M. Miura, Nanostructured oxide superconducting films prepared by metal organic deposition, *Chapter 5 in "Oxide Thin Films, Multilayers, and Nanocomposites"*, ed. by P. Mele, T. Endo, S. Arisawa, C. Li, T. Tsuchiya (Springer, Caham, Heidelberg, 2015), pp. 3–26

26. A. Ignatiev, MOCVD growth of YBCO films for coated conductor applications, *Chapter 15 in "Second-Generation HTS Conductors"*, ed. by A. Goyal (Kluwer Academic Publishers, Dordrecht, 2005), pp. 245–259
27. D.B. Chrisey, G.K. Hubler (eds.), *Pulsed Laser Deposition of Thin Films* (Wiley, New York, 1994), Chapter 8
28. R.K. Singh, D. Kumar, Mat. Sci. Eng. R **22**, 113 (1998)
29. J.M. Hujibregtse, F.C. Klaassen, A. Sziepielow, J.H. Rector, B. Dam, R. Griessen, B.J. Kooi, J.Th.M. De Hosson, Supercond. Sci. Tehnol. **15,** 395 (2002)
30. R.M. Schalk, R. Kundzins, H.W. Weber, E. Stangl, S. Proyer, D. Baüerle, Physica C **257**, 341 (1996)
31. B. Dam, N.J. Koeman, J.H. Rector, B. Stäuble-Pümpin, U. Poppe, R. Griessen, Physica C **261**, 1 (1996)
32. B. Dam, J.M. Hujibregtse, F.C. Klaassen, R.C.F. van der Geest, G. Doornbos, J.H. Rector, A.M. Testa, S. Freisem, J.C. Martinez, B. Stäuble-Pümpin, R. Griessen, Nature **399**, 439 (1999)
33. J.M. Hujibregtse, B. Dam, R.C.F. van der Geest, F.C. Klaassen, R. Elberse, J.H. Rector, R. Griessen, Phys. Rev. B **62**, 1338 (2000)
34. F.C. Klaassen, G. Doornbos, J.M. Hujibregtse, R.C.F. van der Geest, B. Dam, R. Griessen, Phys. Rev. B **64**, 184523 (2001)
35. B. Dam, J.M. Hujibregtse, J.H. Rector, Phys. Rev. B **65**, 064528 (2002)
36. V. Pan, Y. Cherpak, V. Komashko, S. Pozigun, C. Tretiachenko, A. Semenov, E. Pashitskii, A.V. Pan, Phys. Rev. B **73**, 054508 (2006)
37. J. Wang, J.H. Kwon, J. Yoon, H. Wang, T.J. Haughan, F.J. Baca, N.A. Pierce, P.N. Barnes, Appl. Phys. Lett. **92**, 082507 (2008)
38. S. Foltyn, L. Civale, J.L. Macmanus-Driscoll, Q.X. Jia, B. Maiorov, H. Wang, M. Maley, Nat. Mater. **6**, 631 (2007)
39. S.H. Wee, A. Goyal, P.M. Martin, L. Heatherley, Supercond. Sci. Tech. **19**, 865 (2006)
40. Y. Yoshida, K. Matsumoto, Y. Ichino, M. Itoh, A. Ichinose, S. Horii, M. Mukaida, Y. Takai, Jpn. J. Appl. Phys. **44**, L129 (2005)
41. D.R. Nelson, V.M. Vinokur, Phys. Rev. B **48**, 13060 (1993)
42. G. Blatter, M.V. Feigel`man, V.B. Genskenbein, A.I. Larkin, V.M. Vinokur, Rev. Mod. Phys. **86,** 1125 (1994)
43. X. Obradors, T. Puig, Supercond. Sci. Technol. **27**, 044003 (2014)
44. T.J. Haugan, T.A. Campbell, N.A. Pierce, M.F. Locke, I. Maartense, P.N. Barnes, Supercond. Sci. Technol. **21**, 012503 (2008)
45. P.N. Barnes, J.W. Kell, B.C. Harrison, T.J. Haugan, C.V. Varanasi, M. Rane, F. Ramos, Appl. Phys. Lett. **89**, 0122503 (2006)
46. H. Zhou, B. Maiorov, H. Wang, J.L. MacManus-Driscoll, T.G. Holesinger, L. Civale, Q.X. Jia, S.R. Foltyn, Supercond. Sci. Technol. **21**, 025001 (2008)
47. A. Radhika Devi, V. Seshu Bai, P.V. Patanjali, R. Pinto, N. Harish Kumar, S.K. Malik, Supercond. Sci. Technol. **13**, 935 (2000)
48. S. Horii, A. Ichinose, Y. Ichino, T. Ozaki, Y. Yoshida, K. Matsumoto, M. Mukaida, J. Shimoyama, K. Kishio, Physica C **463–465**, 922 (2007)
49. M. Muralidhar, N. Sakai, M. Jirsa, M. Murakami, I. Hirabayashi, Appl. Phys. Lett. **92**, 162512 (2008)
50. M. Muralidhar, N. Sakai, T. Nishiyama, M. Jirsa, T. Machi, M. Murakami, Appl. Phys. Lett. **82**, 943 (2003)
51. M. Muralidhar, N. Sakai, M. Jirsa, N. Koshizuka, M. Murakami, Appl. Phys. Lett. **83**, 5005 (2003)
52. C. Cai, B. Holzapfel, J. Hänisch, L. Fernández, L. Schultz, Applied. Phys. Lett. **84**, 0377 (2004)
53. C. Cai, B. Holzapfel, J. Hänisch, L. Fernández, L. Schultz, Phys. Rev. B **69**, 104531 (2004)
54. R.-K. Ko, S.-H. Bae, M.-J. Jung, S.-H. Jang, K.-J. Song, C. Park, M.-H. Sohn, S.-I. Kang, S.-S. Oh, D.-W. Ha, H.-S. Kim, Y.-C. Kim, I.E.E.E. Trans, Appl. Superc. **19**, 3136 (2009)

55. L. Civale, A.D. Marwick, T.K. Worthington, M.K. Kirk, J.R. Thompson, L. Krusin-Elbaum, Y. Sun, J.R. Clem, F. Holtzberg, Phys. Rev. Lett. **67**, 648 (1991)
56. L. Civale, Supercond. Sci. Technol. **10**, A11 (1997)
57. B. Holzapfel, G. Kreiselmeyer, M. Kraus, G. Saemann-Ishenko, S. Bouffard, S. Klaumünzer, L. Schultz, Phys. Rev. B **48**, 600 (1993)
58. L.M. Paulius, J.A. Fendrich, W.-K. Kwok, A.E. Koshelev, V.M. Vinokur, G.W. Crabtree, B.G. Glagola, Phys. Rev. B **56**, 913 (1997)
59. N. Chikumoto, Enhancement of flux pinning through irradiation, *Chapter 6 in "Critical Currents in Superconductors"*, ed. by T. Takizawa, M. Murakami (Fuzambo International, 2005), pp. 91–103
60. Q. Li, *Presented at CCA2014*, Jeju, South Korea, 2 Dec 2014
61. M. Leroux, K.J. Kihlstrom, S. Holleis, M.W. Rupich, S. Sathyamurthy, S. Fleshler, H. Sheng, D.J. Miller, S. Eley, L. Civale, A. Kayani, P.M. Niraula, U. Welp, W.-K. Kwok, Appl. Phys. Lett. **107**, 192601 (2015)
62. K. Develos-Bagarinao, S.C. Wimbush, H. Matsui, I. Yamaguchi, J.L. MacManus-Driscoll, Supercond. Sci. Technol. **25**, 065005 (2012)
63. T. Sueyoshi, Y. Furuki, T. Kai, T. Fujiyoshi, N. Ishikawa, Physica C **504**, 53 (2014)
64. A. Crisan, S. Fujiwara, J.C. Nie, A. Sundaresan, H. Ihara, App. Phys. Lett. **79**, 4547 (2001)
65. A. Crisan, P. Badica, S. Fujiwara, J.C. Nie, A. Sundaresan, Y. Tanaka, H. Ihara, App. Phys. Lett. **80**, 3566 (2002)
66. A. Crisan, P. Badica, S. Fujiwara, J.C. Nie, A. Sundaresan, A. Iyo, Y. Tanaka, I.E.E.E. Trans, Appl. Supercond. **13**, 3726 (2003)
67. J.C. Nie, H. Yamasaki, H. Yamada, Y. Nakagawa, K. Develos-Bagarinao, Y. Mawatari, Supercond. Sci. Tech. **17**, 845 (2004)
68. X.M. Cui, G.Q. Liu, J. Wang, Z.C. Huang, Y.T. Zhao, B.W. Tao, Y.R. Li, Physica C **466**, 1 (2007)
69. T. Aytug, M. Paranthaman, A.A. Gapud, S. Kang, H.M. Christen, K.J. Leonard, P.M. Martin, J.R. Thompson, D.K. Christen, R. Meng, I. Rusakova, C.W. Chu, T.H. Johansen, J. Appl. Phys. **98**, 114309 (2005)
70. P. Mikheenko, A. Sarkar, V.-S. Dang, J.L. Tanner, J.S. Abell, A. Crisan, Physica C **469**, 798 (2009)
71. A. Sarkar, P. Mikheenko, V.-S. Dang, J.S. Abell, A. Crisan, Physica C **469**, 1550 (2009)
72. T. Aytug, M. Paranthaman, K.J. Leonard, K. Kim, A.O. Ijadoula, Y. Zhang, E. Tuncer, J.R. Thompson, D.K. Christen, J. Appl. Phys. **104**, 043906 (2008)
73. M. Sparing, E. Backen, T. Freudenberg, R. Hühne, B. Rellinghaus, L. Schultz, B. Holzapfel, Supercond. Sci. Tech. **20**, S239 (2007)
74. F.J. Baca, D. Fisher, R.L.S. Emergo, J.Z. Wu, Supercond. Sci. Technol. **20**, 554 (2007)
75. K. Matsumoto, T. Horide, K. Osamura, M. Mukaida, Y. Yoshida, A. Ichinose, S. Horii, Physica C **412–414**, 1267 (2004)
76. K. Matsumoto, T. Horide, P. Mele, A. Ichinose, S. Horii, Y. Yoshida, M. Mukaida, K. Osamura, IEEE Trans. Appl. Supercond. **13,** 3374 (2003)
77. K. Matsumoto, T. Horide, P. Mele, Y. Yoshida, M. Mukaida, A. Ichinose, S. Horii, Physica C **426–431**, 1091 (2005)
78. K. Matsumoto, T. Horide, A. Ichinose, S. Horii, Y. Yoshida, M. Mukaida, Jpn. J. Appl. Phys. **7**, L246 (2005)
79. P. Mele, K. Matsumoto, T. Horide, O. Miura, A. Ichinose, M. Mukaida, Y. Yoshida, S. Horii, Physica C **648**, 445–448 (2006)
80. P. Mele, K. Matsumoto, T. Horide, O. Miura, A. Ichinose, M. Mukaida, Y. Yoshida, S. Horii, Supercond. Sci. Technol. **19**, 44 (2006)
81. Y. Yoshida, K. Matsumoto, M. Miura, Y. Ichino, Y. Takai, A. Ichinose, S. Horii, M. Mukaida, Jpn. J. Appl. Phys. **18**, L546 (2005)
82. C.-F. Tsai, J. Huang, J.-H. Lee, F. Khatkhatay, L. Chen, A. Chen, Q. Su, H. Wang, Physica C **510**, 13 (2015)

83. T. Horide, K. Taguchi, K. Matsumoto, N. Matsukida, M. Ishimaru, P. Mele, R. Kita, Appl. Phys. Let. **108**, 082601 (2016)
84. A. Goyal, S. Kang, K.J. Leonard, P.M. Martin, A.A. Gapud, M. Varela, M. Paranthaman, A.O. Ijaduola, E.D. Specht, J.R. Thompson, D.K. Christen, S.J. Pennycook, F.A. List, Supercond. Sci. Technol. **18**, 1533 (2005)
85. Y. Yamada, K. Takahashi, H. Kobayashi, M. Konishi, T. Watanabe, A. Ibi, T. Muroga, S. Miyata, T. Kato, T. Hirayama, Y. Shiohara, Appl. Phys. Lett. **87**, 132502 (2005)
86. P. Mele, K. Matsumoto, T. Horide, A. Ichinose, M. Mukaida, Y. Yoshida. S. Horii, Supercond. Sci. Technol. **20,** 244 (2007)
87. S. Kang, A. Goyal, J. Li, A.A. Gapud, P.M. Martin, L. Heatherly, J.R. Thompson, D.K. Christen, F.A. List, M. Paranthaman, D.F. Lee, Science **311**, 1911 (2006)
88. A. Ichinose, K. Naoe, T. Horide, K. Matsumoto, R. Kita, M. Mukaida, Y. Yoshida, S. Horii, Supercond. Sci. Technol. **20**, 1144 (2007)
89. S. Kang, A. Goyal, J. Li, P. Martin, A. Ijaduola, J.R. Thompson, M. Paranthaman, Physica C **457**, 41 (2007)
90. M. Inoue, T. Kiss, K. Motoyama, T. Nakamura, T. Fujiwara, D. Mitsui, S. Awaji, K. Watanabe, A. Ibi, K. Takahashi, S. Miyata, Y. Yamada, Y. Shiohara, Physica C **463–465**, 674 (2007)
91. K.J. Song, R.K. Ko, Y.S. Lee, J.S. Yang, Y.M. Park, M. Dixit, H.S. Kim, H.S. Ha, D.W. Ha, S.S. Oh, D.J. Kim, C. Park, S.I. Yoo, Physica C **445–448**, 656 (2006)
92. P. Paturi, M. Irjala, H. Huhtinen, J. Appl. Phys. **103,** 123907 (2008)
93. M. Peurla, H. Huhtinen, M.A. Shakhov, K. Traito, Yu.P. Stepanov, M. Safonchik, P. Paturi, Y. Y. Tse, R. Palai, R. Laiho, Phys. Rev. B **75,** 184524 (2007)
94. M. Peurla, P. Paturi, Yu.P. Stepanov, H. Huhtinen, Y.Y. Tse, A.C. Bódi, J. Raittila, R. Laiho, Supercond. Sci. Technol. **19,** 767 (2006)
95. P. Paturi, M. Irjala, H. Huhtinen, A.B. Abrahamsen, J. Appl. Phys. **105,** 23904 (2009)
96. M. Safonchik, K. Traito, S. Tuominen, P. Paturi, H. Huhtinen, R. Laiho, Supercond. Sci. Technol. **22**, 065006 (2009)
97. K. Traito, M. Peurla, H. Huhtinen, Yu.P. Stepanov, M. Safonchik, Y.Y. Tse, P. Paturi, R. Laiho, Phys. Rev. B **73,** 224522 (2006)
98. F.J. Baca, P.N. Barnes, R.L.S. Emergo, T.J. Haugan, J.N. Reichart, J.Z. Wu, Appl. Phys. Lett. **94**, 102512 (2009)
99. S.H. Wee, A. Goyal, Y.L. Zuev, C. Cantoni, Supercond. Sci. Technol. **21**, 092001 (2008)
100. B. Maiorov, S.A. Baily, H. Zhou, O. Ugurlu, J.A. Kennison, P.C. Dowden, T.G. Holesinger, S.R. Foltyn, L. Civale, Nat. Mater. **8**, 398 (2009)
101. S.H. Wee, A. Goyal, J. Li, Y.L. Zuev, S. Cook, L. Heatherly, Supercond. Sci. Technol. **20**, 789 (2007)
102. S.H. Wee, A. Goyal, J. Li, Y.L. Zuev, S. Cook, J. Appl. Phys. **102**, 063906 (2007)
103. Y.L. Zuev, D.K. Christen, S.H. Wee, A. Goyal, S.W. Cook, Appl. Phys. Lett. **93**, 172512 (2008)
104. S.H. Wee, A. Goyal, P.M. Martin, J. Li, M. Paranthaman, L. Heatherly, Supercond. Sci. Technol. **19**, L42 (2006)
105. M. Mukaida, T. Horide, R. Kita, S. Horii, A. Ichinose, Y. Yoshida, O. Miura, K. Matsumoto, K. Yamada, N. Mori, Jpn. J. Appl. Phys. **44**, L952 (2005)
106. M. Ito, M. Mukaida, T. Ohazama, R. Kita, A. Ichinose, S. Horii, A. Saito, K. Matsumoto, Y. Yoshida, K. Koike, F. Hirose, S. Ohshima, Physica C **426–431**, 1415 (2005)
107. M. Haruta, T. Fujiyoshi, T. Sueyoshi, K. Dezaki, D. Ichigosaki, K. Miyahara, R. Miyagawa, M. Mukaida, K. Matsumoto, Y. Yoshida, A. Ichinose, S. Horii, Supercond. Sci. Technol. **19**, 803 (2006)
108. T. Fujiyoshi, M. Haruta, T. Sueyoshi, K. Yonekura, M. Watanabe, M. Mukaida, R. Teranishi, K. Matsumoto, Y. Yoshida, A. Ichinose, S. Horii, S. Awaji, K. Watanabe, Physica C **468**, 1635 (2008)
109. M. Miura, M. Mukaida, K. Matsumoto, Y. Yoshida, A. Ichinose, S. Horii, R. Kita, A. Saito, K. Kaneko, K. Yamada, N. Mori, Physica C **445–448**, 845 (2006)

110. Y. Yoshida, T. Ozaki, Y. Ichino, Y. Takai, K. Matsumoto, A. Ichinose, M. Mukaida, S. Horii, Physica C **468**, 1606 (2008)
111. V. Selvamanickam, M. Heydari Gharahcheshmeh, A. Xu, E. Galstyan, L. Delgado, C. Cantoni, Appl. Phys. Lett. **106**, 032601 (2015)
112. A. Xu, L. Delgado, M. Heydari Gharahcheshmeh, N. Kahtri, Y. Liu, V. Selvamanickam, Supercond. Sci. Technol. **28,** 082001 (2015)
113. V. Selvamanickam, M. Heydari Gharahcheshmeh, A. Xu, Y. Zhang, E. Galstyan, Supercond. Sci. Technol. **28,** 104003 (2015)
114. Ch.V. Varanasi, P.N. Barnes, J. Burke, L. Brunke, I. Maartense, T.J. Haugan, E.A. Stinzianni, K.A. Dunn, P. Haldar, Supercond. Sci. Technol. **19**, L37 (2006)
115. C.V. Varanasi, J. Burke, L. Brunke, H. Wang, M. Sumption, P.N. Barnes, J. Appl. Phys **102**, 063909 (2007)
116. C.V. Varanasi, J. Burke, L. Brunke, H. Wang, J.H. Lee. P.N. Barnes, J. Mater. Res. **23,** 3363 (2008)
117. P. Mele, K. Matsumoto, T. Horide, A. Ichinose, M. Mukaida, Y. Yoshida. S. Horii, R. Kita, Supercond. Sci. Technol. **21,** 125017 (2008)
118. C.V. Varanasi, J. Burke, H. Wang, J.H. Lee, P.N. Barnes, Appl. Phys. Lett. **93**, 092501 (2008)
119. M. Namba, S. Awaji, K. Watanabe, S. Ito, E. Aoyagi, H. Kai, M. Mukaida, R. Kita, Appl. Phys. Express **2**, 073001 (2009)
120. S. Yasunaga, M. Mukaida, A. Ichinose, S. Horii, R. Teranishi, K. Yamada, K. Matsumoto, R. Kita, Y. Yoshida, N. Mori, Physica C **468**, 1858 (2008)
121. Y. Tanaka, M. Mukaida, R. Teranishi, K. Yamada, A. Ichinose, K. Matsumoto, S. Horii, Y. Yoshida, R. Kita, T. Fujiyosh, N. Mori, Physica C **468**, 1864 (2008)
122. L. Civale, B. Maiorov, A. Serquis, J.O. Willis, J.Y. Coulter, H. Wang, Q.X. Jia, P.N. Arendt, J.L. MacManus-Driscoll, M.P. Maley, S.R. Foltyn, Appl. Phys. Lett. **84**, 2121 (2004)
123. H. Tobita, K. Notoh, K. Higashikawa, M. Inoue, T. Kiss, T. Kato, T. Hirayama, M. Yoshizumi, T. Izumi, Y. Shiohara, Supercond. Sci. Technol. **25**, 062002 (2012)
124. A. Tsuruta, Y. Yoshida, Y. Ichino, A. Ichinose, K. Matsumoto, S. Awaji, J. Appl. Phys. **52**, 010201 (2013)
125. A. Tsuruta, Y. Yoshida, Y. Ichino, A. Ichinose, K. Matsumoto, S. Awaji, Supercond. Sci. Technol. **27**, 065001 (2014)
126. A. Tsuruta, Y. Yoshida, Y. Ichino, A. Ichinose, S. Watanabe, T. Horide, K. Matsumoto, S. Awaji, Appl. Phys. Expr. **8**, 033101 (2015)
127. S.A. Harrington, J.H. Durrell, B. Maiorov, H. Wang, S.C. Wimbush, A. Kursumovic, J.H. Lee, J.L. MacManus-Driscoll, Supercond. Sci. Technol. **22**, 022001 (2009)
128. D.M. Feldmann, T.G. Holesinger, B. Maiorov, S.R. Foltyn, J.Y. Coulter, I. Apodaca, Supercond. Sci. Technol. **23**, 095004 (2010)
129. J.J. Shi, J.Z. Wu, Phil. Mag. **92**, 2911 (2012)
130. J.J. Shi, J.Z. Wu, Phil. Mag. **92**, 4205 (2012)
131. F.J. Baca, T.J. Haugan, P.N. Barnes, T.G. Holesinger, B. Maiorov, R. Lu, X. Wang, J.N. Reichart, J.Z. Wu, Adv. Funct. Mater. **23**, 4826 (2013)
132. J.Z. Wu, J.J. Shi, J.F. Baca, R. Emergo, T.J. Haugan, B. Maiorov, T. Holesinger, Supercond. Sci. Technol. **27**, 044010 (2014)
133. J. Wu, Three-dimensional pinning landscape based on orientation phase diagram of BaZrO3 nanostructures in YBa2Cu3O7 films, in *Presented at IUMRS-ICA 2014*, Fukuoka, Japan, C9-I25-007
134. J.J. Shi, J.Z. Wu, J. Appl. Phys. **118**, 164301 (2015)
135. S.R. Foltyn, P. Tiwari, R.C. Dye, M.Q. Le, X.D. Wu, Appl. Phys. Lett. **63**, 1848 (1993)
136. S.R. Foltyn, Q.X. Jia, P.N. Arendt, L. Kinder, Y. Fan, J.F. Smith, Appl. Phys. Lett. **75**, 3692 (1999)
137. S.R. Foltyn, P.N. Arendt, Q.X. Jia, H. Wang, J.L. MacManus-Driscoll, S. Kreiskott, R.F. DePaula, L. Stan, J.R. Groves, P.C. Dowden, Appl. Phys. Lett. **82**, 4519 (2003)

138. A.O. Ijaduola, J.R. Thompson, R. Feenstra, D.K. Christen, A.A. Gapud, X. Song, Phys. Rev. B **73**, 134502 (2006)
139. S.I. Kim, A. Gurevich, X. Song, X. Li, W. Zhang, T. Kodenkandath, M.W. Rupich, T.G. Holesinger, D.C. Larbalestier, Supercond. Sci. Technol. **19**, 968 (2006)
140. Z.J. Chen, D.M. Feldmann, D.C. Larbalestier, T.G. Holesinger, X. Li, W. Zhang, M.W. Rupich, Appl. Phys. Lett. **91,** 052508 (2007)
141. Yu.V. Cherpak, V.O. Moskaliuk, A.V. Semenov, V.L. Svetchnikov, C.G. Tretiatchenko,V. M. Pan, Supercond. Sci. Technol. **20,** 1159 (2007)
142. Z. Chen, F. Kametani, S.I. Kim, D.C. Larbalestier, H.W. Jang, K.J. Choi, C.B. Eom, J. Appl. Phys. **108**, 043913 (2008)
143. S.R. Foltyn, H. Wang, L. Civale, Q.X. Jia, P.N. Arendt, B. Maiorov, Y. Li, M.P. Maley, J. MacManus-Driscoll, Appl. Phys. Lett. **87**, 162505 (2005)
144. A.A. Gapud, D. Kumar, S.K. Viswanathan, C. Cantoni, M. Varela, J. Abiade, S.J. Pennycook, D.K. Christen, Supercond. Sci. Technol. **18**, 1502 (2005)
145. C.B. Cai, J.C. Zhang, S.X. Cao, J. Hänisch, R. Hühne, B. Holzapfel, Physica C **460–462**, 1355 (2007)
146. K. Develos-Bagarinao, H. Yamasaki, K. Ohki, Y. Nakagawa, Supercond. Sci. Technol. **20**, L25 (2007)
147. K. Develos-Bagarinao, H. Yamasaki, K. Ohki, J. Appl. Phys. **104**, 063907 (2008)
148. T.G. Holesinger, Q. Jia, B. Maiorov, L. Civale, P.C. Dowden, B.J. Gibbons, Adv. Mater. **19**, 1917 (2007)
149. A.V. Pan, S. Pysarenko, S.X. Dou, Appl. Phys. Lett. **98**, 232506 (2006)
150. T. Haugan, P.N. Barnes, I. Maartense, C.B. Cobb, E.J. Lee, M. Sumption, J. Mater. Res. **18**, 2618 (2003)
151. L. Peng, C. Cai, C. Chen, Z. Liu, R. Hühne, B. Holzapfel, J. Appl. Phys. **104**, 033920 (2008)
152. C. Cai, J. Liu, Z. Liu, L. Ying, B. Gao, L. Peng, C. Chen, J. Appl. Phys. **104**, 023913 (2008)
153. S. Kang, K.J. Leonard, P.M. Martin, J. Li, A. Goyal, Supercond. Sci. Technol. **20**, 11 (2007)
154. H. Huhtinen, K. Schlesier, P. Paturi, Supercond. Sci. Technol. **22**, 075019 (2009)
155. P.K. Petrov, Z.G. Ivanov, S.S. Gevorgyan, Mat. Sci. Eng. B **A288**, 231 (2000)
156. W.S. Tan, H.L. Cai, J.S. Liu, X.S. Wu, S.S. Jiang, Z.H. Wu, Q.J. Jia, J. Gao, J. Appl. Phys. **99**, 08M509 (2006)
157. T. Sueyoshi, M. Watanabe, M. Haruta, T. Fujiyoshi, K. Miyahara, T. Ikegami, K. Ebihara, R. Miyagawa, Physica C **468**, 1266 (2008)
158. T. Ozaki, Y. Yoshida, M. Miura, Y. Ichino, Y. Takai, K. Matsumoto, A. Ichinose, S. Horii, M. Mukaida, Physica C **463–465**, 649 (2007)
159. S. Horii, M. Takamura, M. Mukaida, A. Ichinose, K. Yamada, R. Teranishi, K. Matsumoto, R. Kita, Y. Yoshida, J. Shimoyama, K. Kishio, Appl. Phys. Lett. **92**, 132502 (2008)
160. M. Mukaida, S. Miyazawa, Appl. Phys. Lett. **63**, 999 (1993)
161. M. Takamura, M. Mukaida, Y. Shingai, R. Teranishi, K. Yamada, N. Mori, S. Horii, A. Ichinose, R. Kita, S. Kato, K. Matsumoto, Y. Yoshida, Physica C **463–465**, 904 (2007)
162. M. Takamura, M. Mukaida, S. Horii, A. Ichinose, R. Kita, S. Kato, K. Matsumoto, Y. Yoshida, M. Namba, S. Awaji, K. Watanabe, T. Fujiyoshi, R. Teranishi, K. Yamada, N. Mori, Physica C **468**, 1851 (2008)
163. K. Matsumoto, D. Takahara, T. Horide, A. Ichinose, S. Horii, Y. Yoshida, M. Mukaida, K. Osamura, IEEE Trans. Appl. Supercond. **15,** 2719 (2005)
164. P. Mikheenko, V.S. Dang, Y.Y. Tse, M.M. Awang Kechik, P. Paturi, H. Huhtinen, Y. Wang, A. Sarkar, J.S. Abell, A. Crisan, Supercond. Sci. Technol. **23,** 125007 (2010)
165. P. Mikheenko, V.S. Dang, M.M. Awang Kechik, A. Sarkar, P. Paturi, H. Huhtinen, J.S. Abell, A. Crisan, IEEE Trans. Appl. Supercond. **21,** 3184 (2011)
166. A. Crisan, V.S. Dang, G. Yearwood, P. Mikheenko, H. Huhtinen, P. Paturi, Physica C **503**, 89 (2014)
167. M. Miura, T. Kato, M. Yoshizumi, Y. Yamada, T. Izumi, Y. Shiohara, T. Hirayama, Appl. Phys. Express **1**, 051701 (2008)

168. J. Gutiérrez, A. Llordes, J. Gázquez, M. Gibert, N. Romà, S. Ricart, A. Pomar, F. Sandiumenge, N. Mestres, T. Puig, X. Obradors, Nat. Mater. **6**, 367 (2007)
169. A. Lordes, A. Palau, J. Gazquez, M. Coll, R. Vlad, A. Pomar, J. Arbiol, R. Guzman, S. Ye, V. Ruoco, F. Sandiumenge, S. Ricart, T. Puig, M. Varela, D. Chateigner, J. Vanacken, J. Gutierrez, V. Moschalkov, G. Deutscher, C. Magen, X. Obradors, Nat. Mater. **11**, 329 (2012)
170. Y. Yoshida, K. Matsumoto, Y. Ichino, M. Itoh, A. Ichinose, S. Horii, M. Mukaida, Y. Takai, Jpn. J. Appl. Phys. **44**, L129 (2005)
171. Y. Yoshida, Y. Ichino, M. Miura, Y. Takai, K. Matsumoto, A. Ichinose, S. Horii, M. Mukaida, IEEE Trans. Appl. Supercond. **15**, 2727 (2005)
172. M. Miura, Y. Yoshida, Y. Ichino, Y. Takai, K. Matsumoto, A. Ichinose, S. Horii, M. Mukaida, Jpn. J. Appl. Phys. **45**, L11 (2006)
173. Y. Ichino, Y. Yoshida, M. Miura, Y. Takai, K. Matsumoto, M. Mukaida, A. Ichinose, S. Horii, S. Awaji, K. Watanabe, Physica C **463–465**, 639 (2007)
174. Y. Yoshida, Y. Ichino, M. Miura, T. Ozaki, Y. Takai, K. Matsumoto, A. Ichinose, M. Mukaida, S. Horii, Physica C **463–465**, 633 (2007)
175. R.L.S. Emergo, J.Z. Wu, T.J. Haugan, P.N. Barnes, Appl. Phys. Lett. **87**, 232503 (2005)
176. C. Varanasi, P.N. Barnes, J. Burke, J. Carpenter, T.J. Haugan, Appl. Phys. Lett. **87**, 262510 (2005)
177. M.D. Sumption, T.J. Haugan, P.N. Barnes, T.A. Campbell, N.A. Pierce, C. Varanasi, Phys. Rev. B **77**, 094506 (2008)
178. S.I. Kim, F. Kametani, Z. Chen, A. Gurevich, D.C. Larbalestier, T. Haugan, P. Barnes, Appl. Phys. Lett. **90**, 252502 (2007)
179. T.A. Campbell, T.J. Haugan, I. Maartense, J. Murphy, L. Brunke, P.N. Barnes, Physica C **423**, 1 (2005)
180. V.F. Solovyov, H.J. Wiesmann, L. Wu., Q. Li., L.D. Cooley, M. Suenaga, B. Maiorov, L. Civale, Supercond. Sci. Technol. **20,** L20 (2007)
181. H. Yamasaki, K. Ohki, H. Yamada, Y. Nakagawa, Y. Mawatari, Supercond. Sci. Technol. **21**, 125011 (2008)
182. S.K. Viswanathan, A.A. Gapud, M. Varela, J.T. Abiade, D.K. Christen, S.J. Pennycook, D. Kumar, Thin Solid Films **515**, 6452 (2007)
183. J. Hänisch, C. Cai, R. Hühne, L. Schultz, B. Holzapfel, Appl. Phys. Lett. **86**, 122508 (2005)
184. S. Engel, T. Thersleff, R. Hühne, L. Schultz, B. Holzapfel, Appl. Phys. Lett. **90**, 102505 (2007)
185. R. Kalyanaraman, S. Oktyabrsky, J. Narayan, J. Appl. Phys. **85**, 6636 (1999)
186. T. Horide, K. Matsumoto, H. Adachi, D. Takahara, K. Osamura, A. Ichinose, M. Mukaida, Y. Yoshida, S. Horii, Physica C **445–448**, 652 (2006)
187. M.M. Awang Kechik, P. Mikheenko, A. Sarkar, V.S. Dang, N. Hari Babu, D.A. Cardwell, J.S. Abell, A. Crisan, Supercond. Sci. Technol. **22,** 034020 (2009)
188. P. Mele, K. Matsumoto, T. Horide, A. Ichinose, M. Mukaida, Y. Yoshida. S. Horii, Supercond. Sci. Technol. **20,** 616 (2007)
189. P. Mele, R. Guzman, J. Gazquez, X. Obradors, T. Puig, S. Saini, Y. Yoshida, M. Mukaida, A. Ichinose, K. Matsumoto, M.I. Adam, Supercond. Sci. Technol. **28**, 024002 (2015)
190. C. Meingast, D.C. Larbalestier, J. Appl. Phys. **66**, 5971 (1989)
191. A. Gurevich, Supercond. Sci. Technol. **20**, S128 (2007)
192. E.H. Brandt, Phys. Rev. Lett. **69**, 1105 (1992)
193. A.E. Koshelev, A.B. Kolton, Phys. Rev. B **84**, 104528 (2011)
194. T. Horide, K. Matsumoto, P. Mele, Y. Yoshida, A. Ichinose, R. Kita, S. Horii, M. Mukaida, Phys. Rev. B **79**, 092504 (2009)
195. K. Matsumoto, I. Tanaka, T. Horide, P. Mele, Y. Yoshida, S. Awaji, J. Appl. Phys. **116**, 163903 (2014)
196. K. Matsumoto, T. Horide, A.K. Jha, P. Mele, Y. Yoshida, S. Awaji, IEEE Trans. Appl. Supercond. **25,** 800106 (2015)

197. T. Horide, T. Murayama, K. Takata, K. Matsumoto, P. Mele, Y. Yoshida, Y. Ichino, S. Awaji, J. Appl. Phys. **114**, 073903 (2013)
198. P. Mele, K. Matsumoto, T. Horide, A. Ichinose, M. Mukaida, Y. Yoshida. S. Horii, Supercond. Sci. Technol. **21**, 015019 (2008)
199. H. Zhou, B. Maiorov, S.A. Baily, P.C. Dowden, J.A. Kennison, L. Stan, T.G. Holesinger, Q.X. Jia, S.R. Foltyn, L. Civale, Supercond. Sci. Technol. **22**, 085013 (2009)
200. A.K. Jha, K. Matsumoto, T. Horide, S. Saini, P. Mele, Y. Yoshida, S. Awaji, I.E.E.E. Trans, Appl. Supercond. **25**, 8000505 (2015)
201. A.K. Jha, K. Matsumoto, T. Horide, S. Saini, P. Mele, A. Ichinose, Y. Yoshida, S. Awaji, Supercond. Sci. Technol. **28**, 114004 (2015)
202. T. Horide, T. Kawamura, K. Matsumoto, A. Ichinose, M. Yoshizumi, T. Izumi, Y. Shiohara, Supercond. Sci. Technol. **26**, 075019 (2013)
203. P. Mele, M.I. Adam, T. Suzuki, Y. Yoshida, S. Awaji, A. Ichinose, S. Saini, A.K. Jha, K. Matsumoto, Sci. Adv. Mater. **9**, 1042 (2017)
204. S.H. Wee, A. Goyal, Y.L. Zuev, C. Cantoni, Appl. Phys. Express **1**, 111702 (2009)
205. G. Ercolano, S.A. Harrington, H. Wang, C.F. Tsai, J.L. MacManus-Driscoll, Supercond. Sci. Technol. **23**, 022003 (2010)
206. G. Ercolano, M. Bianchetti, S.C. Wimbush, S.A. Harrington, H. Wang, J.H. lee, J.L. MacManus-Driscoll, Supercond. Sci. Technol. **24**, 095012 (2011)

# Chapter 3
# Chemically and Mechanically Engineered Flux Pinning for Enhanced Electromagnetic Properties of $MgB_2$

**Soo Kien Chen, Minoru Maeda, Akiyasu Yamamoto and Shi Xue Dou**

## 3.1 Impurities as Flux Pinning Centers

### *3.1.1 Introduction*

Formation of nanoprecipitates and their distribution within the samples play an important role in influencing the flux pinning properties of $MgB_2$. It is highly desirable that reaction between the chemical additives and Mg and B should not lead to severe degradation of critical transition temperature, $T_c$, especially when substitution into the crystal structure of $MgB_2$ takes place. Depending on the starting raw powders and sample preparation procedures, the impurities in the undoped $MgB_2$ samples are mainly oxide phases such as MgO and Mg-B-O (Table 3.1) as shown by X-ray diffraction (XRD) and transmission electron

S.K. Chen (✉)
Department of Physics, Faculty of Science, Universiti Putra Malaysia,
43400 Serdang, Selangor, Malaysia
e-mail: chensk@upm.edu.my

M. Maeda
Department of Physics, College of Science and Technology, Nihon University,
1-8-14 Kanda-Surugadai, Chiyoda-ku, 101-8308 Tokyo, Japan
e-mail: mm723@uowmail.edu.au

A. Yamamoto
Department of Applied Physics, Tokyo University of Agriculture and Technology,
2-24-16 Nakacho, 184-8588 Koganei, Tokyo, Japan
e-mail: akiyasu@cc.tuat.ac.jp

S.X. Dou
Institute for Superconducting & Electronic Materials,
Australian Institute of Innovative Materials, University of Wollongong,
Innovation Campus, Squires Way, North Wollongong, 2500 Wollongong,
NSW, Australia
e-mail: shi@uow.edu.au



A. Crisan (ed.), *Vortices and Nanostructured Superconductors*, Springer Series in Materials Science 261, DOI 10.1007/978-3-319-59355-5_3

microscopy (TEM) techniques. For the $MgB_2$ samples doped with oxides, the impurities are mainly MgO and other reaction products between the oxide dopants and Mg and B (Tables 3.2, 3.3, and 3.4). The sizes of such oxide phases are in a wide range. To pin magnetic flux lines effectively from moving in order to enhance critical current density, $J_c$, the size of the precipitates and defects must be comparable to the coherent length, $\xi$ of $MgB_2$. Due to the anisotropic electronic structure of $MgB_2$, the coherent length in the *ab* plane and *c*-axis has been measured to be in the range of 3.7–12.8 nm ($\xi_{ab}(0)$) and 1.6–5.0 nm ($\xi_c(0)$), respectively [1]. For single crystals, $\xi_{ab}(0) = 6.1 - 6.5$ nm and $\xi_c(0) = 2.5 - 3.7$ nm [1]. Nevertheless, the presence of excessive oxide phases would dirty the grain boundaries. Subsequently, the effective cross-sectional area for current transport is reduced leading to a decrease in $J_c$ [2–4]. Because of the formation of MgO or Mg-B-O-related phases, this suggests that Mg deficiency exists in the samples which further enhances upper critical field, $H_{c2}$ [5].

**Table 3.1** Oxides and other precipitates as imaged by TEM technique in various types of polycrystalline $MgB_2$

| Type of samples | Sample areas where the precipitates were observed | Type of precipitates (size) | References |
|---|---|---|---|
| In situ | Within $MgB_2$ grains | • MgO ($\geq$40 nm)<br>• $Mg(BO)_x$ (~5–15 nm and 2–5 nm in the longest axis and shortest axis, respectively) | [13, 25] |
| In situ | Within $MgB_2$ grains | Mg-B-O (5–100 nm) | [21, 26, 35] |
| In situ | Within $MgB_2$ grains and grain boundaries | • Within grains: $MgB_7$ (1–5 nm), Mg-B-O (100–400 nm)<br>• At grain boundaries: Amorphous B-rich phases with crystalline MgO particles | [27, 28] |
| Ex situ | Grain boundaries | MgO (10–500 nm) | [30] |
| HIP Ex situ | Grain boundaries | • amorphous-like $BO_x$ with a typical thickness ~1–3 nm<br>• $BO_x$-$MgO_y$(B)-$BO_z$ in which the thickness of B-O phase is ~2 nm while the typical thickness of MgO is ~10 nm | [22, 29] |
| HIP Ex situ | Within $MgB_2$ grains | Mg-B-O (10–200 nm) | [36] |
| Hot pressing and spark plasma synthesis (both in situ and ex situ) | – | Mg-B-O (15–20 nm or smaller) | [19] |

**Table 3.2** Dopants and their sizes, impurities, lattice parameters, and superconducting properties of the oxides-doped $MgB_2$. These samples were prepared using in situ reaction method at ambient pressure

| Rare earth oxides (REO) | | | | | | | | |
|---|---|---|---|---|---|---|---|---|
| Dopant | Size of dopant | Impurities | Lattice parameters | $T_c$ | $J_c$ | $H_{c2}$ | $H_{irr}$ | References |
| $Pr_6O_{11}$ | – | MgO, $PrB_6$ | No significant change | Decrease by ~2 K for x up to 0.05 in $Mg_{1-x}(Pr_6O_{11})_{x/6}B_2$ | $1.15 \times 10^4$ A/cm$^2$ (10 K, 5 T) for x = 0.03 | Improve | Improve | [55] |
| | 100 nm | MgO, $PrB_6$ | *a*- and *c*-axes increase slightly | Decrease by ~0.2 K for 1 wt% $Pr_6O_{11}$ | $4.5 \times 10^4$ A/cm$^2$ (10 K, 2 T) for 1 wt% $Pr_6O_{11}$ | – | 7.2 T (10 K) | [50, 51] |
| $Eu_2O_3$ | | MgO, $EuB_6$ | No significant change | Decrease by ~2 K for x up to 0.05 in $Mg_{1-x}(Eu_2O_3)_{x/2}B_2$ | $1.10 \times 10^4$ A/cm$^2$ (10 K, 5 T) for x = 0.04 | Improve | Improve | [55] |
| | | MgO, $EuB_6$ | *a*- and *c*-axes increase and then decrease slightly | Decrease by ~2 K for x up to 0.05 in $Mg_{1-x}(Eu_2O_3)_{x/2}B_2$ | $6.05 \times 10^4$ A/cm$^2$ (10 K, 2 T) $4.86 \times 10^3$ A/cm$^2$ (20 K, 2 T) for x = 0.04 | – | 4.46 T (10 K) for x = 0.04 | [52] |
| $CeO_2$ | 100 nm | MgO, $CeB_4$ | No significant change | No significant change up to 10 wt% | Not significant enhancement | – | – | [50] |
| | 10–20 nm | MgO, $CeO_2$ | *a*- and *c*-axes increase slightly | Decrease by ~1 K for 2 wt % $CeO_2$ | $0.45 \times 10^6$ A/cm$^2$ (20 K, 1 T) for 2 wt% $CeO_2$ | – | – | [48] |
| $Lu_2O_3$ | 100 nm | $LuB_4$, MgO | No significant change | No significant change up to 5 wt% | Not significant enhancement | – | – | [50] |

(continued)

**Table 3.2** (continued)

| Rare earth oxides (REO) | | | | | | | | |
|---|---|---|---|---|---|---|---|---|
| Dopant | Size of dopant | Impurities | Lattice parameters | $T_c$ | $J_c$ | $H_{c2}$ | $H_{irr}$ | References |
| $Ho_2O_3$ | 50 nm | $HoB_4$, MgO | No significant change | Decrease by 0.2 K for x up to 10% in $Mg_{1-x}(Ho_2O_3)_xB_2$ | $1 \times 10^3$ A/cm$^2$ (20 K, 5 T) $2.0 \times 10^4$ A/cm$^2$ (10 K, 5 T) $1.2 \times 10^5$ (5 K, 5 T) for x = 3% $Ho_2O_3$ | No significant change | Improve | [56] |
| $Dy_2O_3$ | 1–3 μm | MgO, $DyB_4$, $Dy_2O_3$ | No significant change | Decrease by 0.5 K for up to 5 wt% $Dy_2O_3$ | $6.5 \times 10^5$ A/cm$^2$ (6 K, 1 T) $3.5 \times 10^5$ A/cm$^2$ (20 K, 1 T) for 0.5 wt % $Dy_2O_3$ | – | – | [53] |
| $Y_2O_3$ | 15–30 nm | MgO, $YB_4$ | No significant change | Decrease by 1 K for 15 wt % $Y_2O_3$ | $8 \times 10^4$ A/cm$^2$ (20 K, 2 T) for 10 wt % $Y_2O_3$ | – | 11.5 T (4.2 K), 5.5 T (20 K) for 10 wt% $Y_2O_3$ | [54] |
| $Nd_2O_3$ | – | MgO, $NdB_4$, $NdB_6$, $Nd_2O_3$ | No significant change in the position of the XRD peaks | Decrease by <2 K for 7 at. % in Mg:$Nd_2O_3$:B according to the ratio (1 −x):(x/2):(2 + x) | $5.27 \times 10^2$ A/cm$^2$ (4.2 K, 12 T) for 5 at. % $Nd_2O_3$ | – | – | [49] |
| $Tb_4O_7$ | <60 nm | MgO, $TbB_4$, $Tb_2O_3$ | No significant change | No significant change up to 10 wt% $Tb_4O_7$ | $2.6 \times 10^3$ A/cm$^2$ (5 K, 6 T) for 10 wt% $Tb_4O_7$ | – | 8.1 T (5 K) for 10 wt% $Tb_4O_7$ | [71] |

(continued)

**Table 3.2** (continued)

| Dopant | Size of dopant | Impurities | Lattice parameters | $T_c$ | $J_c$ | $H_{c2}$ | $H_{irr}$ | References |
|---|---|---|---|---|---|---|---|---|
| Rare earth oxides (REO) | | | | | | | | |
| Other oxide dopants | | | | | | | | |
| $Al_2O_3$ | 10–20 nm | MgO, $Al_2O_3$ | *a*- and *c*-axes decrease | Decrease by ~5 K for x = 20% in $MgB_2(Al_2O_3)_x$ | $J_c$ higher than the pure samples at field <3.5 T, 10 K for x = 15% | – | Decrease of $H_{irr}$ is not that great compared to Al doping | [57] |
| $TiO_2$ | 25-50 nm | MgO | *a*- and *c*-axes decrease slightly | Decrease by ~0.5 K for 15 wt% $TiO_2$ | 4250 A/cm$^2$ (20 K, 4.2 T) for 10 wt% $TiO_2$ | – | 4.9 T (20 K) for 10 wt% $TiO_2$ | [60] |
| $SiO_2$ | 10–20 nm | MgO, $Mg_2Si$ | *a*-axis remains unchanged, *c*-axis increases slightly | Decrease by ~0.8 K for x = 15% in $MgB_2(SiO_2)_x$ | $2.5 \times 10^5$ A/cm$^2$ (20 K, 1 T) for x = 7% | – | – | [64] |
| $TiO_2$/ $SiO_2$ | 15 nm | MgO, $Mg_2Si$, $MgB_4$ | *a*-axis remains unchanged, *c*-axis increases slightly | Decrease by ~3 K for up to 15 wt% $TiO_2/SiO_2$ | $3.9 \times 10^3$ A/cm$^2$ (5 K, 8 T) for 10 wt% $TiO_2/SiO_2$ | – | – | [62] |
| $Sb_2O_3$ | 90–210 nm | MgO, $Mg_3Sb_2$ | No significant change | Decrease by <1 K for up to 15 wt% of $Sb_2O_3$ | $2.4 \times 10^3$ A/cm$^2$ (5 K, 8 T) for 2.5 wt% $Sb_2O_3$ | Improve | Improve slightly | [66] |
| $Co_3O_4$ | 25 nm, | MgO, $Co_3O_4$ | – | Decrease by ~4–5 K for 1 mol% $Co_3O_4$, depending on sintering temperature | Best $J_c$ for 1 mol% $Co_3O_4$ sample, $J_c$ increases with sintering temperature | – | $H_{irr}$ is enhanced at ≤ 20 K | [69] |
| MgO | 10 nm | MgO | – | Decrease by <1 K for up to 5 wt% MgO | $3.59 \times 10^3$ A/cm$^2$ (4.2 K, 10 T) for 1.25 wt% MgO | – | – | [70] |

**Table 3.3** Dopant, impurities, evolution of lattice parameters, $T_c$ and $J_c$ of the oxides-doped $MgB_2$ processed by spark plasma sintering [72, 73] and mechanical alloying [74] methods

| Dopant (size) | Impurities | Lattice parameters | $T_c$ | $J_c$ | References |
|---|---|---|---|---|---|
| $Ho_2O_3$ (0.9–15 μm, 0.95–48 μm) | MgO, $HoB_4$, $MgB_4$ | No significant change | remains at ~38 K | – | [72] |
| $GeO_2$ (14.3 μm) | MgO, $MgB_4$, $Mg_2Ge$ | No significant change | Decrease by <1 K for x = 0.005–0.03 in $MgB_2(GeO_2)_x$ | ~100 A/cm$^2$ (20 K, 5.1 T) for the sample $MgB_2(GeO_2)_{0.005}$ | [73] |
| MgO (45 μm) | Mg, MgO, WC | No significant change | Increase ~1 K for 10 wt% MgO | 5 wt% MgO: 1.3 × $10^6$ A/cm$^2$ (7.5 K, self-field) | [74] |
| $SiO_2$ (45 μm) | Mg and $SiO_2$, MgO, $Mg_2Si$, SiC | No significant change | Decrease by ~2 K for 20 wt% $SiO_2$ | 2 wt% $SiO_2$: 1.2 × $10^6$ A/cm$^2$ (7.5 K, self-field) | [74] |

## 3.1.2 *Oxide Phases in Undoped $MgB_2$ Bulks*

Oxygen doping and thus the formation of oxides represents one natural means of introducing flux pinning centers in $MgB_2$. The formation of oxide phases depends on several factors such as the source of starting raw powders and the environment of heat treatment. Very often, the type of oxides and their distribution within the samples are greatly influenced by the sample preparation procedure. In the following subsections, we discuss the sources for oxidation and compare the properties of the samples prepared using in situ, ex situ, high-pressure, and high-temperature techniques.

### 3.1.2.1 Sources of Oxidation

#### Oxidized Mg Powder

It is common that MgO is contained in the commercial magnesium (Mg) powder [6]. The presence of MgO could be a residue of the processing of Mg such as using the thermal reduction process [7]. Mg is a reactive metal and oxidation of Mg can take place even at room temperature if it is exposed to air for a sufficient period of time. MgO has a very negative Gibbs energy of formation (−568.945 kJ · mol$^{-1}$ and −492.952 kJ · mol$^{-1}$, at 298.15 K and 1000 K, respectively, for crystalline phase) compared with $MgB_2$ (−89.473 kJ · mol$^{-1}$ and −80.376 kJ · mol$^{-1}$, at 298.15 K and 1000 K, respectively, for crystalline phase) [8]. In attempts to reduce oxidation,

**Table 3.4** Dopant, impurities, evolution of lattice parameters and superconducting properties of the oxides and different carbon sources (C) co-doped $MgB_2$. These samples were prepared using in situ reaction method at ambient pressure

| Dopant (size) | Impurities | Lattice parameters | $T_c$ | $J_c$ | $H_{c2}$ | $H_{irr}$ | References |
|---|---|---|---|---|---|---|---|
| Nanodiamond (<10 nm) + nano-$SiO_2$ (10 nm) | Mg, MgO, $Mg_2Si$ | *a*-axis decreases; *c*-axis unchanged | Decrease by ~2 K for $MgB_{1.9}C_{0.1}$ + 5 wt% $SiO_2$ | 2.1 × $10^3$ A/cm$^2$ (5 K, 8 T) for $MgB_{1.9}C_{0.1}$ + 5 wt% $SiO_2$ | – | 7.6 T (15 K) for $MgB_{1.9}C_{0.1}$ + 5 wt% $SiO_2$ | [77] |
| Nanocarbon (<50 nm) + nano-$SiO_2$ (10 nm) | MgO, $Mg_2Si$ | *a*-axis decreases; *c*-axis unchanged | Decrease by ~3 K for $MgB_{1.9}C_{0.1}$ + 2.5 wt% $SiO_2$ | 9 × $10^3$ A/cm$^2$ (5 K, 8 T) for $MgB_{1.9}C_{0.1}$ + 2.5 wt% $SiO_2$ | – | – | [78] |
| 3 wt% GO + 1 wt% REO (submicron) (REO = La, Sm, Eu, Gd, Tb, Ho oxides) | MgO, $REB_4$, $REB_6$ | No significant change | No significant change | For GO + Tb 46.6 × $10^3$ A/cm$^2$ (10 K, 5 T) 14.2 × $10^3$ A/cm$^2$ (20 K, 4 T) | 22.4 T (0 K) for Tb doping | – | [79] |
| Graphite + $Eu_2O_3$ (–) | MgO, $EuB_6$, $EuC_2$ | *a*- and *c*-axes decrease | Decrease by ~1 K for $Mg_{0.97}(Eu_2O_3)_{0.03}(B_{0.95}C_{0.05})_2$ | 1.57 × $10^3$ A/cm$^2$ (20 K, 4 T) for $Mg_{0.97}(Eu_2O_3)_{0.03}(B_{0.95}C_{0.05})_2$ | – | – | [76] |
| Graphite + $Pr_6O_{11}$ (–) | MgO, $PrB_4$, $PrB_6$ | *a*- and *c*-axes decrease | Decrease by ~3 K for $Mg_{0.99}(Pr_6O_{11})_{0.01}(B_{0.95}C_{0.05})_2$ | 2.83 × $10^2$ A/cm$^2$ (20 K, 4 T) for $Mg_{0.99}(Pr_6O_{11})_{0.01}(B_{0.95}C_{0.05})_2$ | | | [76] |

some researchers make use of $MgH_2$ instead of Mg powder to prepare $MgB_2$ via in situ reaction method [9]. It was reported that the use of $MgH_2$ increases $J_c$ significantly probably by reducing the formation of MgO especially at the grain boundaries leading to improved grain connectivity [9, 10]. Even some commercial $MgB_2$ powders were found to contain MgO [2] possibly due to surface oxidation of $MgB_2$ [11]. Moreover, MgO could be a residue of the manufacturing process.

**Impurities in Boron Powder**

The presence of $B_2O_3$ in commercial boron (B) powder has been reported widely [12– 14]. The type of residual impurities in B powder depends on the manufacturing processes [15]. One way to produce elemental B powder is using magnesiothermic process in which magnesium is reacted with $B_2O_3$ [16]. Hence, the presence of $B_2O_3$ as a residue is inevitable after the reduction process. As shown by Jiang et al. [13], $B_2O_3$ can be removed from the starting B powder by sintering in reducing $H_2$-Ar gas atmosphere. Reaction of B with oxygen in air at room temperature is also possible [17]. In addition, some commercial B powders have been found to contain impurities such as C, Mg, and other metallic elements [18–20].

**Residual Oxygen in the Heating Environment**

The presence of residual oxygen in a variety of atmosphere during heat treatment is inherent despite the heat treatment was conducted in reducing or inert gas atmosphere [21, 22] and using evacuated quartz ampoule [23]. This is due to the presence of ppm oxygen in the commercial gas cylinders. All these lead to the formation of oxide phases in $MgB_2$ samples. As shown by Kim et al. [24], the formation of MgO fraction in the $MgB_2$ samples does not depend on the sintering temperature implying that the sources for oxidation are mainly from the starting Mg and B powders as well as the heat treatment atmosphere.

#### 3.1.2.2 Types of $MgB_2$ Samples

**In Situ $MgB_2$**

By removing $B_2O_3$ from the precursor boron (B) powder, the amount of MgO in the synthesized $MgB_2$ was reduced from 5.1 to 1.5 mol% as estimated based on the XRD data [13]. This has resulted in an increase of effective cross-sectional area for current transport, $A_F$ from 0.25 to 0.48 indicating that the MgO phase is mostly formed at the intergranular areas [13]. The resistivity at 40 K, $\rho_{40K}$, was reduced from 5.4 $\mu\Omega \cdot cm$ in the $MgB_2$ synthesized using the as-received B powder to 1.7 $\mu\Omega \cdot cm$ in the samples synthesized using the purified B powder. Hence, the estimated residual resistivity ratio, *RRR*, increased from 6.5 to 10. The $J_c$ of the samples synthesized using the purified B powder is higher compared with the samples synthesized using the as-received B powder as a result of improved grain connectivity [13, 14]. In those samples, TEM imaging showed the presence of spherical MgO (size ~40 nm or larger) within $MgB_2$ grains [25]. In addition, platelet-shape nanoprecipitates were found within $MgB_2$ grains (size ~5–15 nm and 2–5 nm in the longest and shortest axes, respectively). These precipitates have a well-defined orientation with $MgB_2$ matrix, i.e., the longest of the platelet lying perpendicular to the *c*-axis of $MgB_2$ matrix, and oxygen ordering in every other unit cell occurs along the *c*-direction of $MgB_2$ lattice [25]. Electron energy loss spectroscopy (EELS) showed that the precipitates contain Mg, B, and O with the composition of $Mg(BO)_x$ [25]. The observation of such an oxygen ordering agrees with previous findings [21, 26] using TEM imaging too showing a broad range of 5–100-nm-sized precipitates consisting of coherently ordered $MgB_{2-x}O_x$ phases in the $MgB_2$ grains. These phases were formed by oxygen substitution onto the boron lattice in an ordered manner [26]. The periodicity of the oxygen ordering is governed by the oxygen concentration in the precipitates and mainly occurs in the (010) plane [26]. Such Mg-B-O phases may not be stable at high temperature as they tend to transform into MgO and reside at the grain boundaries [21]. In addition, it is possible that point defects present due to oxygen doping and the formation of other impurities. Together, they contribute toward flux pinning.

For in situ samples which were synthesized using high-purity B and Mg in a well-controlled environment [27], no obvious MgO phase was detected by both XRD and TEM [28]. TEM imaging showed various types of intragrain nanoprecipitates. Some of these are spherical ones (1–5 nm) which can be indexed to $MgB_7$. These precipitates have no specific orientation with respect to $MgB_2$ matrix [28] while the larger ones (100–400 nm) could be Mg-B-O as reported elsewhere [25]. The grain boundaries are dominated by amorphous B-rich phases. In these amorphous phases, there are crystalline particles which could be identified as MgO [29]. It is worth mentioning here that the clean samples reported in [28] have *RRR* as high as ~15. By annealing the samples in Mg vapor, the *RRR* value was reduced to ~3 with no significant change in the inter- and intragrain microstructure [28]. The Mg vapor-treated samples showed enormous enhanced $J_c$, $H_{c2}$, and irreversibility, $H_{irr}$, compared with the clean samples though the $T_c$ of the former is about 2 K lower than the latter. The reduced *RRR* and increased resistivity at 40 K, $\rho_{40K}$, indicate enhanced impurity scattering in those samples. In addition, the nanoprecipitates could act as flux pinning centers.

**Ex Situ $MgB_2$**

By sintering the pellets made of commercial $MgB_2$ powder in Mg vapor, Zhu et al. found that MgO is the second major phase in the samples as shown by XRD and TEM [30]. However, the sintered pellets appeared to be crumbly suggesting poor connectivity among the grains. The size of MgO was found to be 10–500 nm. The MgO was found to be in the amorphous grain boundaries as a result of oxidation of Mg vapor. Due to the difference of thermal expansion and compressibility between $MgB_2$ and MgO [31], coexistence of the both phases may induce the formation of defects and related interfacial strain [30, 32]. Under TEM imaging, the diffraction patterns of MgO (cubic) and $MgB_2$ (hexagonal) can be quite similar in some orientations in spite of their different crystal symmetries [30].

### High-Pressure and High-Temperature Sintered $MgB_2$ Samples

For the homemade hot-pressed ex situ samples [22], XRD data showed the presence of 5% MgO [29]. TEM did not reveal any appreciable oxygen content within the $MgB_2$ grains [29]. Two distinct types of oxides were found at the grain boundaries: (i) amorphous-like $BO_x$ with a typical thickness ~1–3 nm and (ii) $BO_x$-$MgO_y$(B)-$BO_z$ in which the thickness of B-O phase is about 2 nm while the typical thickness of the MgO is about 10 nm. The formation of such oxide phase could be due to increased oxygen uptake resulting in oxidation in the $MgB_2$ regions. This in turn assists MgO to nucleate and hence forming MgO to be sandwiched between the $BO_x$ layers [29].

In the $MgB_2$ samples prepared using hot isostatic pressing (HIP) technique, better grain connectivity and high density of defects are the essential factors for increasing $J_c$ considerably [33, 34]. The enhanced intergrain connectivity is shown by reduced resistivity at room temperature compared with the un-HIPed samples [34]. HIP also reduces *RRR* from ~9 (un-HIPed) to ~3. The lower *RRR* suggests higher strain level probably because of a higher density of defects and dislocation resulted from the HIP process [34, 35]. It was found that MgO, which resides in the grain boundaries of the un-HIPed samples, was well dispersed within $MgB_2$ grains after HIP [34]. The size of these fine MgO particles is about 10–50 nm [34]. In addition, Mg(B, O) also presents [36]. As mentioned before, the Mg(B, O) phase is unstable at high temperature and transforms into MgO with the presence of trace amount of oxygen [21]. On a separate finding, Yates et al. reported the formation of $Mg_x(B_yO_z)$ in their $MgB_2$ thin films annealed in oxygen flow [37]. Such oxide phase is unstable at room temperature and decays into MgO and $B_2O_3$. Surprisingly, $T_c$ of the films was restored as the oxide phase decays.

Prikhna et al. reported the preparation of a series of $MgB_2$ samples using hot pressing and spark plasma sintering methods [19]. SEM imaging and Auger spectroscopy showed the presence of nanolayers (~15–20 nm thick) with oxygen-enriched areas Mg-B-O. As the processing temperature increases, such layers transform into smaller sized Mg-B-O inclusions. The pinning mechanism shifts from grain boundary to point pinning as the processing temperature increases probably because of increased density of defects.

## *3.1.3 Oxides Doping*

To date, various oxides have been attempted as dopants for reaction with $MgB_2$ including graphene oxide, rare earth oxides as well as other magnetic and non-magnetic oxides. Mostly these samples were prepared by in situ process whereby the dopants are mixed with Mg and B and subsequently reacted at elevated temperature in Ar or Ar/$H_2$ gas flow (Sects. 3.1.3.1–3.1.3.3). However, some of the oxides-doped samples were prepared using spark plasma sintering method at high temperature (Sect. 3.1.3.4) and mechanical alloying process (Sect. 3.1.3.5). Some studies on co-addition of oxides with a variety of carbon sources have also been undertaken (Sect. 3.1.3.6). This section discusses the influence of oxides additions and sample preparation procedure on the lattice and superconducting properties of $MgB_2$.

#### 3.1.3.1 Graphene Oxide

In $MgB_2$ samples doped with graphene oxide (GO) [38], the volume of MgO was found to increase from ~3 to 5% for doping up to 10 wt% as estimated based on the XRD data. Both *a*- and *c*-axes decrease with the amount of GO additions. However, the reduction in lattice parameters is relatively smaller compared with doping using other carbon (C) sources [39, 40], suggesting C substitution is smaller in GO-doped $MgB_2$ samples. Full width at half maximum (FWHM) of the XRD peaks increases with GO doping indicating a decrease in the crystallinity. SEM images showed denser grain structure and the grain size decreases with GO doping. The calculated effective cross-sectional area for current transport, $A_F$, increases with GO doping. This is in agreement with the results obtained from atom probe microscopy (APM) showing that the presence of GO between the grains may lead to improved grain connectivity [41]. This is because GO has a good electrical and thermal conductivity. Apart from this, the aligned GO also serves as a base for texturing. Nanoscale oxide phases such as MgO and Mg-B-O were also observed in the GO-doped samples [41]. In particular, $Mg(B, O)_2$ was observed to be surrounded by MgO (~5–7 nm) consistent with the previous finding by Liao et al. that the former is unstable and transforms into MgO [21]. Such a distribution of oxygen within the samples is considered as interstitial. These oxide precipitates serve as effective flux pinning centers.

The $T_c$ remains largely unchanged ~38 K for doping up to 10 wt% of GO. This is quite similar to graphene (G) doping alone whereby 5 mol% of G doping only results in 1 K decrease of $T_c$ [42]. This is consistent with a small reduction in the lattice parameters reinforcing the fact that low C substitution takes place in GO-doped samples probably due to strong C–C bonding in graphene [43]. The improvement of both $H_{c2}$ and $H_{irr}$ in the GO-doped samples over the undoped samples is insignificant due to a low C substitution resulting in a small degree of charge carrier scattering. This is different from other sources of carbon doping [39, 40] in which C substitution level is a lot higher leading to a strong impurity scattering which is known to enhance $H_{c2}$ [44, 45]. The enhancement of $J_c$ due to GO doping is so enormous and even better than graphene alone [41] or nanocarbon [46] doping. The superior $J_c$ performance of GO-doped samples is believed to be the effect of a combination of grain texturing and improved grain connectivity apart from effective flux pinning due to the presence of defects and nanoprecipitates as well as C substitution [41]. In graphene-doped $MgB_2$ samples, it was reported that the spatial fluctuation in $T_c$ is responsible for the pinning mechanism [46].

#### 3.1.3.2 Rare Earth Oxides

In situ reaction of rare earth oxides (REOs) with Mg and B leads to the formation of $REB_4$, $REB_6$, and MgO as impurities [47] (Table 3.2). These phases increase with the increased amount of rare earth oxide additions. Unreacted rare earth oxides can be found in the XRD patterns if the dopant addition is excessive [48, 49]. The lattice parameters of the $MgB_2$ samples doped with rare earth oxides do not change much indicating that substitution of rare earth element into the lattice of $MgB_2$ is insignificant. This also shows limited solubility of RE in $MgB_2$ which may due to the large difference between the ionic size of RE and Mg [47, 50].

As shown by TEM imaging, the formed nanoprecipitates are distributed throughout the samples leading to two distinct effects: (i) degraded grain connectivity when the nanoprecipitates are segregated at the grain boundaries [49–51]. This is evidenced in the decrease of effective cross-sectional area for current transport, $A_F$ [49]. However, in $Eu_2O_3$-doped samples, it was found that $A_F$ increases with doping level indicating improved grain connectivity [52] (ii) enhanced flux pinning capability as the nanoprecipitates are mainly distributed within the $MgB_2$ matrix (intragrain) if the effect from (i) is not overwhelmed [50, 53, 54]. TEM images showed that the size of some nanoprecipitates is comparable or even smaller than the coherent length of $MgB_2$, thus acting as effective flux pinning centers.

The $T_c$ remains unchanged largely with rare earth oxide doping (Table 3.2). For example, $T_c$ decreases by about 1 K only for 15 wt% of doping with $Y_2O_3$ [54]. This can be understood as the oxides do not dope into the lattice structure of $MgB_2$ as shown by insignificant change of lattice parameters. This shows that the nanoprecipitates are mostly distributed within the matrix without changing much the crystal structure of $MgB_2$. The formation of some RE-boride phases leads to paramagnetism in the samples as shown in the temperature-dependent magnetization measurement as well as the magnetization hysteresis loops [50, 55, 56].

The improvement in $J_c$ for REO-doped samples is believed to be due to enhanced pinning properties arising from the nanoprecipitates which act as effective flux pinners. As shown by the reduced *RRR*, the improved $H_{c2}$ is a result of enhanced impurity scattering [44] due to the presence of various types of nanoprecipitates in addition to defects. It has been shown that the enhanced $H_{irr}$ is correlated to the increased FWHM indicating that lattice distortion may play an important role in this regards [52]. This is also supported by the increase in strain because of lattice distortion [52].

Table 3.2 summarizes the dopant size, impurities, lattice parameters, and superconducting properties of the rare earth oxides-doped $MgB_2$ samples.

#### 3.1.3.3 Other Oxide Dopants

***$Al_2O_3$***

In nano$Al_2O_3$-doped $MgB_2$ samples [57], the decrease in lattice parameters is due to substitution of Al for Mg, similar to aluminum doping alone [58]. Such change in lattice parameters is rather small compared with Al doping alone indicating the more reactive form of pure Al in substituting Mg during chemical reaction. While the $T_c$ does not decrease much, $J_c$ at low field (<3.5 T, 10 K) is enhanced compared with the pure sample. This behavior is quite similar to Al doping [59]. However, $J_c$ at 20 and 30 K for the $Al_2O_3$-doped samples is lower than the pure samples. This could be attributed to the fact that pinning capability is not great as compared to that of rare earth oxide doping in which substitution is insignificant and the nanoprecipitates are distributed within the matrix of $MgB_2$ acting as magnetic flux pinners. It was found that $H_{irr}$ of the $Al_2O_3$-doped samples is poorer than the pure sample. The particle size of the starting nanoalumina powder is 10–20 nm. After reaction, some fine particles with the size 10–50 nm were found to be dispersed within $MgB_2$ grains. EDS showed the presence of elements Mg, B, Al, and O in the grains.

***$TiO_2$***

For $MgB_2$ samples doped with $TiO_2$ [60], MgO is the only impurity detected by XRD. Compared with Ti-doped $MgB_2$ samples [61], Ti, $TiB_2$, and $TiB_4$ were found in addition to MgO showing the more reactive form of Ti compared with $TiO_2$ in chemical reaction. As shown by the small change in lattice parameters and $T_c$, Ti does not substitute into the $MgB_2$ lattice. Similar finding was reported in Ti-doped $MgB_2$ [61]. The small change in the lattice parameters is attributed to the presence of MgO and Mg deficiency [60]. The enhanced $J_c$ and $H_{irr}$ are believed to be due to a number of factors such as the presence of nanoprecipitates and Mg deficiency. In Ti-doped samples [61], formation of thin $TiB_2$ layers at the grain boundaries is beneficial to connecting $MgB_2$ grains. Thus, the grain connectivity and area for grain boundary pinning are increased leading to enhancement in $J_c$.

Zhang et al. reported their study on $MgB_2$ bulks doped with $TiO_2$ particles [62]. The surfaces of the starting $TiO_2$ particles have been modified by 5–10% $SiO_2$. XRD showed the presence of MgO, $MgB_4$, and $Mg_2Si$ indicating the presence of a significant amount of $SiO_2$ in the dopant. The *a*-axis does not change much while the *c*-axis decreases slightly with the amount of dopant consistent with the fact that neither Ti [61] nor Si [63] substitutes into the lattice of $MgB_2$. 15 wt% of dopant addition causes only a reduction of $T_c$ by about 3 K. $J_c$ versus field is enhanced significantly when the samples were sintered at the optimal temperature. Such enhancement is probably due to improved flux pinning by nanoprecipitates which are formed as a result of reaction between $TiO_2$ and $SiO_2$ with Mg and B during the in situ reaction in addition to Mg deficiency as Mg reacts with Si to form $Mg_2Si$.

### *SiO₂*

In nano$SiO_2$-doped $MgB_2$ samples [64], the *c*-axis lattice parameter increases slightly while that of the *a*-axis remains unchanged. Consequently, $T_c$ drops by 0.8 K only for 15% of dopant additions due to insignificant Si substitution into the $MgB_2$ lattice similar to Si doping alone [63]. Some fine particles (10–50 nm) which are identified as $Mg_2Si$ using EDS were found to distribute within the $MgB_2$ grains. However, improvement of $J_c$ is not as great as nanoSi doping [65]. $J_c$ of $SiO_2$-doped sample shows improvement at low applied magnetic field only. However, $H_{irr}$ is not improved much. The reasons for this could be the following: (i) the presence of significant amount of impurities such as $Mg_2Si$ and MgO that degrade the supercurrent path and (ii) ineffective flux pinning due to the large size of nanoprecipitates (some as large as 50 nm) compared with the coherent length of $MgB_2$.

### *Sb₂O₃*

By doping nanosized $Sb_2O_3$ particles into $MgB_2$, both *a*- and *c*-axes lattice parameters decrease slightly suggesting the possibility of oxygen doping [66]. XRD showed the presence of MgO and $Mg_3Sb_2$ phases as the doping level increases. The decrease in the estimated value $A_F$ indicates the increase of insulating phases MgO and $Mg_3Sb_2$ which reduce the supercurrent path. SEM images showed smaller grain size of the doped samples compared with the pure one. $T_c$ of the samples decreases by less than 1 K only even for doping up to 15 wt%. Both $H_{c2}$ and $H_{irr}$ are improved probably because of enhanced impurity scattering and lattice distortion as indicated by a lower value of *RRR* and increased FWHM of the (110) and (002) XRD reflections, respectively. In addition, the increase of $H_{c2}$ is also enhanced by Mg deficiency as indicated by the formation of $Mg_3Sb_2$. The great improvement of $J_c$ versus field especially at high fields is believed to be contributed by the enhanced $H_{c2}$ and grain boundary pinning.

### *Co₃O₄*

For the samples doped with $Co_3O_4$, XRD showed the presence of MgO and unreacted $Co_2O_3$ as the doping level increases [67]. No significant change in the lattice parameters was found. The crystallinity of the samples is degraded as shown by the increased FWHM values of the XRD peaks. $T_c$ does not change much (37–38 K) for up to 6% of $Co_3O_4$ additions, suggesting that no Co substitution into $MgB_2$ lattice has taken place. On the other hand, Co may substitute for Mg in Co-doped $MgB_2$ as concluded from Mössbauer spectroscopy measurement [68]. This may be ascribed to the more reactive form of pure Co element compared to $Co_3O_4$ during chemical reaction. Field-cooled (FC) and zero-field-cooled (ZFC) temperature-dependent magnetic susceptibility measurements showed the presence of paramagnetic Meissner effect (PME). Such phenomenon can be related to the stacked $MgB_2/Co_3O_4/MgB_2$ layers giving rise to superconductor-insulator-superconductor (SIS) or superconductor-normal metal-superconductor (SNS) junctions. However, Kuroda et al. reported that PME was observed in both their undoped and $Co_3O_4$-doped samples, suggesting that such phenomenon may not originate from the stacked junctions [69]. It should be noted that the samples reported in [69] show greater degradation in $T_c$ compared to that in [67]. $J_c$ is enhanced almost by an order of magnitude compared with the undoped $MgB_2$ samples [67]. $J_c$ and $H_{irr}$ were found to be influenced greatly by the sintering temperature indicating that sample preparation procedure is crucial in influencing the superconducting properties of these samples [69].

***MgO***

Addition of MgO into $MgB_2$ gives rise to a higher volume fraction of MgO as shown in the XRD patterns [70]. As expected, the change of lattice parameters is negligible small and hence $T_c$ does not degrade much either [70]. $T_c$ decreases less than 1 K for addition of up to 5 wt% MgO. To enhance $J_c$, optimal amount of MgO additions is crucial for increasing flux pinning without degrading grain connectivity due to the presence of excessive insulating MgO, especially when they reside at the grain boundaries [2, 70].

The dopant size, impurities, lattice parameters, and superconducting properties of the various oxides-doped $MgB_2$ samples are given in Table 3.2.

#### 3.1.3.4 Samples Prepared Using Spark Plasma Sintering (SPS)

The microstructure and superconducting properties of $Ho_2O_3$ [72]- and $GeO_2$ [73]-doped $MgB_2$ synthesized using ex situ SPS method have been studied. The advantage of using such material processing route is that samples with high density (>90% of the theoretical density) can be obtained. In addition, the samples have nanograin structure and high density of defects.

***$Ho_2O_3$***

The physical properties of the $Ho_2O_3$-doped samples prepared using SPS method [72] are quite similar to those prepared at ambient pressure [56]. XRD data indicated the presence of MgO, $HoB_4$, and $MgB_4$ (Table 3.3). C doping is possible due to the C containing synthesis environment. No obvious change of lattice constants is noticeable as also reflected in a small change of $T_c$. SEM images showed that the samples are composed of regions with relatively free of Ho which are embedded in regions containing Ho. It is believed that the starting powder of $Ho_2O_3$ is very crucial in defining the microstructure which will affect $J_c$. Two different $Ho_2O_3$ particles were used as dopant in the study reported in [72]. The first type consists mainly of bar-shaped particles (0.9–15 μm) while the second type has sphere-like shape (0.95–48 μm). In the samples prepared using a combination of the two types of dopants, $J_c$ could be controlled so that enhancement of $J_c$ occurs at all fields. TEM images showed the presence of nanoinclusions with the size of 8–60 nm at the grain boundaries. Most of these nanoinclusions are found to have a size of even less than 20 nm. EDS showed that they are Ho-B-O phases. It is considered that $\delta T_c$

pinning is the main pinning mechanism due to defects, grain boundaries, and nanoscale impurities.

***$GeO_2$ doping***

XRD patterns of the $GeO_2$-doped samples showed the presence of impurities MgO, $MgB_4$, and $Mg_2Ge$ [73] (Table 3.3). The *a*-axis lattice parameter decreases slightly (may be due to accidental C doping) while that of the *c*-axis remains constant. Hence, substitution effect is insignificant in agreement with the very little change of $T_c$ (<1.0 K). Moreover, the residual strain in the samples does not change with $GeO_2$ additions. SEM showed that the grain size of $MgB_2$ decreases after reaction with $GeO_2$, thus giving rise to more grain boundary pinning. TEM images showed 25–100-nm C-rich magnesium boride phases. In addition, about 100-nm-sized Mg-O and Mg-Ge phases were also found. The size of these phases is too big to act as effective pinning centers. Therefore, the pinning mechanism is attributed to grain boundaries. $GeO_2$ doping shifts the pinning mechanism from point pinning to grain boundary pinning. $J_c$ at 20 K, 5.1 T, is about 100 A/cm$^2$ for the sample with nominal composition $MgB_2(GeO_2)_{0.005}$.

#### 3.1.3.5 Samples Prepared Using Mechanical Alloying and Hot Pressing

Perner et al. [74] reported the preparation $MgB_2$ samples doped with MgO and $SiO_2$ using mechanical alloying method. The starting Mg and B were mixed with either MgO (45 μm) or $SiO_2$ (45 μm) and mechanical milled using a planetary ball mill. The mixture was then hot pressed. XRD patterns showed the presence of unreacted Mg, MgO, and debris from the milling tool (WC) for the MgO-doped samples while unreacted Mg, $SiO_2$, MgO, $Mg_2Si$, and SiC were found in the $SiO_2$-doped samples (Table 3.3). It is unclear whether SiC appears as a reaction product during the mechanical alloying or impurity in the $SiO_2$ nanopowders. No significant change in the lattice parameters in both samples was found. For the $SiO_2$-doped samples, the *c*-axis lattice parameter increases slightly only which is attributed to the possibility of more effective substitution of Si into $MgB_2$ lattice assisted by mechanical alloying process. The relatively lower $T_c \sim 34$ K for the undoped $MgB_2$ implies that the samples are much distorted due to high-energy milling. The change of $T_c$ is only ~1 and 2 K, for additions up to 10 wt% MgO and 20 wt% of $SiO_2$, respectively. The $J_c$, $H_{c2}$, and $H_{irr}$ increase significantly with appropriate amount of MgO or $SiO_2$ addition. TEM images showed 10–20-nm-sized MgO and 20–40-nm-sized $MgB_2$ grains for the MgO-doped samples. For the $SiO_2$-doped samples, TEM images showed the presence of 10–30 nm MgO, 20–160 nm $Mg_2Si$ as well as about 20 nm $MgB_2$ grains. These nanoprecipitates should act effectively as flux pinning centers. It should be noted that the mechanically milled and subsequently hot-pressed samples have much higher density than those prepared in ambient pressure [75] suggesting better grain connectivity of the former. The processing method also leads to a high density of defects and grain boundaries (as a result of nanosize grains) as well as nanoprecipitates which are crucial for increasing $J_c$, $H_{c2}$, and $H_{irr}$.

#### 3.1.3.6 Co-additions

Studies of superconducting properties of $MgB_2$ with co-additions of different sources of C and $SiO_2$ as well as rare earth oxides have been reported (Table 3.4). XRD patterns showed the presence of impurities which are MgO and reaction products between the dopant with Mg and B. Due to C, doping effect is obvious as evident in the change of the $a$-axis lattice parameter while that of $c$-axis remains unchanged largely. However, the decrease in both $a$- and $c$-axes was also noticed in graphite and rare earth oxides co-doped samples [76]. Due to C doping and impurity phases, the sample crystallinity is degraded as shown by the increased FWHM and strain [76–79]. The grain connectivity is degraded because of the presence of impurity phases as indicated by the decrease in $A_F$ [76]. However, the graphene oxide- and rare earth oxide co-doped samples showed increased $A_F$ indicating improved grain connectivity [79] probably due to grain alignment induced by graphene oxide as reported elsewhere [38]. The decrease in $T_c$ is expected mainly because of C doping. The enhancement in $J_c$, $H_{c2}$, and $H_{irr}$ is attributed to the increased pinning and enhancement of impurity scattering because of the presence of nanoprecipitates and C doping.

## 3.2 Carbon Impurity Doping

### *3.2.1 Why Carbon Doping?*

Chemical doping by introducing appropriate impurity atoms into materials is a simple and conventional but effective way of modifying the structures, especially at the atomic scale and the nanoscale. Such structural modifications have strong influences on many properties, for example, electrical and magnetic properties, and it is essential to understand the nature of materials and to enhance the physical properties for their applications. The approach by chemical doping has been of course applied for binary $MgB_2$ materials, such as $MgB_2$ superconducting wires, bulks, and films. Many attempts to dope various impurities and to find active third and fourth elements for changing nanostructure parameters that characterize the crystal system, and thus, tuning the superconducting properties has been made since the discovery of the superconductivity in 2001 [80]. The main objective of these trials is to specifically improve two superconducting properties, namely the upper critical field and the in-field critical current density, which are the most important criterions for technical applications such as the next generation of superconducting magnets for magnetic resonance imaging (MRI) scanner. The performance of the in-field transport current without dissipation is particularly important and it is mainly determined by several factors, for example, the upper critical field, the irreversibility field, flux pinning, anisotropy, and the connectivity, which are depended on nanostructures in the superconducting materials.

Today, carbon is known to be the most appropriate impurity element for high-field applications using $MgB_2$ materials, especially the wires and films. The chemical doping reduces the number of holes, introduces scattering centers for the charge carriers, and thus decreases the critical transition temperature, but reduces the anisotropy and drastically improves the upper critical field, the irreversibility field, and the in-field critical current property when an appropriate dopant as a carbon source is doped by adding an appropriate amount and using an effective method [81–86]. The pronounced effect was firstly discovered in 2002 by Dou et al. through silicon carbide (SiC) doping [87], which is being studied even now and considered as one of the most effective inorganic dopants. Numerous carbon-containing compounds including organic matter as well as inorganic one can be available as dopants, and thus, the first result has attracted two main questions of what the most appropriate dopant as a carbon source is and how the superconducting properties change upon incorporating the dopant. A considerable number of studies have been done so far and the intense research efforts have elucidated the mechanism of the improvement in the in-field critical current density, especially for $MgB_2$ wires and bulks. Many reviews [81–86] published on this subject have already described the mechanism, and they have led to a better understanding of the effect of carbon doping. The nanoscopic roles for $MgB_2$ wires and bulks are, however, not shown in these reviews, and this is because the origin has recently been observed at the atomic-resolution scale by using advanced microscopic techniques [88]. Different nanostructures formed by carbon doping have also been observed in $MgB_2$ thin film, and the unique structures have been investigated to elucidate the origin and the mechanism [89–91]. The studies of the doping effect besides the investigation of the nanoscopic origin have been conducted on many carbon dopants for wires and bulks, and it means that the first question regarding the most appropriate carbon dopant is still open. Here, we therefore compare various dopants, describe the impact of the recent doping methods, and discuss the mechanism, especially in terms of the nanostructural results of thin films, wires, and bulks.

### 3.2.2 Carbon Doping Effect

Recent progress of nanostructural studies has clearly revealed that the origin of carbon doping into $MgB_2$ wires and bulks, which is indispensable for understanding the mechanism of the improvement in the in-field critical current density. The nanoscopic origin was investigated on in situ processed $MgB_2$ wires, which are doped with malic acid ($C_4H_6O_5$) as a carbon source and sintered at a low temperature of 650 °C [88]. The carbon compound is known to be one of the most effective organic dopants [92], and the doped wire was thus selected as a research target for examining the carbon role on the nanostructure. The wire cores were treated by a wedge polishing method and then by an ion milling process with a low acceleration voltage of 100 eV. The ion-milled sample was analyzed using a 300-keV field-emission transmission electron microscope, and a resulting image is shown in Fig. 3.1a. Many structural defects (marked with red arrows in the figure) are found to arise in the carbon-doped sample sintered at the low temperature, and they are identified as stacking fault which is a type of plane defect. The planar defects are assumed to be formed, for example, by condensation of vacancies or interstitials introduced in the hexagonal structure consisting of alternatively stacked magnesium and boron planes in $MgB_2$ crystals, and the structural change can be caused by substitutional or interstitial effect of the chemical doping. The transmission electron microscope image obtained at the range of a few hundreds of nanometers (as shown in Fig. 3.1a) is, however, not determined which planes, atoms, and atomic positions are associated with and responsible for such defects. The nanoscopic origin to be determined for elucidating the fundamental mechanism was further investigated by magnifying and analyzing the region where the stacking faults are present, and it was performed by using a Cs-corrected scanning transmission electron microscope equipped with electron energy loss spectrum analysis as well as the transmission electron microscope [88]. The resulting images and analyses can distinguish the two different layers of magnesium and boron and prove clearly that partial boron planes are missing from the original crystal structure. It is thus confirmed that carbon doping with the low-temperature sintering introduces stacking faults on boron planes of the $MgB_2$ crystal structure in the case of the polycrystalline materials.

The atomic arrangement and structural defects caused by carbon doping with the low-temperature sintering were also studied by using the scanning transmission electron microscope [88, 93], and a resulting image is shown in Fig. 3.1b. A magnesium atom and two boron atoms can be distinguished (and some of them

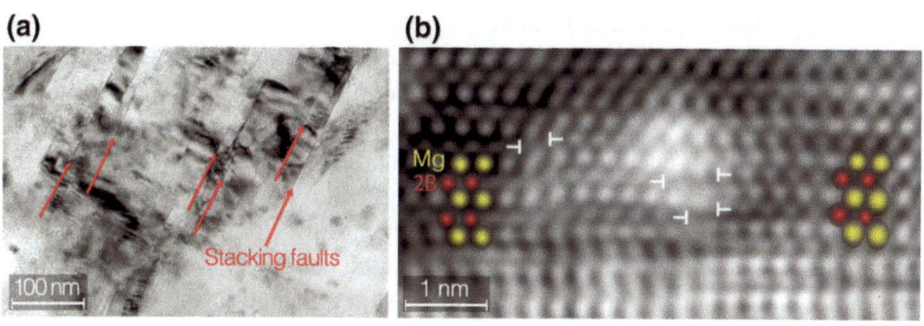


**Fig. 3.1** Nanostructures of the core region of the malic acid-doped $MgB_2$ wires. **a** Low-magnification transmission electron microscope image (reproduced from [88]). *Red arrows* denote stacking faults. **b** Annular dark-field scanning transmission electron microscope image (reproduced from [93]). *Yellow and red balls* denote a magnesium atom and two boron atoms, respectively. *White 'T' marks* denote edge dislocations and areas sandwiched between the marks correspond to boron vacancies

are marked with yellow and red balls as a guide for eyes in the figure), and they are confirmed to form a stacking structure of magnesium and boron layers in $MgB_2$ crystals. Such a high-resolution image can also capture atomic-scale behaviors of planer, line, and point defects such as stacking faults, dislocations, and vacancies. For example, lattice defects (marked with white 'T' marks in the figure) are found to arise in the layer structure, and they are identified as edge dislocation which indicates a terminal point of incomplete lattice rows. The dislocations are accompanied by local stress fields, and the sandwiched areas on boron layers are found to be missing. The missing areas correspond to intrinsic stacking faults formed by condensation of boron vacancies, and the generation mechanism may be explained by dislocation climb [88] underlying the maintenance of equilibrium between the stress and the vacancy concentration and the carbon substitutional effect suggested by first-principle calculations [94, 95, 96]. A study of the vacancy formation behavior as well as the influence on lattice parameters was performed by the calculations using the projector-augmented wave method within the generalized gradient approximation for the exchange-correlation interaction, and the simulations adopted a model, in which supercell geometry in size of 8 formula units was constructed [96]. The calculation results indicate that a single carbon substitution into a boron site decreases the *a*-axis lattice parameter of the basal plane by 0.73%, while the *c*-axis parameter remains almost equal to the original value. The calculated lattice volume is contracted 1.43% by carbon substitution, and as a consequence, the calculated vacancy formation energy is 0.37 eV lower with carbon substitution than without carbon. This fact makes it clear that more vacancies are introduced by carbon doping into $MgB_2$, which is in agreement with the experimental observations (as shown in Fig. 3.1). It is thus clear that these structural defects, especially the boron vacancies, play a central role in the structural change caused by carbon doping with the low-temperature sintering in the case of $MgB_2$ polycrystalline wires and bulks.

The structural change due to carbon doping may also have a strong influence on lattice parameters that characterize the $MgB_2$ crystal system reflecting the superconductivity. The experimental studies to evaluate changes in the nanostructure parameters as well as the first-principle calculations to simulate the effect [94–96] were made, and they were performed, for example, by X-ray powder diffraction [97–100], four-circle X-ray diffraction [63, 91, 101–103], neutron powder diffraction [104], and high-energy synchrotron radiation powder diffraction [88, 105–107]. These structural analyses based on X-ray, neutron, or synchrotron radiation crystallography revealed that carbon doping into $MgB_2$ single crystals, polycrystalline bulks, and wires decreases the *a*-axis lattice parameter while leaving the *c*-axis lattice parameter unaffected or causing a slight change in the lattice spacing. The change in the nanostructure parameters, especially the shrinkage of the *a*-axis lattice parameter, may arise from the formation of boron vacancies and associated stacking faults (as shown in Fig. 3.1) and that of other substitutional defects in the case of $MgB_2$ wires and bulks. On the other hand, the doping into $MgB_2$ thin films leads to a distinctly different behavior compared with those of the single crystals, bulks, and wires, and in fact, the *a*- and *c*-axes lattice parameters increase with increasing carbon content except heavily doped films [91, 102]. The lattice expansion may not be simply explained by an intrinsic model of only the interstitial (or substitutional) incorporation [95], and extrinsic factors, such as the secondary phase of $MgB_2C_2$ in the grain boundaries and the thick layer, have been suggested to be mainly responsible for the effect [89, 91, 108]. It is thus obvious that the changes in lattice parameters are inherently different between the intrinsic and extrinsic effects of carbon doping.

The lattice changes in both cases of the intrinsic and extrinsic effects may cause lattice disorder, and an evidence for this is shown in Fig. 3.1b. The atomic-scale image of the carbon-doped sample proves clearly that the atomic arrangement of magnesium and boron atoms is not perfectly ordered in crystal layers and the lattice structure is distorted by the intrinsic effect. This is consistent with the results of other structural analyses by X-ray powder diffraction [109, 110], high-energy synchrotron radiation powder diffraction [88, 107], and Raman scattering spectra [111, 112]. Different nanostructures most likely reflecting the extrinsic effect have also been studied, and they are found to have extensive lattice disorder [89–91]. The structural disorder induced by the intrinsic or extrinsic effect of carbon doping increases the scattering rate of charge carriers, shortens the mean free paths, decreases the superconducting coherence length, and thus improves the irreversibility field and the upper critical field. Experimental results showing the drastic improvement by carbon doping have been reported so far, and for example, the upper critical field at 4.2 K exceeds 33 T in round wires by the intrinsic effect [113] and may reach over 60 or 70 T for the parallel field in thin films by the extrinsic effect [90, 91, 103, 114]. The exceptional high values shown in the thin films approach or exceed the paramagnet limit of 65 T, which is calculated from the conventional BCS theory for single-gap superconductors [115]. This is because $MgB_2$ is theoretically and experimentally identified as the first superconductor where two superconducting gaps exist, and the two-band and two-gap structure may be responsible for such anomalous behaviors of the upper critical field. A two-band dirty-limit theory to explain the nature has been proposed by Gurevich, and the exceptional upper critical field can be achieved by tuning the scattering within the two bands [45, 116]. The theory predicts further that the paramagnet limit of $MgB_2$ is estimated to be 130 T, and it means that there is still large room for a further increase in the upper critical field. The interesting prediction has been controversial, but in any case, an increase in the upper critical field (and the irreversibility field) improves the field dependence of superconducting properties including the transport critical current density, as shown in many reviews [81–86]. The anisotropy of the upper critical field is also known to be crucial for the in-field properties, and the effect has been studied on $MgB_2$ materials [81, 117–119]. The anisotropic property confirmed in thin films and single crystals arises from the nature of the superconducting gap on the two-dimensional $\sigma$ bands, which is the larger of the two gaps on the two bands [39, 120, 121]. The anisotropy exerts a strong influence on the in-field critical current property of untextured polycrystalline materials such as wires and bulks, and in fact the randomly oriented grains having different properties in fields cause field-induced inhomogeneity in the materials, leading to rapid degradation of the in-field critical current property [81, 118]. A method to reduce the upper critical field anisotropy is carbon doping [81, 110], which introduces the lattice disorder including boron vacancies (as shown in Fig. 3.1) and thus may affect the two-dimensional σ band conduction localized on the boron planes [39]. It is therefore evident that carbon doping into $MgB_2$ materials introduces structural disorder and reduces the anisotropy, resulting in the improvement of the upper critical field and the in-field critical current property.

### *3.2.3 Carbon Doping Methodology*

Carbon doping can improve the in-field superconducting properties of $MgB_2$ wires, bulks, and films, but it strongly depends on how carbon is doped into the materials. Numerous carbon-containing compounds as well as the carbon element itself like pure carbon nanoparticles can be used as carbon sources, and thus, various doping methods have been proposed to effectively improve the upper critical field and the in-field transport critical current density. Active carbon sources, for example, bis (methylcyclopentadienyl)magnesium ($(MeCp)_2Mg$) [103, 114], methane ($CH_4$) [90, 122], and trimethylborane (TMB) [91], for $MgB_2$ thin films fabricated by hybrid physical–chemical vapor deposition have been studied, and the carbon-containing gases seem to be promising sources for realizing the exceptional high upper critical field over 60 or 70 T. On the other hand, numerous kinds of carbon dopants have been examined on wires and bulks, and it raises a general question of which sources are the most effective for the polycrystalline materials. The answer, of course, depends on how the criteria are chosen in experimental results obtained so far by various measurement methods and conditions. A criterion to compare the effects of various carbon dopants is given here by the in-field transport critical current density, which is mainly determined by several factors including the flux pinning and universally used as an important criterion for in-field technical applications. Many studies of the doping through various carbon sources have reported transport results at 4.2 K and around 10 T, and thus, the best values achieved by each dopant are compared and listed in Table 3.5. The comparison is carried out among results reported for round mono-core wires [85, 88, 93, 96, 110, 123–144], which were fabricated by a conventional powder-in-tube method and prepared from metal sheaths of Fe or Nb/Monel, so that the strong influences of wire fabrication methods [145, 146] except for carbon doping methods are reduced on the transport property. The nature of starting boron powder is also known to have a strong effect on the critical current property [12], and hence, conventional round wires prepared from high purity of the amorphous powder (at least over ~99%) are selected here for the comparison. (The boron powder sizes were not mentioned in some selected results, but to our knowledge, the kinds of high-purity powder of amorphous boron are limited and the sizes are known to be submicron and under ~1 micron. It is thus thought that the round wires selected here were prepared from the similar fine powders except for a result [133].) A different way for carbon doping without mixing or treating directly with pure boron powder has recently shown a drastic improvement of the in-field transport critical current density for round mono-core wires fabricated by a conventional powder-in-tube method, and thus, the results [138, 140] are also added for the comparison. Carbon sources finally compared here are TiC [136], $B_4C$ [85, 127, 132], silicon oil [142], pyrene ($C_{16}H_{10}$) [96, 135], single-walled carbon nanotubes (SWCNTs) [126, 131], carbon nanoparticles [129], multi-walled carbon nanotubes (MWCNTs) [126, 128, 144], coronene ($C_{24}H_{12}$) [143], SiC [85, 88, 123–125, 129, 130, 133, 134, 136], malic acid ($C_4H_6O_5$) [88, 93, 110, 136, 137, 139, 141], and methane ($CH_4$) [138, 140]. Some of the sources have also been evaluated by magnetization measurements, and the doping effects have been confirmed by the Bean critical-state model [87, 92, 107, 109, 147–151]. Although the effects of these carbon sources cannot be compared at entirely the same condition in terms of other factors including sintering conditions, this comparison of the best values achieved by each dopant would be a guide for the evaluation. The most effective carbon sources for the high-field transport property seem to be SiC, malic acid, and methane (as listed in Table 3.5), and in particular, the latter doping by radio frequency (RF) plasma injecting methane gas [140] reaches a high value of ~$4.0 \times 10^4$ A $cm^{-2}$ at 4.2 K and 10 T [138] among the round mono-core wires (as given in Table 3.5). It is thus found that methane gas is an active carbon source for improving the high-field superconducting properties not only of $MgB_2$ thin films but also of polycrystalline wires.

**Table 3.5** Comparison of the best values of transport critical current densities ($J_c$) at 4.2 K and 10 T, which are obtained so far by various carbon sources. Results reported for round mono-core wires fabricated by an in situ powder-in-tube process [85, 88, 93, 96, 110, 123–144] (without densification methods to address void formation) are selected and compared, and the best values achieved by each dopant are extracted and listed here. Metal sheaths of the listed wires are either Fe or Nb/Monel

| Transport $J_c$ (A $cm^{-2}$) at 4.2 K, 10 T | Carbon source | Carbon doping method |
|---|---|---|
| ~$4.0 \times 10^4$ [138] | Methane ($CH_4$) | RF plasma injecting $CH_4$ gas and solid mixing |
| ~$2.9 \times 10^4$ [136] | Malic acid ($C_4H_6O_5$) | Liquid mixing and solid mixing |
| ~$2.4 \times 10^4$ [129] | SiC | Solid mixing |
| ~$1.8 \times 10^4$ [143] | Coronene ($C_{24}H_{12}$) | Liquid process and solid mixing |
| ~$1.3 \times 10^4$ [144] | MWCNTs | Solid mixing |
| ~$1.3 \times 10^4$ [129] | Carbon | Solid mixing |
| ~$1.0 \times 10^4$ [126] | SWCNTs | Solid mixing |
| ~$1.0 \times 10^4$ [135] | Pyrene ($C_{16}H_{10}$) | Solid mixing |
| ~$0.6 \times 10^4$ [142] | Silicon oil | Liquid mixing |
| ~$0.6 \times 10^4$ [127] | $B_4C$ | Solid mixing |
| ~$0.2 \times 10^4$ [136] | TiC | Solid mixing |

Similar transport critical current densities at 4.2 K and 10 T to the highest performance of round wires have been reported on densified $MgB_2$ tapes doped by different carbon dopants that are $C_{60}$ [152], carbon nanoparticles [153, 154], hollow carbon nanoboxes [155], and cumene ($C_9H_{12}$) [156]. Further improvement of the high-field critical current densities is known to require appropriate combinations of densification methods and several carbon dopants, and the transport results [143, 157–169] are compared here and listed in Table 3.6. The transport critical current densities (as listed in Table 3.6) are found to be significantly higher than those of conventional round wires (as listed in Table 3.5). The high-field transport properties have been achieved by the carbon sources of malic acid, nanostructured carbon, coronene, SiC, and methane, and the densification methods of cold high-pressure densification (CHPD), mechanical allowing (MA), internal magnesium diffusion (IMD), and advanced internal magnesium infiltration (AIMI) [143, 158, 160, 162, 167]. These densification methods have been proposed to address voids formed within cores of $MgB_2$ polycrystalline conductors, which are caused by the large volume reduction of the precursor powders when magnesium fully reacts with boron to form $MgB_2$ [88]. Hardness of sheath metals besides the densification methods may also exert a strong influence on the transport property, and it is thus hard to evaluate the most effective sources from only the comparison (as listed in Table 3.6). It is, however, interesting to note that the carbon doping by using methane gas achieves the highest transport critical current density at 4.2 K and 10 T among all the $MgB_2$ polycrystalline conductors including round wires and tapes.

The doping method by using a carbon-containing gas, for example, methane gas [138, 140, 167, 170], ethylene ($C_2H_4$) gas [171], or pyrene gas [107], has been studied and proposed on $MgB_2$ bulks and wires. The basic concept is to homogeneously incorporate carbon into the polycrystalline materials without any severe deterioration of grain connectivity, which is caused by agglomeration of unreacted carbon due to the inhomogeneity [107]. The carbon incorporation methods to address the agglomeration have been established through chemical vapor deposition

**Table 3.6** Comparison of the best values of transport critical current densities ($J_c$) at 4.2 K and 10 T, which are obtained so far by densification methods and several carbon sources. Results of mono-core conductors reported for the $J_c$ over ~5.0 × $10^4$ A $cm^{-2}$ [143, 157–169] are selected and compared, and the best values achieved by each dopant are extracted and listed here

| Transport $J_c$ (A $cm^{-2}$) at 4.2 K, 10 T | Carbon source | Densification method | Sheath, barrier |
|---|---|---|---|
| ~15.0 × $10^4$ [167] | Methane ($CH_4$) | AIMI | Monel, Nb |
| ~13.0 × $10^4$ [162] | SiC | IMD | Ta |
| ~10.7 × $10^4$ [143] | Coronene ($C_{24}H_{12}$) | IMD | Fe |
| ~6.0 × $10^4$ [158] | Nanostructured carbon | MA | $Cu_{70}Ni_{30}$, Nb |
| ~5.0 × $10^4$ [160] | Malic acid ($C_4H_6O_5$) | CHPD | Monel, Nb |

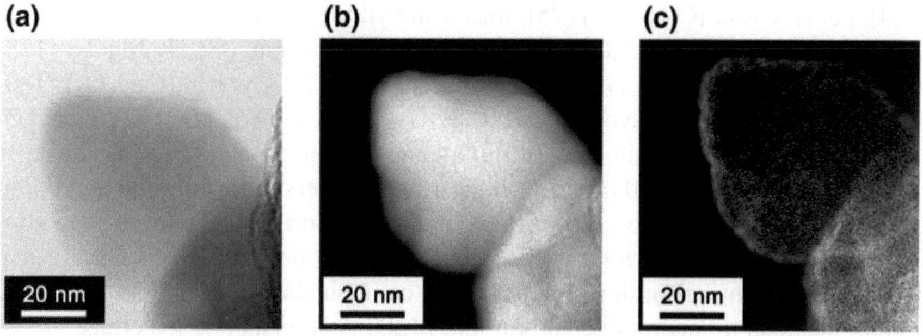


**Fig. 3.2** Carbon-treated boron nanopowder made by injecting trichloride ($BCl_3$), hydrogen ($H_2$), and methane ($CH_4$) into radio frequency (RF) argon plasma. **a** Zero-loss energy-filtering transmission electron microscope image, **b** the corresponding boron *K*, and **c** carbon *K* maps. White regions shown in boron *K* and carbon *K* maps denote boron and carbon, respectively, and it is clear that a very thin carbon layer surrounds the boron nanoparticle (reproduced from [140])

or diffusion [107, 171] and RF plasma synthesis [138, 140, 167, 170], and the latter is also known to be one of the synthetic methods for pure boron nanopowder [170]. The plasma synthesis can be applied for carbon doping, and in this process, carbon-treated boron nanopowder, called carbon-doped boron, are made from boron trichloride ($BCl_3$), hydrogen ($H_2$), and methane as a carbon source. The nature of the nanopowder has been investigated by using 300-keV field-emission and zero-loss energy-filtering transmission electron microscopes [140], and resulting images are shown in Fig. 3.2. A nanoparticle located at the center of Fig. 3.2a is captured, and from boron *K* and carbon *K* maps shown in Fig. 3.2b and c, a very thin carbon layer, about 1–2 nm, is found to almost homogeneously surround the boron nanoparticle. The carbon-encapsulated boron nanopowder can be used as a stating material for bulks and wires, and thus, the thin carbon layer can be easily incorporated into $MgB_2$ polycrystalline materials, leading to the introduction of lattice disorder (as shown in Fig. 3.1) and the drastic improvement of the high-field property (as given in Tables 3.5 and 3.6) compared with $MgB_2$ wires prepared from pure boron nanopowder which is synthesized by RF plasm injecting boron trichloride and hydrogen without using methane gas as a carbon source [138, 140]. Similar nanostructured particles (as shown in Fig. 3.2) for carbon incorporation into $MgB_2$ materials can be formed even by liquid and chemical solution processes, which are used with pure boron powder and organic carbon sources, for example, coronene [172] and malic acid [88, 92], and thus, the doping through the two sources may result in high values of the high-field transport property (as given in Tables 3.5 and 3.6). These processes besides the doping effects can decrease the void density and the void size in the polycrystalline materials, and thus, the carbon doping itself is known to be one of the densification methods [88]. Other liquid and chemical solution processes have also been studied extensively for homogeneously carbon incorporation into the polycrystalline materials by using different carbon sources, for example, benzene ($C_6H_6$) [173], thiophene ($C_4H_4S$) [173], ethyltoluene ($C_9H_{12}$)

[174], benzoic acid ($C_7H_6O_2$) [175], maleic anhydride ($C_4H_2O_3$) [176], maleic acid ($C_4H_4O_4$) [177], stearic acid ($C_{18}H_{36}O_2$) [178], Mg stearate ($C_{36}H_{70}MgO_4$) [178], Zn stearate ($C_{36}H_{70}ZnO_4$) [178], sugar ($C_6H_{12}O_6$) [179], glycerin ($C_3H_8O_3$) [180], acetone ($C_3H_6O$) [181], glycolic acid ($C_2H_4O_3$) [182], mineral oil [183], oleic acid ($C_{18}H_{34}O_2$) [184], and chitosan [185]. Organic carbon dopants can, of course, be used for conventional solid mixing with starting powders of magnesium and boron, and the doping effect has been confirmed on the transport critical current properties, for example, by using pyrene powder as a carbon source [96, 135] (as given in Table 3.5). It is thus clear that organic carbon compounds can be applied for various doping methods to incorporate carbon into $MgB_2$ polycrystalline materials.

Different doping methods by using inorganic carbon sources, for example, nanodiamond [186], graphite [187, 188], $Al_4C_3$ [189], ZrC [189], NbC [189, 190], $Mo_2C$ [189], HfC [189], WC [189] hollow carbon spheres [191], graphene [46, 155], and carbon black [183], have also been studied on $MgB_2$ bulks and wires. The most effective inorganic dopants on the high-field transport properties seem to be nanostructured carbon [158] made from high-purity graphite [192–194] and SiC [129, 162] (as given in Tables 3.5 and 3.6), and in particular, the latter doping has been widely investigated on various $MgB_2$ conductors, for example, round wires [85, 88, 123–125, 129, 130, 133, 134], square wires [146, 195], tapes [156, 173, 174, 196, 197], and IMD-processed wires [143, 157, 159, 162, 163, 165, 166, 168]. The nature of the doping effect has been well understood by a dual reaction mechanism among SiC, magnesium, and boron [129], and thus, co-doping of separate silicon and carbon may not be optimum for $MgB_2$ polycrystalline materials doped by the two elements [198]. The SiC-doped $MgB_2$ wires fabricated by the IMD process show high values of high-field critical current densities at 4.2 K (as given in Table 3.6), but coarse impurity phases of $Mg_2Si$ as well as the nanometer-sized ones are formed within the wire cores [199, 200]. Such large-sized nonsuperconducting phases may strongly limit the effective cross-sectional area for supercurrent flow, and thus, there is still room for improvement on the homogeneity of SiC-doped $MgB_2$ cores. It is therefore evident that the homogeneity is a key requirement on carbon doping even by using either of these sources including inorganic and organic matter, especially for $MgB_2$ polycrystalline materials, and further tailoring approaches for carbon homogenous incorporation could open up the way to enhancing the in-field critical current density for their technical applications.

## 3.3 Ball Milling

### *3.3.1 Introduction*

Ball milling process is a widely used technique to grind and produce fine powders. The scalability of this process makes it a preferable choice when powders should be uniformly mixed and its grain size be reduced in production of metallic and ceramic materials [201]. The attainable particle size reaches as low as ~10 nm depending on materials and milling conditions, i.e., milling balls and vial, ball to powder ratio, rotation speed, and time. Moreover, it is capable of synthesizing both equilibrium and non-equilibrium alloy phases when two or more elemental powders were milled.

In the case of $MgB_2$, it is now well established that grain size plays an important role since grain boundaries act as flux pinning centers [81, 85, 109, 202–206], in similar manner with $Nb_3Sn$ [207, 208]. Hence, reduction of effective grain size $D_g$ by ball milling contributes to improvement of flux pinning force $F_P$ and $J_c$ through increasing grain boundaries density according to the following equation,

$$J_c = \frac{F_P}{B} = \frac{f_P}{D_g a_f}, \tag{3.1}$$

where $B$ is flux density, $f_P$ is the elementary pinning force, and $a_f$ is the flux line spacing. Such relationship between grain size and flux pinning strength can be clearly shown in Fig. 3.3. The reciprocal grain size dependence of $J_c$ shown in (3.1) drives investigation on ball milling process in synthesis of $MgB_2$ materials which requires high critical current and irreversibility field, $H_{irr}$, such as tapes, wires, and more recently bulk magnets.

In situ and ex situ are the primarily employed two techniques to fabricate $MgB_2$ materials [209]. Perhaps the most commonly studied is the in situ method, formation of $MgB_2$ from mixed powders of Mg + 2B, since relatively high $J_c$ is easily attained owing to its reasonably strong intergrain bonding. In the in situ reaction process, magnesium grains melt and diffuse into boron grains during heat treatment and transform into voids resulting in a low bulk density (packing factor, $P \sim 50\%$) and low connectivity. On the other hand, ex situ method, using prereacted $MgB_2$

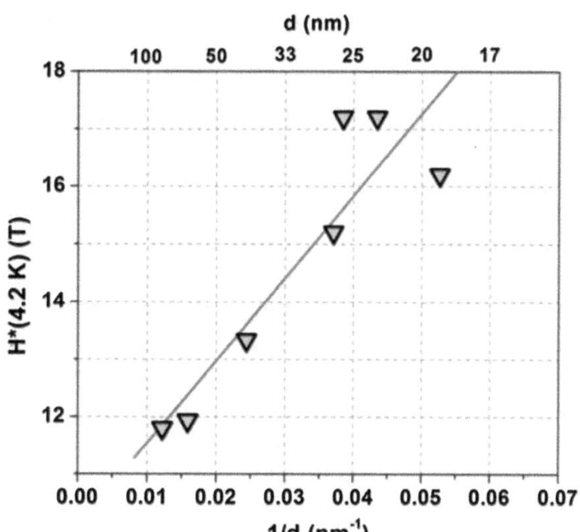


**Fig. 3.3** Linear relationship between irreversibility field (4.2 K) and inverse grain size (reproduced from [205])

powder, is favorable in terms of bulk density. Packing factor close to ~75% (the close-packing of spheres) can be expected. Even unsintered, as-pressed ex situ $MgB_2$ tapes show relatively high transport $J_c$ of ~$10^4$ A $cm^{-2}$ at 20 K [210]. Heat treatment after cold working is effective to improve $J_c$ through strengthening of intergrain bonding [211–215]. Ball milling has been used for both techniques.

In this sub-chapter, the influences of ball milling on phase formation, grain boundaries, $J_c$, $H_{c2}$, $H_{irr}$, and anisotropy of $MgB_2$ samples prepared by in situ and ex situ processes are introduced.

### 3.3.2 Ex Situ Processed Materials

Suo and Flukiger et al. firstly studied the influence of initial $MgB_2$ powder grain size on transport $J_c$ in ex situ $MgB_2$ tapes by using ball milling [209, 216, 217]. The commercial $MgB_2$ powder with medium particle size of ~60 μm was ball milled using an agate mortar and agate balls with different milling time and energy. 2, 3, and 100 h of ball milling yield fine $MgB_2$ powders with particle size distribution peaks of ~3/30, ~1.5/10, and ~1 μm, respectively. Transport $J_c$ at 4.2 K under 6.5 T for the annealed $MgB_2$ tapes based on the powders milled for 0, 2, 3, and 100 h showed 2.5, 11, 6.7, and 3.5 kA/$cm^2$, respectively. It is clearly seen that fine grain size contributed to high transport $J_c$. The decrease of $J_c$ for the latter samples is attributed to generation of impurity phases at the grain surfaces. Ball milling with the option of carbon was reported by Kario et al. [218] using high-energy milling. Long multifilament tapes with 20-h-milled powder with 5 wt% carbon showed high $J_c$ of $10^4$ A/$cm^2$ at 4.2 K under 8.8 T.

Effects of ball milling on the irreversibility field $H_{irr}$ and anisotropy of ex situ $MgB_2$ tapes were reported by Malagoli et al. [212]. Significant grain refinement produced by milling of $MgB_2$ powder enhanced grain boundary pinning resulting in high $J_c$ of $10^4$ A/$cm^2$ at 4.2 K under 10 T and high $H_{irr}$ of 14 T at 4.2 K. Anisotropy in $H_{irr}$ versus $T$ diagram is commonly observed in ex situ $MgB_2$ tapes owing to intrinsic mass anisotropy 4–5 [39] and partial alignment of $MgB_2$ grains during cold working. $H_{irr}$ values at 4.2 K obtained by $I$–$V$ measurement perpendicular and parallel to tape surface for the unmilled tape were 5.4 and 8.2 T, respectively, while those for the milled tape were 14.0 and 14.4 T, respectively, showing that macroscopic anisotropy almost diminished by milling due to fine grain size.

Evolution of $MgB_2$ phase by mechanical milling was studied in detail by Chen et al. [219]. After 4 h of milling with $ZrO_2$ media, a detectable amount of $ZrO_2$ was present in $MgB_2$ and its content increased with milling time. Partially amorphous $MgB_2$ was observed in the 400-h-milled powder together with the formation of a minority $ZrB_2$ phase. The expansion of lattice parameters in both $a$- and $c$-axes was observed which may be due to possible alloying of zirconium.

Formation of grain boundaries during sintering is an important issue to achieve high transport current in ex situ $MgB_2$ materials. Tanaka and Mizutani et al. studied microstructural characteristics of ex situ $MgB_2$ bulks prepared from laboratory-made $MgB_2$ powder with different grain size by ball milling [220–222]. Secondary electron images of the polished surfaces of ex situ $MgB_2$ bulks sintered at 900 °C for 24 h using different starting $MgB_2$ powders ball milled for 0, 2, and 20 h showed that the intergranular coupling was seen to improve with longer ball milling period (Fig. 3.4). In addition, the shape of the grains became plate-like and they were connected with many neighboring grains. Compared to the green compacts before sintering and the sintered sample without ball milling, the numbers of isolated grains in the two-dimensional view were observed to decrease due to mass transfer and the formation of grain boundaries. Since temperature-dependent part of resistivity due to electron phonon scattering is not affected by impurity scattering (the cleanness of a sample), connectivity as a barometer of macroscopic effective conductivity for transport current can be described as,

$$K = \frac{\rho\,(300\,\mathrm{K}) - \rho\,(40\,\mathrm{K})}{\Delta\rho_{\mathrm{single}}}, \tag{3.2}$$

where $\rho$(300 K) and $\rho$(40 K) are resistivity at 300 and 40 K, respectively, and $\Delta\rho_{\mathrm{single}}$ is the difference in resistivity between 300 and 40 K for an ideal single crystals (4.3 μΩ cm [4, 223], 6.32 μΩ cm [206, 220, 221, 224], 7.3 μΩ cm [13], and 8.1 μΩ cm [205, 225]). Connectivity values calculated from (3.2) using $\Delta\rho_{\mathrm{single}}$ = 6.32 μΩ cm for the ex situ $MgB_2$ bulks sintered at 900 °C for 24 h using different starting $MgB_2$ powders ball milled for 0, 2, and 20 h were 25.9, 38.2, and 38.2%, respectively. The network structure with an array of grain boundaries observed in the ball-milled samples is expected to provide superior electrical current paths. Ball milling is considered to contribute to the formation of intergranular necks and grain boundaries initial point contacts between neighboring $MgB_2$ grains, and generation of new intergranular point contacts and subsequent formation of a network structure through shape optimization of the grains through reducing grain size.

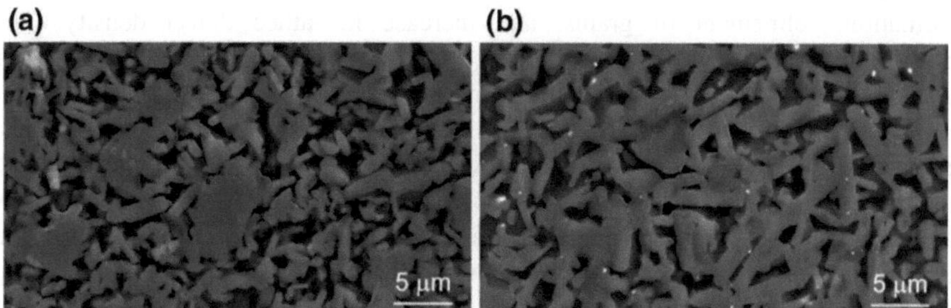


**Fig. 3.4** Secondary electron images of the polished surfaces of ex situ $MgB_2$ bulks sintered at 900 °C for 24 h using different starting $MgB_2$ powders ball milled for **a** 0 h and **b** 20 h (reproduced from [220])

### 3.3.3 In Situ Processed Materials

Preparation of iron-clad $MgB_2$ tapes using an ultrafine starting precursor (Mg, B) was reported by Fang et al. [226]. Commercially available magnesium and boron powder (−325 mesh in size) with stoichiometric ratio was milled by high-energy ball milling for 2 h using stainless steel balls and vial. After milling, the powder with reduced particle size (200–300 nm) was filled in 1020 iron tube followed by groove rolling and flat rolling to a tape with about 2.5 mm by 0.2 mm in cross section. The tape after annealing at 850 °C for 30 min showed transport $J_c$ of $2.0 \times 10^4$ A/cm$^2$ at 30 K under self-field. Matsumoto et al. [227] investigated transport $J_c$ properties of $MgB_2$ tapes prepared from various ball-milled powders, such as Mg + B, $MgH_2$ + B, and Mg + B + SiC. The ball milling showed almost no positive effects on $J_c$ for $MgH_2$ + B; however, significant enhancement of $J_c$ was observed in tapes prepared from the ball-milled powders of Mg + B and Mg + B + SiC. Transport $J_c$ of $1.2 \times 10^4$ A/cm$^2$ was obtained at 4.2 K under 10 T. The origin of high $J_c$ may be attributed to that fine boron grains abraded magnesium grains during ball milling, resulting in fresh magnesium and boron grain surfaces. Jiang et al. [228] reported ball milling effects on the $MgB_2$ samples prepared with milled original powders of Mg, B, and 3 at.% graphite. It was revealed that ball milling of original powders promoted substitution of carbon for boron in $MgB_2$ lattice, resulting in generation of lattice distortion and enhancement of $J_c$ under high field.

Influence of milling media was extensively studied by Xu et al. [229, 230]. Among acetone, ethanol, and toluene, it was found that using toluene was most effective to enhance $J_c$. $J_c$ obtained at 5 K under 8 T reached $5 \times 10^3$ A/cm$^2$ owing to smaller $MgB_2$ grains with carbon substitution.

Influence of particle size refinement on phase crystallinity was studied and reported by Lee and Jun et al. [231, 232]. With increasing ball milling period, semicrystalline boron peaks in XRD pattern became broader indicative of loss of crystalline response of the boron particles. Influence of milling of low-purity boron powder ($\beta$-rhombohedral crystalline phase) on $MgB_2$ bulks and tapes was examined by Wang et al. [233–235]. It was found that ball milling facilitated $MgB_2$ phase formation, refinement of grains, and increase in lattice defect density. The 80-h-milled $MgB_2$ bulk sample showed $J_c$ of $2.8 \times 10^4$ A/cm$^2$ which was an order of magnitude higher than that of the unmilled sample.

Ball milling is effective to enhance trapped field properties in bulk magnets. Sugino et al. [233–236] studied the effects of the grain size, lattice strain, and microstructure on the trapped field properties of ball-milled $MgB_2$ bulks. A decrease in the *in-plane* lattice parameters and transition temperature and an increase in the *in-plane* XRD peak widths occurred with milling as a result of grain refinement and carbon substitution. Microstructural analysis revealed that submicron-size fine $MgB_2$ grains were well connected, resulting in increased grain boundary density. The trapped field of the magnetized bulk samples exhibited a large improvement with milling and reached 3.72 T at 5 K as shown in Fig. 3.5,

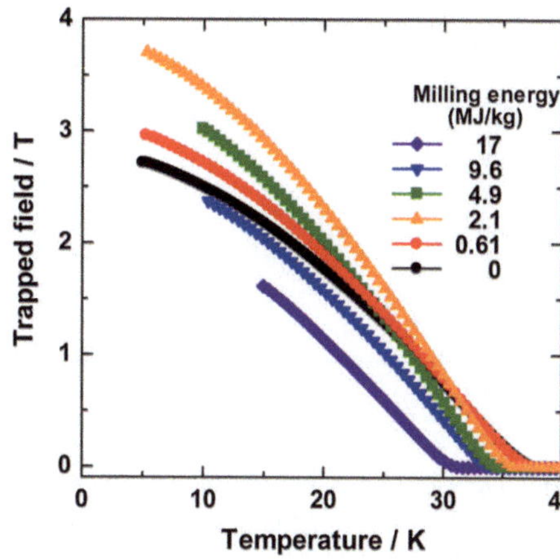


**Fig. 3.5** The trapped magnetic field as a function of increasing temperature for the magnetized, disk-shaped $MgB_2$ bulks with 30 mm diameters and 10 mm thickness synthesized from powders under different milling conditions (reproduced from ref [236]). The milling energy was estimated according to Häßler's model [237] (3.3)

which is the highest among $MgB_2$ bulks prepared using pressureless synthesis. The increased grain boundary density and electron scattering are considered to contribute to the high trapped field through quantitatively and qualitatively enhancing the grain boundary flux pinning.

### 3.3.4 Mechanical Alloying

The influence of the milling energy of precursor powder (Mg and B) on the mechanical $MgB_2$ formation (alloying), microstructure, and the superconducting properties of $MgB_2$ bulk samples and wires were quantitatively elaborated by Häßler et al. [194, 218, 237–240]. They proposed an explicit approximate formula for the energy $E_t$ transferred to the powder sample with mass $m$ during milling:

$$\frac{E_t}{m} = c\beta \frac{\left(\omega_p r_p\right)^3}{r_v} t, \tag{3.3}$$

where $\beta$ is the mass ratio of the balls to the sample, $\omega_p$ is the angular frequency, $r_p$ is the radius of the sun disk, $r_v$ is the radius of the vial, $t$ is the milling time, and $c$ is the coefficient. The range of the milling energies varied from $4 \times 10^4$ (mixing) to $1 \times 10^{10}$ J/kg (high-energy milling). Figure 3.6 shows the phase fraction of $MgB_2$ in the powder in dependence on the normalized milling energy based on Rietveld analysis of the X-ray diffraction data. It can be seen that a transition from pure milling to mechanical alloying occurs at a milling energy of about $10^8$ J/kg. Electron microscopy analysis suggested that crush of Mg particles occurs during the first stage of milling followed by formation of $MgB_2$ phase. The transport critical current density of the mono-filamentary tapes systematically increased with an

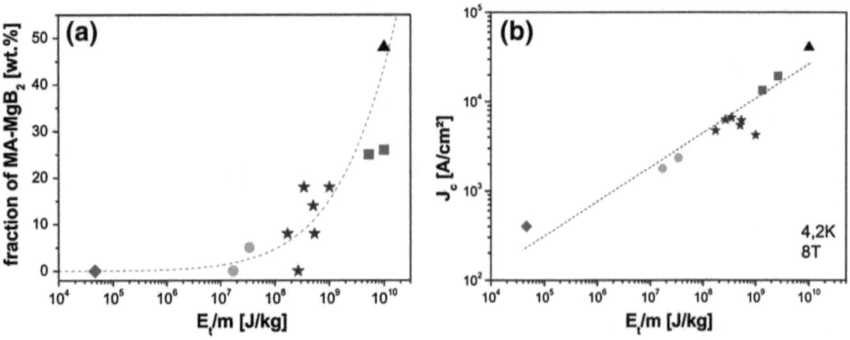


**Fig. 3.6** *Left* The phase fraction of $MgB_2$ in the powder in dependence of the normalized energy based on Rietveld analysis of the diffraction data. *Right* The transport critical current density at 4.2 K under 8 T in dependence on the milling energy of the mono-filamentary tapes (reproduced from [237])

increase in milling energy (Fig. 3.6). The $J_c$ values at 4.2 K under 8 T varied over two orders of magnitude from $4 \times 10^2$ to $4 \times 10^4$ A/cm$^2$.

### 3.3.5 Mechanical Alloying with Dopants

High-energy milling of $MgB_2$ prereacted powder renders the material largely amorphous through extreme mechanical deformation and is suitable for mechanical alloying with dopants including carbon. Senkowicz et al. reported high-energy ball milling of the mixtures of Alfa Aesar prereacted $MgB_2$ powder and graphite (0.9 $MgB_2$ + 0.1C) in a Spex 8000 M Mixer/Mill [205, 225, 241, 242]. The milled powders were hot isostatic pressed at 1000 °C for 200 min under 30 ksi. Then, the obtained $MgB_2$ pellets were annealed with Mg at 900 °C for 5 h to compensate Mg deficiency. The *a*-axis lattice parameter of the sample was shorter than the literature and estimated carbon content was $x = 0.051$ in $Mg(B_{1-x}C_x)_2$, indicative of successful carbon alloying. Upper critical field at 0 K reached 33 T, which is equal to that of the carbon substituted single crystal [39]. Critical current density had a maximum approaching $10^6$ A/cm$^2$ at 4 K. Increased lattice disorder induced in the milling process is indicated by weakened XRD patterns, high normal-state resistivity, and a low-temperature upturn in $H_{c2}(T)$ and considered to be the origin of improved $H_{c2}$, and $J_c$. Senkowicz et al. further investigated the influences of the mechanical alloying by systematically changing milling time up to 3000 min and found that long milling time can be responsible for very fine <30 nm grains, which yield high irreversibility field. With increasing milling time, $T_c$, *RRR*, *a*-axis lattice parameter, and grain size systematically decreased while residual resistivity increased. $J_c$ and irreversibility field showed a nearly 50% increase at medium milling time (600–1200 min). The $J_c$ values at 4.2 K under 12 T varied from $1 \times 10^2$ to $2 \times 10^4$ A/cm$^2$.

Flux pinning efficiency in ball-milled $MgB_2$ bulks with the addition of $ZrB_2$ and $TaB_2$ was reported by Rodrigues and Silva et al. [243–245]. $MgB_2$ bulk samples prepared by milling prereacted $MgB_2$ and $ZrB_2$ or $TaB_2$ powders using a Spex 8000 M Mill with WC jars and balls followed by heating in a hot isostatic press at 1000 °C under a pressure of 30 ksi for 24 h showed improved $J_c$ under high magnetic fields and flux pinning. The sample added with 5 at.% $TaB_2$ and milled for 5 h showed high $J_c$ of $7 \times 10^5$ A/cm$^2$ and macroscopic flux pinning force of 14 GN/m$^3$ at 4.2 K under 2 T. The effect of the additive, acting as doping or not, is not still well understood. Since the milled and mixed samples showed higher values of $B_{irr}$ than the unmilled and unmixed samples, the milling increased electron scattering and resistivity, increasing $H_{irr}$ of the samples, resulting improvement of $J_c$ under fields.

## 3.4 Concluding Remarks

It is clear that formation of nanoprecipitates is an effective way to introduce flux pinning centers in $MgB_2$ in order to enhance the pinning strength of this material. This can be done through oxygen doping, chemical dopant addition, and mechanical ball milling. For undoped $MgB_2$, pinning capability is strengthened as a result of oxygen doping leading to the formation of oxide phases. By doping $MgB_2$ with oxides, impurities arise from reaction between the dopant with Mg and B are the sources of pinning centers. For carbon doping, carbon-related nanoprecipitates can be effective pinning centers in addition to substitution-induced enhanced impurity scattering which is beneficial to increasing $H_{c2}$ and high-field $J_c$. Flux pinners in the form of defects and grain boundary (because of grain refinement) can be introduced via mechanical ball milling. Hence, the electromagnetic properties of $MgB_2$ can be tailored by using a combinatorial approach to introduce various forms of flux pinners in the material. In line with this, minimizing the degree of decrease in $T_c$ and degradation in grain connectivity is equally important.

**Acknowledgements**

**Soo Kien Chen**

This work was supported by The Ministry of Education, Malaysia, and Universiti Putra Malaysia.

**Minoru Maeda**

This work was supported by JSPS KAKENHI Grant Number 26709021 and Nihon University Multidisciplinary Research Grant for 2015–2016. The author expresses his gratitude to Prof. Yoshiki Takano, Nihon University, Japan, for his help and support.

**Akiyasu Yamamoto**
This work was supported by JSPS KAKENHI Grant Numbers 22860019, 23246110, and 24656368 and by JST-PRESTO. The authors gratefully acknowledge H. Tanaka, A. Ito, S. Sugino, S. Mizutani, J. Shimoyama, K. Kishio (Univ. Tokyo), and T. Matsushita (KIT) for fruitful discussion.

**Shi Xue Dou**
This work was supported by Hyper Tech Research, Inc., and Australian Research Council.

## References

1. C. Buzea, T. Yamashita, Review of the superconducting properties of $MgB_2$. Supercond. Sci. Technol. **14**(11), R115 (2001)
2. P. Kováč et al., The role of MgO content in ex situ $MgB_2$ wires. Supercond. Sci. Technol. **17** (10), L41 (2004)
3. P. Kováč et al., Properties of $MgB_2$ wires made of oxidized powders. Physica C **477**, 20 (2012)
4. J.M. Rowell, The widely variable resistivity of $MgB_2$ samples. Supercond. Sci. Technol. **16** (6), R17 (2003)
5. S.K. Chen et al., Structural and superconducting property variations with nominal Mg non-stoichiometry in $Mg_xB_2$ and its enhancement of upper critical field. Adv. Func. Mater. **18**(1), 113 (2008)
6. H. Yamada et al., Critical current densities of powder-in-tube $MgB_2$ tapes fabricated with nanometer-size Mg powder. Appl. Phys. Lett. **84**(10), 1728 (2004)
7. H.E. Friedrich, B.L. Mordike, in *Magnesium Technology*, vol. 788 (Springer, 2006)
8. M.W. Chase, Jr., NIST-JANAF Thermochemical Tables, 4th Edition. J. Phys. Chem. Ref. Data, Monograph. **9**, 1–1951 (1998)
9. H. Fujii, K. Togano, H. Kumakura, Enhancement of critical current densities of powder-in-tube processed $MgB_2$ tapes by using $MgH_2$ as a precursor powder. Supercond. Sci. Technol. **15**(11), 1571 (2002)
10. C.H. Jiang, H. Hatakeyama, H. Kumakura, Preparation of $MgB_2$/Fe tapes with improved $J_c$ property using $MgH_2$ powder and a short pre-annealing and intermediate rolling process. Supercond. Sci. Technol. **18**(5), L17 (2005)
11. A. Talapatra et al., X-ray photoelectron spectroscopy studies of $MgB_2$ for valence state of Mg. Physica C **419**(3–4), 141 (2005)
12. S.K. Chen et al., Strong influence of boron precursor powder on the critical current density of $MgB_2$. Supercond. Sci. Technol. **18**(11), 1473 (2005)
13. J. Jiang et al., Influence of boron powder purification on the connectivity of bulk $MgB_2$. Supercond. Sci. Technol. **19**(8), L33 (2006)
14. S. Zhou, Y. Zhang, S. Dou, Effects of $B_2O_3$ in precursor B powder on $MgB_2$ critical current density. Physica C **470**, S635 (2010)
15. J. Zhou, P. Bai, A review on the methods of preparation of elemental boron. Asia-Pac. J. Chem. Eng. **10**(3), 325 (2015)
16. L.Y. Markovskii, Chemistry of magnesiothermal preparation of boron. Electron Technol. **3**, 95 (1970)
17. Y. Wang, J. Fan, M. Trenary, Surface chemistry of boron oxidation. 1. Reactions of oxygen and water with boron films grown on tantalum (110). Chem. Mater. **5**(2), 192 (1993)
18. G.V. Tsagareishvili et al., On the interaction of hydrogen and oxygen with boron powder and the kinetics of removal of the gases. J. Less Common Metals **82**, 131 (1981)

19. T.A. Prikhna et al., Nanostructural inhomogeneities acting as pinning centers in bulk $MgB_2$ with low and enhanced grain connectivity. Supercond. Sci. Technol. **27**(4), 044013 (2014)
20. R.A. Ribeiro et al., Effects of stoichiometry, purity, etching and distilling on resistance of $MgB_2$ pellets and wire segments. Physica C **382**(2–3), 194 (2002)
21. X.Z. Liao et al., $Mg(B, O)_2$ precipitation in $MgB_2$. J. Appl. Phys. **93**(10), 6208 (2003)
22. D.C. Larbalestier et al., Strongly linked current flow in polycrystalline forms of the superconductor $MgB_2$. Nature **410**(6825), 186 (2001)
23. Y. Akiyasu et al., Improved critical current properties observed in $MgB_2$ bulks synthesized by low-temperature solid-state reaction. Supercond. Sci. Technol. **18**(1), 116 (2005)
24. J.H. Kim et al., Study of MgO formation and structural defects in in situ processed $MgB_2$/Fe wires. Supercond. Sci. Technol. **20**(10), 1026 (2007)
25. X. Song, Atomic structure and chemistry of dense nano-precipitates in $MgB_2$ ceramic. Ceram. Int. **39**(4), 4299 (2013)
26. R.F. Klie et al., Observation of coherent oxide precipitates in polycrystalline $MgB_2$. Appl. Phys. Lett. **80**(21), 3970 (2002)
27. V. Braccini et al., Significant enhancement of irreversibility field in clean-limit bulk $MgB_2$. Appl. Phys. Lett. **81**(24), 4577 (2002)
28. X. Song, V. Braccini, D.C. Larbalestier, Inter- and intragranular nanostructure and possible spinodal decomposition in low-resistivity bulk $MgB_2$ with varying critical fields. J. Mater. Res. **19**(08), 2245 (2004)
29. R.F. Klie et al., Direct observation of nanometer-scale Mg- and B-oxide phases at grain boundaries in $MgB_2$. Appl. Phys. Lett. **79**(12), 1837 (2001)
30. Y. Zhu et al., Microstructure and structural defects in $MgB_2$ superconductor. Physica C **356** (4), 239 (2001)
31. J.D. Jorgensen, D.G. Hinks, S. Short, Lattice properties of $MgB_2$ versus temperature and pressure. Phys. Rev. B **63**(22), 224522 (2001)
32. Y.Y. Xue et al., Nonstoichiometry, defects and transport properties in $MgB_2$. Physica C **377** (1–2), 7 (2002)
33. A. Serquis et al., Hot isostatic pressing of powder in tube $MgB_2$ wires. Appl. Phys. Lett. **82** (17), 2847 (2003)
34. A. Serquis et al., Influence of microstructures and crystalline defects on the superconductivity of $MgB_2$. J. Appl. Phys. **92**(1), 351 (2002)
35. A. Serquis et al., Effect of lattice strain and defects on the superconductivity of $MgB_2$. Appl. Phys. Lett. **79**(26), 4399 (2001)
36. X.Z. Liao et al., Controlling flux pinning precipitates during $MgB_2$ synthesis. Appl. Phys. Lett. **80**(23), 4398 (2002)
37. K.A. Yates et al., The effect of oxygenation on the superconducting properties of $MgB_2$ thin films. Appl. Phys. Lett. **86**(2), 022502 (2005)
38. Sudesh et al., Effect of graphene oxide doping on superconducting properties of bulk $MgB_2$. Supercond. Sci. Technol. **26**(9), 095008 (2013)
39. T. Masui, S. Lee, S. Tajima, Carbon-substitution effect on the electronic properties of $MgB_2$ single crystals. Phys. Rev. B **70**(2), 024504 (2004)
40. R.H.T. Wilke et al., Systematic effects of carbon doping on the superconducting properties of $MgB_{(1-x}C_{x)2}$. Phys. Rev. Lett. **92**(21), 217003 (2004)
41. W.K. Yeoh et al., On the roles of graphene oxide doping for enhanced supercurrent in $MgB_2$ based superconductors. Nanoscale **6**(11), 6166 (2014)
42. K.S.B. De Silva et al., Flux pinning mechanisms in graphene-doped $MgB_2$ superconductors. Scripta Mater. **65**(7), 634 (2011)
43. W.X. Li et al., Graphene micro-substrate-induced $\pi$ gap expansion in $MgB_2$. Acta Mater. **59** (19), 7268 (2011)
44. A. Gurevich et al., Very high upper critical fields in $MgB_2$ produced by selective tuning of impurity scattering. Supercond. Sci. Technol. **17**(2), 278 (2004)
45. A. Gurevich, Enhancement of the upper critical field by nonmagnetic impurities in dirty two-gap superconductors. Phys. Rev. B **67**(18), 184515 (2003)

46. X. Xu et al., Graphene doping to enhance the flux pinning and supercurrent carrying ability of a magnesium diboride superconductor. Supercond. Sci. Technol. **23**(8), 085003 (2010)
47. Y. Katsura et al., Effects of rare earth doping on the superconducting properties of $MgB_2$. Physica C **463–465**, 225 (2007)
48. M. Gharaibeh et al., Effective incorporation of nanoceria into polycrystalline $MgB_2$. J. Appl. Phys. **107**(6), 063908 (2010)
49. Y. Chao et al., Doping effects of $Nd_2O_3$ on the superconducting properties of powder-in-tube $MgB_2$ tapes. Supercond. Sci. Technol. **24**(5), 055016 (2011)
50. X.F. Pan, C.H. Cheng, Y. Zhao, Effect of rare-earth oxides doping on the superconductivity and flux pinning of $MgB_2$ superconductor. J. Supercond. Novel Magn. **24**(5), 1611 (2011)
51. X.F. Pan et al., Doping effect of $Pr_6O_{11}$ on superconductivity and flux pinning of $MgB_2$ bulk. Phys. status solidi (a) **204**(5), 1555 (2007)
52. N. Ojha et al., Enhanced superconducting properties of $Eu_2O_3$-doped $MgB_2$. Physica C **469** (14), 846 (2009)
53. S.K. Chen, M. Wei, J.L. MacManus-Driscoll, Strong pinning enhancement in $MgB_2$ using very small $Dy_2O_3$ additions. Appl. Phys. Lett. **88**(19), 192512 (2006)
54. J. Wang et al., High critical current density and improved irreversibility field in bulk $MgB_2$ made by a scalable, nanoparticle addition route. Appl. Phys. Lett. **81**(11), 2026 (2002)
55. N. Ojha et al., Effect of rare-earth doping on the superconducting properties of $MgB_2$. J. Appl. Phys. **105**(7), 07E315 (2009)
56. C. Cheng, Y. Zhao, Enhancement of critical current density of $MgB_2$ by doping $Ho_2O_3$. Appl. Phys. Lett. **89**(25), 252501 (2006)
57. X.F. Rui et al., Doping effect of nano-alumina on $MgB_2$. Physica C **412–414, Part 1**, 312 (2004)
58. J.S. Slusky et al., Loss of superconductivity with the addition of Al to $MgB_2$ and a structural transition in $Mg_{1-x}Al_xB_2$. Nature **410**(6826), 343 (2001)
59. A. Berenov et al., Enhancement of critical current density in low level Al-doped $MgB_2$. Supercond. Sci. Technol. **17**(10), 1093 (2004)
60. G.J. Xu et al., Enhancement of the irreversibility field in bulk $MgB_2$ by $TiO_2$ nanoparticle addition. Physica C **406**(1–2), 95 (2004)
61. Y. Zhao et al., Nanoparticle structure of $MgB_2$ with ultrathin $TiB_2$ grain boundaries. Appl. Phys. Lett. **80**(9), 1640 (2002)
62. Y. Zhang et al., Effect of addition of nanoparticle $TiO_2/SiO_2$ on the superconducting properties of $MgB_2$. Physica C **468**(15–20), 1383 (2008)
63. S.M. Kazakov et al., Carbon substitution in $MgB_2$ single crystals: Structural and superconducting properties. Phys. Rev. B **71**(2), 024533 (2005)
64. X.F. Rui et al., Improved flux pinning behaviour in bulk $MgB_2$ achieved by nano-$SiO_2$ addition. Supercond. Sci. Technol. **17**(4), 689 (2004)
65. X.L. Wang et al., Significant enhancement of flux pinning in $MgB_2$ superconductor through nano-Si addition. Physica C **385**(4), 461 (2003)
66. Y. Zhang, S.X. Dou, Influence of antimony trioxide nanoparticle doping on superconductivity in $MgB_2$ bulk. J. Mater. Res. **26**(21), 2701 (2011)
67. V.P.S. Awana et al., Fluxoid jump coupled high critical current density of nano-$Co_3O_4$ doped $MgB_2$. Supercond. Sci. Technol. **19**(6), 551 (2006)
68. E. Kuzmann et al., Local environments of iron and cobalt in doped $MgB_2$ superconductors. Supercond. Sci. Technol. **15**(11), 1479 (2002)
69. T. Kuroda, T. Nakane, H. Kumakura, Effects of doping with nanoscale $Co_3O_4$ particles on the superconducting properties of powder-in-tube processed $MgB_2$ tapes. Physica C **469**(1), 9 (2009)
70. C.H. Jiang, H. Hatakeyama, H. Kumakura, Effect of nanometer MgO addition on the in situ PIT processed $MgB_2$/Fe tapes. Physica C **423**(1–2), 45 (2005)
71. K. Vinod et al., Structural and superconducting properties of bulk $MgB_2$ with added nano $Tb_4O_7$. Supercond. Sci. Technol. **21**(2), 025003 (2008)

72. G. Aldica et al., Addition of $Ho_2O_3$ of different types to $MgB_2$ in the ex-situ Spark Plasma Sintering: Simultaneous control of the critical current density at low and high magnetic fields. Mater. Chem. Phys. **146**(3), 313 (2014)
73. D. Batalu et al., $GeO_2$-added $MgB_2$ superconductor obtained by Spark Plasma Sintering. Solid State Sci. **48**, 23 (2015)
74. O. Perner et al., Effects of oxide particle addition on superconductivity in nanocrystalline $MgB_2$ bulk samples. Physica C **432**(1–2), 15 (2005)
75. A. Gumbel et al., High density nanocrystalline $MgB_2$ bulk superconductors with improved pinning. IEEE Trans. Appl. Supercond. **13**(2), 3064 (2003)
76. N. Ojha et al., The effect of carbon and rare earth oxide co-doping on the structural and superconducting properties of $MgB_2$. Supercond. Sci. Technol. **23**(4), 045005 (2010)
77. S. Rahul et al., Combined addition of nano diamond and nano $SiO_2$, an effective method to improve the in-field critical current density of $MgB_2$ superconductor. Mater. Res. Bull. **46** (11), 2036 (2011)
78. S. Thomas et al., Co-addition of nano-carbon and nano-silica: an effective method for improving the in-field properties of magnesium diboride superconductor. Mater. Chem. Phys. **148**(1–2), 190 (2014)
79. Sudesh, et al., Enhanced superconducting properties of rare-earth oxides and graphene oxide added $MgB_2$. Physica C **505**, 32 (2014)
80. J. Nagamatsu et al., Superconductivity at 39 K in magnesium diboride. Nature **410**(6824), 63 (2001)
81. M. Eisterer, Magnetic properties and critical currents of $MgB_2$. Supercond. Sci. Technol. **20** (12), R47 (2007)
82. W.K. Yeoh, S.X. Dou, Enhancement of $H_{c2}$ and $J_c$ by carbon-based chemical doping. Physica C **456**(1–2), 170 (2007)
83. K. Vinod, R.G.A. Kumar, U. Syamaprasad, Prospects for $MgB_2$ superconductors for magnet application. Supercond. Sci. Technol. **20**(1), R1 (2007)
84. K. Vinod, N. Varghese, U. Syamaprasad, Superconductivity of $MgB_2$ in the BCS framework with emphasis on extrinsic effects on critical temperature. Supercond. Sci. Technol. **20**(10), R31 (2007)
85. E.W. Collings et al., Prospects for improving the intrinsic and extrinsic properties of magnesium diboride superconducting strands. Supercond. Sci. Technol. **21**(10), 103001 (2008)
86. X.X. Xi, Two-band superconductor magnesium diboride. Rep. Prog. Phys. **71**(11), 116501 (2008)
87. S.X. Dou et al., Enhancement of the critical current density and flux pinning of $MgB_2$ superconductor by nanoparticle SiC doping. Appl. Phys. Lett. **81**(18), 3419 (2002)
88. J.H. Kim et al., Microscopic role of carbon on $MgB_2$ wire for critical current density comparable to NbTi. NPG Asia Mater. **4**, e3: 1 (2012)
89. Y. Zhu et al., Nanoscale disorder in high critical field, carbon-doped $MgB_2$ hybrid physical-chemical vapor deposition thin films. Appl. Phys. Lett. **91**(8), 082513 (2007)
90. Y. Zhu et al., MgO platelets and high critical field in $MgB_2$ thin films doped with carbon from methane. Supercond. Sci. Technol. **22**(12), 125001 (2009)
91. W. Dai et al., High-field properties of carbon-doped $MgB_2$ thin films by hybrid physical-chemical vapor deposition using different carbon sources. Supercond. Sci. Technol. **24**(12), 125014 (2011)
92. J.H. Kim et al., Carbohydrate doping to enhance electromagnetic properties of $MgB_2$ superconductors. Appl. Phys. Lett. **89**(14), 142505 (2006)
93. A. Motaman et al., Power-law relationship between critical current density, microstructure, and the n-value in $MgB_2$ superconductor wires. J. Supercond. Novel Magn. **27**(7), 1643 (2014)
94. Y.F. Yan, M.M. Al-Jassim, Carbon impurities in $MgB_2$. J. Appl. Phys. **92**(12), 7687 (2002)
95. A.K. Bengtson et al., Impact of substitutional and interstitial carbon defects on lattice parameters in $MgB_2$. J. Appl. Phys. **107**(2), 023902 (2010)

96. M. Maeda et al., Superior $MgB_2$ superconducting wire performance through oxygen-free pyrene additive. Appl. Phys. Express **5**(1), 013101 (2012)
97. T. Takenobu et al., Intralayer carbon substitution in the $MgB_2$ superconductor. Phys. Rev. B **64**(13), 134513 (2001)
98. Z.H. Cheng et al., Superconductivity of $Mg(B_{1-x}C_x)_2$ ternary compounds. J. Appl. Phys. **91** (10), 7125 (2002)
99. W. Mickelson et al., Effects of carbon doping on superconductivity in magnesium diboride. Phys. Rev. B **65**(5), 052505 (2002)
100. A. Bharathi et al., Carbon solubility and superconductivity in $MgB_2$. Physica C **370**(4), 211 (2002)
101. S. Lee et al., Carbon-substituted $MgB_2$ single crystals. Physica C **397**(1–2), 7 (2003)
102. A.V. Pogrebnyakov et al., Properties of $MgB_2$ thin films with carbon doping. Appl. Phys. Lett. **85**(11), 2017 (2004)
103. V. Braccini et al., High-field superconductivity in alloyed $MgB_2$ thin films. Phys. Rev. B **71** (1), 012504 (2005)
104. M. Avdeev et al., Crystal chemistry of carbon-substituted $MgB_2$. Physica C **387**(3–4), 301 (2003)
105. I. Maurin et al., Phase separation in carbon-doped $MgB_2$ superconductors. Phys. B **318**(4), 392 (2002)
106. I. Maurin et al., Carbon miscibility in the boron layers of the $MgB_2$ superconductor. Chem. Mater. **14**(9), 3894 (2002)
107. M. Maeda et al., In-field $J_c$ improvement by oxygen-free pyrene gas diffusion into highly dense $MgB_2$ superconductor. J. Appl. Phys. **109**(2), 023904 (2011)
108. A. Saengdeejing et al., Effects of carbon in $MgB_2$ thin films: Intrinsic or extrinsic. Appl. Phys. Lett. **90**(15), 151920 (2007)
109. A. Yamamoto et al., Universal relationship between crystallinity and irreversibility field of $MgB_2$. Appl. Phys. Lett. **86**(21), 212502 (2005)
110. J.H. Kim et al., Correlation between doping induced disorder and superconducting properties in carbohydrate doped $MgB_2$. J. Appl. Phys. **104**(6), 063911 (2008)
111. W.X. Li et al., Electron-phonon coupling properties in $MgB_2$ observed by Raman scattering. J. Phys.: Condens. Matter **20**(25), 255235 (2008)
112. W.K. Yeoh et al., Evaluation of carbon incorporation and strain of doped $MgB_2$ superconductor by Raman spectroscopy. Scripta Mater. **64**(4), 323 (2011)
113. M.D. Sumption et al., Large upper critical field and irreversibility field in $MgB_2$ wires with SiC additions. Appl. Phys. Lett. **86**(9), 092507 (2005)
114. C. Ferdeghini et al., Upper critical fields up to 60T in dirty magnesium diboride thin films. IEEE Trans. Appl. Supercond. **15**(2), 3234 (2005)
115. G. Sarma, On the influence of a uniform exchange field acting on the spins of the conduction electrons in a superconductor. J. Phys. Chem. Solids **24**(8), 1029 (1963)
116. A. Gurevich, Limits of the upper critical field in dirty two-gap superconductors. Physica C **456**(1–2), 160 (2007)
117. M. Eisterer, M. Zehetmayer, H.W. Weber, Current percolation and anisotropy in polycrystalline $MgB_2$. Phys. Rev. Lett. **90**(24), 247002 (2003)
118. M. Eisterer, C. Krutzler, H.W. Weber, Influence of the upper critical-field anisotropy on the transport properties of polycrystalline $MgB_2$. J. Appl. Phys. **98**(3), 033906 (2005)
119. A.A. Polyanskii et al., Roles of intrinsic anisotropy and pi-band pairbreaking effects on critical currents in tilted-c-axis $MgB_2$ films probed by magneto-optical and transport measurements. Phys. Rev. B **90**(21), 214509 (2014)
120. S. Patnaik et al., Electronic anisotropy, magnetic field-temperature phase diagram and their dependence on resistivity in c-axis oriented $MgB_2$ thin films. Supercond. Sci. Technol. **14** (6), 315 (2001)
121. L. Lyard et al., Anisotropy of the upper critical field and critical current in single crystal $MgB_2$. Phys. Rev. B **66**(18), 180502(R) (2002)

122. C.G. Zhuang et al., Significant improvements of the high-field properties of carbon-doped $MgB_2$ films by hot-filament-assisted hybrid physical-chemical vapor deposition using methane as the doping source. Supercond. Sci. Technol. **21**(8), 082002 (2008)
123. S. Soltanian et al., High transport critical current density and large $H_{c2}$ and $H_{irr}$ in nanoscale SiC doped $MgB_2$ wires sintered at low temperature. Supercond. Sci. Technol. **18**(5), 658 (2005)
124. C.H. Jiang et al., Enhanced $J_c$ property in nano-SiC doped thin $MgB_2$/Fe wires by a modified in situ PIT process. Physica C **422**(3–4), 127 (2005)
125. O. Shcherbakova et al., The effect of doping level and sintering temperature on $J_c(H)$ performance in nano-SiC doped and pure $MgB_2$ wires. J. Appl. Phys. **99**(8), 08M510 (2006)
126. J.H. Kim et al., Enhancement of in-field $J_c$ in $MgB_2$/Fe wire using single- and multiwalled carbon nanotubes. Appl. Phys. Lett. **89**(12), 122510 (2006)
127. P. Lezza, C. Senatore, R. Flukiger, Improved critical current densities in $B_4C$ doped $MgB_2$ based wires. Supercond. Sci. Technol. **19**(10), 1030 (2006)
128. J.H. Kim et al., Superconductivity of $MgB_2$ with embedded multiwall carbon nanotube. Physica C **449**(2), 133 (2006)
129. S.X. Dou et al., Mechanism of enhancement in electromagnetic properties of $MgB_2$ by nano SiC doping. Phys. Rev. Lett. **98**(9), 097002 (2007)
130. M.A. Susner et al., Influence of Mg/B ratio and SiC doping on microstructure and high field transport $J_c$ in $MgB_2$ strands. Physica C **456**(1–2), 180 (2007)
131. J.H. Kim et al., Improvement of upper critical field and critical current density in single walled CNT doped $MgB_2$/Fe wires. IEEE Trans. Appl. Supercond. **17**(2), 2907 (2007)
132. R. Flukiger et al., Simultaneous addition of $B_4C$ + SiC to $MgB_2$ wires and consequences for $J_c$ and $B_{irr}$. IEEE Trans. Appl. Supercond. **17**(2), 2846 (2007)
133. H.L. Suo et al., High critical current densities in SiC doped in-situ $MgB_2$ wires prepared by continuous tube forming and filling technique. IEEE Trans. Appl. Supercond. **17**(2), 2822 (2007)
134. M. Bhatia et al., Superconducting properties of SiC doped $MgB_2$ formed below and above Mg's melting point. IEEE Trans. Appl. Supercond. **17**(2), 2750 (2007)
135. J.H. Kim et al., Influence of disorder on the in-field $J_c$ of $MgB_2$ wires using highly active pyrene. Appl. Phys. Lett. **92**(4), 042506 (2008)
136. M.A.A. Mahmud et al., Comparison of critical current density in $MgB_2$ with different boron sources and nano-particle dopant additions. IEEE Trans. Appl. Supercond. **19**(3), 2756 (2009)
137. J.H. Kim et al., Comparative study of mono- and multi-filament $MgB_2$ wires with different boron powders and malic acid addition. Supercond. Sci. Technol. **23**(7), 075014 (2010)
138. M.A. Susner et al., Enhanced critical fields and superconducting properties of pre-doped B powder-type $MgB_2$ strands. Supercond. Sci. Technol. **24**(1), 012001 (2011)
139. M. Maeda et al., Influence of hydrogen-containing argon gas on the structural parameters and superconducting properties of malic acid-doped $MgB_2$ wires. Scripta Mater. **64**(11), 1059 (2011)
140. J.H. Kim et al., Tailored materials for high-performance $MgB_2$ wire. Adv. Mater. **23**(42), 4942 (2011)
141. M.A. Susner et al., Drawing induced texture and the evolution of superconductive properties with heat treatment time in powder-in-tube in situ processed $MgB_2$ strands. Supercond. Sci. Technol. **25**(6), 065002 (2012)
142. M.S.A. Hossain et al., The effects of annealing temperature on the in-field $J_c$ and surface pinning in silicone oil doped $MgB_2$ bulks and wires. Cryogenics **52**(12), 755 (2012)
143. S.J. Ye et al., Strong enhancement of high-field critical current properties and irreversibility field of $MgB_2$ superconducting wires by coronene active carbon source addition via the new B powder carbon-coating method. Supercond. Sci. Technol. **27**(8), 085012 (2014)
144. D. Patel et al., Multiwalled carbon nanotube-derived superior electrical, mechanical and thermal properties in $MgB_2$ wires. Scripta Mater. **88**, 13 (2014)

145. M.S.A. Hossain et al., The roles of CHPD: superior critical current density and n-value obtained in binary in situ $MgB_2$ cables. Supercond. Sci. Technol. **27**(9), 095016 (2014)
146. P. Kovac et al., Stainless steel reinforced multi-core $MgB_2$ wire subjected to variable deformations, heat treatments and mechanical stressing. Supercond. Sci. Technol. **23**(6), 065010 (2010)
147. S.X. Dou et al., Effect of carbon nanotube doping on critical current density of $MgB_2$ superconductor. Appl. Phys. Lett. **83**(24), 4996 (2003)
148. A. Yamamoto et al., Effects of $B_4C$ doping on critical current properties of $MgB_2$ superconductor. Supercond. Sci. Technol. **18**(10), 1323 (2005)
149. S.K. Chen et al., Effect of heating rates on superconducting properties of pure $MgB_2$, carbon nanotube- and nano-SiC-doped in situ $MgB_2/Fe$ wires. Appl. Phys. Lett. **87**(18), 182504 (2005)
150. A. Yamamoto et al., Doping effects of TiC and $Mo_2C$ on critical current properties of $MgB_2$ tapes. IEEE Trans. Appl. Supercond. **16**(2), 1411 (2006)
151. X.L. Wang, Z.X. Cheng, S.X. Dou, Silicon oil: A cheap liquid additive for enhancing in-field critical current density in $MgB_2$. Appl. Phys. Lett. **90**(4), 042501 (2007)
152. X.P. Zhang et al., Strongly enhanced current-carrying performance in $MgB_2$ tape conductors by $C_{60}$ doping. J. Appl. Phys. **103**(10), 103915 (2008)
153. V. Braccini et al., Improvement of magnetic field behavior of ex-situ processed magnesium diboride tapes. IEEE Trans. Appl. Supercond. **17**(2), 2766 (2007)
154. C.D. Wang et al., Significant improvement in critical current densities of C-doped $MgB_2$ tapes made by high-energy ball milling. Supercond. Sci. Technol. **25**(7), 075010 (2012)
155. D.L. Wang et al., Effects of three different homemade nanocarbons doping on the superconducting properties of $MgB_2$ tapes. Physica C **508**, 49 (2015)
156. H. Yamada et al., Improvement of the critical current properties of in situ powder-in-tube-processed $MgB_2$ tapes by hot pressing. Supercond. Sci. Technol. **23**(4), 045030 (2010)
157. J.M. Hur et al., Fabrication of high-performance $MgB_2$ wires by an internal Mg diffusion process. Supercond. Sci. Technol. **21**(3), 032001 (2008)
158. W. Häßler et al., Further increase of the critical current density of $MgB_2$ tapes with nanocarbon-doped mechanically alloyed precursor. Supercond. Sci. Technol. **21**(6), 062001 (2008)
159. K. Togano et al., Microstructures and critical currents of single- and multi-filamentary $MgB_2$ superconducting wires fabricated by an internal Mg diffusion process. Supercond. Sci. Technol. **23**(8), 085002 (2010)
160. R. Flukiger et al., A new generation of in situ $MgB_2$ wires with improved $J_c$ and $B_{irr}$ values obtained by cold densification (CHPD). IEEE Trans. Appl. Supercond. **21**(3), 2649 (2011)
161. R. Flukiger et al., Improved transport properties and connectivity of in situ $MgB_2$ wires obtained by Cold High Pressure Densification (CHPD). Physica C **471**(21–22), 1119 (2011)
162. H. Kumakura et al., Superconducting properties of diffusion-processed multifilamentary $MgB_2$ Wires. IEEE Trans. Appl. Supercond. **21**(3), 2643 (2011)
163. S.J. Ye et al., Enhancement of the critical current density of internal Mg diffusion processed $MgB_2$ wires by the addition of both SiC and liquid aromatic hydrocarbon. Physica C **471** (21–22), 1133 (2011)
164. G.Z. Li et al., The critical current density of advanced internal-Mg-diffusion-processed $MgB_2$ wires. Supercond. Sci. Technol. **25**(11), 115023 (2012)
165. S.J. Ye et al., Enhancing the critical current properties of internal Mg diffusion-processed $MgB_2$ wires by Mg addition. Supercond. Sci. Technol. **25**(12), 125014 (2012)
166. S.J. Ye et al., Comparison of SiC and/or toluene additives to the critical current density of internal Mg diffusion-processed $MgB_2$ wires. Physica C **484**, 167 (2013)
167. G.Z. Li et al., Effects of carbon concentration and filament number on advanced internal Mg infiltration-processed $MgB_2$ strands. Supercond. Sci. Technol. **26**(9), 095007 (2013)

168. S.J. Ye et al., High-performance MgB2 superconducting wires for use under liquid-helium-free conditions fabricated using an internal Mg diffusion process. Supercond. Sci. Technol. **26**(12) (2013)
169. S.J. Ye et al., Uniformity of coronene-added $MgB_2$ superconducting wires fabricated using an internal Mg diffusion process. IEEE Trans. Appl. Supercond. **25**(5), 6200807 (2015)
170. J.V. Marzik et al., Plasma synthesized doped B powders for $MgB_2$ superconductors. Physica C **423**(3–4), 83 (2005)
171. E.A. Young, Y. Yang, Preparation and properties of $Mg(B_{1-x}C_x)_2$ using carbon chemical vapor coated boron. IEEE Trans. Appl. Supercond. **17**(2), 2794 (2007)
172. S.J. Ye, M.H. Song, H. Kumakura, Novel nanometer-level uniform amorphous carbon coating for boron powders by direct pyrolysis of coronene without solvent. Nanotechnology **26**(4), 045602 (2015)
173. H. Yamada et al., Effect of aromatic hydrocarbon addition on in situ powder-in-tube processed $MgB_2$ tapes. Supercond. Sci. Technol. **19**(2), 175 (2006)
174. H. Yamada et al., The excellent superconducting properties of in situ powder-in-tube processed $MgB_2$ tapes with both ethyltoluene and SiC powder added. Supercond. Sci. Technol. **20**(6), L30 (2007)
175. H. Fujii et al., The effect of SiC addition on the superconducting properties of $MgB_2$ tapes fabricated through an ex situ process using chemically treated powder. Supercond. Sci. Technol. **20**(6), 579 (2007)
176. Z.S. Gao et al., Enhancement of the critical current density and the irreversibility field in maleic anhydride doped $MgB_2$ based tapes. J. Appl. Phys. **102**(1), 013914 (2007)
177. Z.S. Gao et al., Influence of oxygen contents of carbohydrate dopants on connectivity and critical current density in $MgB_2$ tapes. Appl. Phys. Lett. **91**(16), 162504 (2007)
178. Z.S. Gao et al., Strongly enhanced critical current density in $MgB_2$/Fe tapes by stearic acid and stearate doping. Supercond. Sci. Technol. **20**(5), 485 (2007)
179. S.H. Zhou et al., Sugar coating of boron powder for efficient carbon doping of $MgB_2$ with enhanced current-carrying performance. Adv. Mater. **19**(10), 1373 (2007)
180. B.H. Jun et al., Effective carbon incorporation in $MgB_2$ by combining mechanical milling and the glycerin treatment of boron powder. Supercond. Sci. Technol. **21**(10), 105006 (2008)
181. D.L. Wang et al., Effect of processing temperature on the superconducting properties of acetone doped $MgB_2$ tapes. Physica C **469**(1), 23 (2009)
182. C.D. Wang et al., Enhanced superconducting properties of $MgB_2$ tapes achieved by treating B powder with glycolic acid. Supercond. Sci. Technol. **24**(10), 105005 (2011)
183. H. Fujii, K. Ozawa, H. Kitaguchi, Effect of the carbon addition to filling powder ball-milled with oils on the critical current density in ex situ processed $MgB_2$ tapes. Supercond. Sci. Technol. **25**(10), 105008 (2012)
184. E. Martinez, R. Navarro, J.M. Andres, Improvement of the critical current density on in situ PIT processed Fe/$MgB_2$ wires by oleic acid addition. Supercond. Sci. Technol. **26**(12), 125017 (2013)
185. G. Bovone et al., An innovative technique to synthesize C-doped $MgB_2$ by using chitosan as the carbon source. Supercond. Sci. Technol. **27**(2), 022001 (2014)
186. Y. Zhao et al., Improved irreversibility behavior and critical current density in $MgB_2$-diamond nanocomposites. Appl. Phys. Lett. **83**(14), 2916 (2003)
187. H.L. Xu et al., Enhancement of critical current density in graphite doped $MgB_2$ wires. Chin. Phys. Lett. **21**(12), 2511 (2004)
188. P. Kováč et al., Transport current improvements of in situ $MgB_2$ tapes by the addition of carbon nanotubes, silicon carbide or graphite. Supercond. Sci. Technol. **20**(1), 105 (2007)
189. A. Yamamoto et al., Reactivity of carbides in synthesis of $MgB_2$ bulks. Physica C **445**, 801 (2006)
190. Z.S. Gao et al., Effects of NbC addition on the critical current density of $MgB_2$ tapes. Supercond. Sci. Technol. **20**(1), 57 (2007)

191. Z.S. Gao et al., Hollow carbon spheres as an efficient dopant for enhancing the critical current density of $MgB_2$-based tapes. Supercond. Sci. Technol. **21**(10), 105020 (2008)
192. M. Francke et al., Modification of carbon nanostructures by high energy ball-milling under argon and hydrogen atmosphere. Carbon **43**(6), 1204 (2005)
193. W. Gruner et al., Reactive nanostructured carbon as an effective doping agent for $MgB_2$. Supercond. Sci. Technol. **20**(7), 601 (2007)
194. M. Herrmann et al., Touching the properties of NbTi by carbon doped tapes with mechanically alloyed $MgB_2$. Appl. Phys. Lett. **91**(8), 082507 (2007)
195. M. Kulich et al., Effect of C and SiC additions into in situ or mechanically alloyed $MgB_2$ deformed in Ti sheath. Physica C **469**(14), 827 (2009)
196. A. Matsumoto et al., Evaluation of connectivity, flux pinning, and upper critical field contributions to the critical current density of bulk pure and SiC-alloyed $MgB_2$. Appl. Phys. Lett. **89**(13), 132508 (2006)
197. Y. Zhu et al., Microstructures of SiC nanoparticle-doped $MgB_2$/Fe tapes. J. Appl. Phys. **102** (1), 013913 (2007)
198. S.K. Chen et al., Reaction method control of impurity scattering in C-doped $MgB_2$: proving the role of defects besides C substitution level. Supercond. Sci. Technol. **26**(12), 125018 (2013)
199. Y. Shimada et al., Electron microscopy observations of $MgB_2$ wire prepared by an internal Mg diffusion method. Physica C **471**(21–22), 1137 (2011)
200. Y. Shimada et al., Microstructure in high-density $MgB_2$ wires prepared by an internal Mg diffusion method. IEEE Trans. Appl. Supercond. **21**(3), 2668 (2011)
201. C. Suryanarayana, Mechanical alloying and milling. Prog. Mater Sci. **46**(1–2), 1 (2001)
202. L. Cooley, X.Y. Song, D. Larbalestier, Improving flux pinning at high fields in intermetallic superconductors: clues from $MgB_2$ and $MgCNi_3$. IEEE Trans. Appl. Supercond. **13**(2), 3280 (2003)
203. H. Kitaguchi et al., $MgB_2$ films with very high critical current densities due to strong grain boundary pinning. Appl. Phys. Lett. **85**(14), 2842 (2004)
204. E. Martinez et al., Flux pinning force in bulk $MgB_2$ with variable grain size. Phys. Rev. B **75** (13), 134515 (2007)
205. B.J. Senkowicz et al., Nanoscale grains, high irreversibility field and large critical current density as a function of high-energy ball milling time in C-doped magnesium diboride. Supercond. Sci. Technol. **21**(3), 035009 (2008)
206. T. Matsushita et al., Essential factors for the critical current density in superconducting $MgB_2$: connectivity and flux pinning by grain boundaries. Supercond. Sci. Technol. **21**(1), 015008 (2008)
207. R.M. Scanlan, W.A. Fietz, E.F. Koch, Flux pining centers in superconducting $Nb_3Sn$. J. Appl. Phys. **46**(5), 2244 (1975)
208. W.E. Yetter, D.A. Thomas, E.J. Kramer, Grain-boundary flux pinning by the electron-scattering mechanism. Philos. Mag. B **46**(5), 523 (1982)
209. R. Flukiger et al., Improved transport critical current and irreversibility fields in mono- and multifilamentary Fe/$MgB_2$ tapes and wires using fine powders. Supercond. Sci. Technol. **16** (2), 264 (2003)
210. G. Grasso et al., Large transport critical currents in unsintered $MgB_2$ superconducting tapes. Appl. Phys. Lett. **79**(2), 230 (2001)
211. R. Flukiger et al., Superconducting properties of $MgB_2$ tapes and wires. Physica C **385**(1–2), 286 (2003)
212. A. Malagoli et al., Effect of grain refinement on enhancing critical current density and upper critical field in undoped $MgB_2$ ex situ tapes. J. Appl. Phys. **104**(10) (2008)
213. C.E.J. Dancer et al., Fabrication and properties of dense ex situ magnesium diboride bulk material synthesized using spark plasma sintering. Supercond. Sci. Technol. **22**(9), 095003 (2009)
214. P. Kovac et al., Progress in electrical and mechanical properties of rectangular $MgB_2$ wires. Supercond. Sci. Technol. **22**(7), 075026 (2009)

215. A. Yamamoto et al., Towards the realization of higher connectivity in $MgB_2$ conductors: In-situ or sintered ex-situ? Jpn. J. Appl. Phys. **51**(1), 010105 (2012)
216. H.L. Suo et al., High transport and inductive critical currents in dense Fe- and Ni- clad $MgB_2$ tapes using fine powder. AIP Conf. Proc. **614**(1), 872 (2002)
217. P. Lezza et al., Transport properties and exponential n-values of Fe/$MgB_2$ tapes with various $MgB_2$ particle sizes. Physica C **401**(1–4), 305 (2004)
218. A. Kario et al., High energy milled ex situ $MgB_2$ as precursor for superconducting tapes without critical current anisotropy. J. Supercond. Novel Magn. **25**(7), 2337 (2012)
219. S.K. Chen et al., Influence of in situ and ex situ $ZrO_2$ addition on the properties of $MgB_2$. Supercond. Sci. Technol. **17**(2), 243 (2004)
220. S. Mizutani et al., Self-sintering-assisted high intergranular connectivity in ball-milled ex situ $MgB_2$ bulks. Supercond. Sci. Technol. **27**(11), 114001 (2014)
221. S. Mizutani et al., Understanding routes for high connectivity in ex situ $MgB_2$ by self-sintering. Supercond. Sci. Technol. **27**(4), 044012 (2014)
222. H. Tanaka et al., Strongly connected ex situ $MgB_2$ polycrystalline bulks fabricated by solid-state self-sintering. Supercond. Sci. Technol. **25**(11), 115022 (2012)
223. Y. Eltsev et al., Anisotropic superconducting properties of $MgB_2$ single crystals probed by in-plane electrical transport measurements. Phys. Rev. B **65**(14), 140501 (2002)
224. A. Yamamoto et al., Limiting factors of normal-state conductivity in superconducting $MgB_2$: an application of mean-field theory for a site percolation problem. Supercond. Sci. Technol. **20**(7), 658 (2007)
225. B.J. Senkowicz et al., Understanding the route to high critical current density in mechanically alloyed $Mg(B_{1-x}C_x)_2$. Supercond. Sci. Technol. **20**(7), 650 (2007)
226. H. Fang et al., High critical current density in iron-clad $MgB_2$ tapes. Appl. Phys. Lett. **82** (23), 4113 (2003)
227. A. Matsumoto et al., The microstructures and superconducting properties of $MgB_2$ tapes processed in-situ by a ball-milling method. IEEE Trans. Appl. Supercond. **15**(2), 3333 (2005)
228. H. Jiang et al., The effect of ball-milling treatment of original powders on the sintering process and critical current density of graphite-doped $MgB_2$ bulks. J. Mater. Sci. **48**(6), 2485 (2013)
229. X. Xu et al., Influence of ball-milled low purity boron powder on the superconductivity of $MgB_2$. IEEE Trans. Appl. Supercond. **17**(2), 2782 (2007)
230. X. Xu et al., Improved $J_c$ of $MgB_2$ superconductor by ball milling using different media. Supercond. Sci. Technol. **19**(11), L47 (2006)
231. J.H. Lee et al., Superconducting properties of $MgB_2$ prepared from attrition ball-milled boron powder. J. Alloy. Compd. **476**(1–2), 919 (2009)
232. B.H. Jun, S.D. Park, C.J. Kim, Refinement and carbon incorporation effects on the superconducting properties of $MgB_2$ through wet milling process of low purity boron powder. J. Alloy. Compd. **535**, 27 (2012)
233. C. Wang et al., Significant improvement in critical current densities of C-doped $MgB_2$ tapes made by high-energy ball milling. Supercond. Sci. Technol. **25**(7), 075010 (2012)
234. C. Wang et al., Effect of high-energy ball milling time on superconducting properties of $MgB_2$ with low purity boron powder. Supercond. Sci. Technol. **25**(3), 035018 (2012)
235. C. Wang et al., Improved $J_c$-B properties of $MgB_2$ multifilamentary wires and tapes. Supercond. Sci. Technol. **25**(12), 125001 (2012)
236. S. Sugino et al., Enhanced trapped field in $MgB_2$ bulk magnets by tuning grain boundary pinning through milling. Supercond. Sci. Technol. **28**(5), 055016 (2015)
237. W. Häßler et al., Influence of the milling energy transferred to the precursor powder on the microstructure and the superconducting properties of $MgB_2$ wires. Supercond. Sci. Technol. **26**(2), 025005 (2013)
238. M. Herrmann et al., Conflicting effects of SiC doping on the properties of mechanically alloyed bulk $MgB_2$. IEEE Trans. Appl. Supercond. **19**(3), 2726 (2009)

239. A. Kario et al., Properties of $MgB_2$ tapes prepared by using MA in-ex situ powder. IEEE Trans. Appl. Supercond. **20**(3), 1521 (2010)
240. A. Kario et al., Critical current density enhancement in strongly reactive ex situ $MgB_2$ bulk and tapes prepared by high energy milling. Supercond. Sci. Technol. **24**(7), 075011 (2011)
241. B.J. Senkowicz et al., Improved upper critical field in bulk-form magnesium diboride by mechanical alloying with carbon. Appl. Phys. Lett. **86**(20), 202502 (2005)
242. B.J. Senkowicz et al., Atmospheric conditions and their effect on ball-milled magnesium diboride. Supercond. Sci. Technol. **19**(11), 1173 (2006)
243. D. Rodrigues et al., Flux pinning optimization of $MgB_2$ bulk samples prepared using high-energy ball milling and addition of $TaB_2$. IEEE Trans. Appl. Supercond. **19**(3), 2797 (2009)
244. L.B.S. Da Silva, E.E. Hellstrom, D. Rodrigues Jr., The influence of milling time and $ZrB_2$ addition on the superconducting properties of $MgB_2$. IEEE Trans. Appl. Supercond. **25**(3), 6800404 (2015)
245. L.B.S. Da Silva et al., Study of $TaB_2$ and SiC additions on the properties of $MgB_2$ superconducting bulks. Supercond. Sci. Technol. **28**(2), 025008 (2015)

# Chapter 4
# Critical Current Anisotropy in Relation to the Pinning Landscape

**Nick J. Long**

## 4.1 Introduction

Raising the critical currents of high-temperature superconductors has the desirable outcome of lowering the barrier to widespread application of these materials. With creative materials engineering, researchers have been successful in improving critical currents over the field and temperature ranges relevant to applications [1–7]. Intrinsic thermodynamic properties and the interaction between the magnetic vortices and the pinning landscape combine to determine how much current is supported. Due to both factors, the resultant critical currents generally vary with the angle of the imposed external field. In this chapter, we will explore what is known about this anisotropy.

The key to improved critical current performance has been optimizing the density and morphology of pinning centers both for native crystal defects, such as ion vacancies, stacking faults, and dislocations, and engineered additions such as randomly distributed normal phase nanoparticles, columnar and transverse arrays of nanoparticles. The improvements made in critical currents have been achieved by trial and error guided by the understanding that it is desirable to have pinning centers which are of similar dimension to the superconductor coherence length, and that defects which are correlated in one direction will mostly enhance pinning when the external magnetic field is also oriented in this direction. To go beyond trial and error and more rationally improve nanoengineering of the pinning landscape, we need to better understand the structure-property relationships between the pinning microstructure and the critical current density.

But analyzing the critical current density to produce structure-property relationships has not proved straightforward or without controversy for the anisotropic

N.J. Long (✉)
Robinson Research Institute, Victoria University of Wellington,
Wellington, New Zealand
e-mail: Nick.Long@vuw.ac.nz



A. Crisan (ed.), *Vortices and Nanostructured Superconductors*, Springer Series
in Materials Science 261, DOI 10.1007/978-3-319-59355-5_4

superconductors. The difficulty stems partially from having two sources of anisotropy, the anisotropy in the vortex cross section, arising from the intrinsic mass anisotropy of the carriers of the vortex current, and the anisotropy of the pinning centers themselves. The effects of these two factors are not easily separated. It is never possible to have a fully isotropic pinning landscape in an anisotropic superconductor. By definition, the modulation of the superconductor order parameter in the *c*-axis direction gives some directional pinning. In addition, particularly for thin films, some correlated native defects in the *c*-axis or *a/b* axis directions are inevitable. The other aspect of the difficulty lies in data analysis, as we will describe in this chapter these two effects can produce similar outcomes.

A better understanding of the structure-property relations has technical value as the anisotropy in critical currents complicates the use of these superconductors. In most applications, wires and tapes will experience magnetic fields at many different angles to the tape. If the critical current ($I_c$) varies strongly with angle, this presents challenges for coil designers who must keep the coil within safe operating margins of $I/I_c$ everywhere within the coil. An unpredictable or only weakly predictable $I_c$ with field and temperature means exploiting the full capacity of these materials is compromised.

Scientifically, the anisotropy of $I_c(\theta)$ highlights that for anisotropic superconductors or even superconductors in general, we do not understand well enough the structure-property relationships between pinning and critical currents. The phenomenon of pinning is not in dispute, but the understanding of how pinning is aggregated across a complex pinning landscape, and the role played by carrier anisotropy, is not complete.

Therefore, in this chapter, we will look at the common methods used to extract information about the pinning landscape and the role of mass anisotropy from measurements of the critical current density anisotropy $J_c(\theta)$. We define $J_c = I_c/A$ as the critical current density in the plane of the tape or thin film, or *ab*-plane in the case of a crystal, and $\theta$ is the angle between the applied magnetic field and the normal to the tape, usually the *c*-axis direction, with the current always perpendicular to the field. This is often referred to as the maximum Lorentz force configuration and is of most relevance to magnet and coil design.

Experimentally, we define the transport critical current as the current when an electric field $E_0 = 1$ μV/cm is present in the current direction. The $E(J)$ relation for transport currents is found to be a power law $E/E_0 = (J/J_c)^n$ so strictly we have an arbitrary engineering criterion rather than a value uniquely determined by experiment. With magnetization measurements of $J_c$, the electric field criterion is effectively much lower although this does not affect the analyses examined here. Theoretically, the critical current is defined by $\vec{J}_c \times \vec{\Phi}_0 + \vec{p}_{max} = 0$ [8, 9], that is, the sum of the forces on the vortex arising from the presence of the transport current is in balance with the restoring force on the vortex due to its spatial position in a potential minimum. Researchers commonly call the force due to the current the Lorentz force although there are good reasons not to do this [9, 10]. We shall persist with the mainstream nomenclature and refer to this as the Lorentz force.

Common methods to analyze the critical current anisotropy are firstly scaling methods, such as the Blatter scaling [8, 11] and other scaling approaches which are a modification of this approach. Secondly, there are more direct methods of calculating the expected response from defects, which examine the pinning forces on vortices from defects under certain assumptions, and finally, we examine the vortex path model or maximum entropy method, which is an information theory or statistical approach for extracting information from $J_c(\theta)$.

## 4.2 Mass Anisotropy Scaling

### 4.2.1 Theory

For high-temperature superconductors, taking account of the fundamental electronic anisotropy of the materials is necessary to understand the thermodynamic and electromagnetic properties. Hao and Clem [12] first discussed how the Gibbs free-energy difference ΔG between the mixed and normal states can be expressed as a function of the reduced field $h(\theta) = \frac{H}{H_{c2}(\theta)} \propto H\epsilon(\theta)$ for $H \gg H_{c1}$, where $\epsilon(\theta) = (cos^2\theta + \gamma^{-2}sin^2\theta)^{1/2}$ and $\gamma = \sqrt{m_c/m_{ab}}$ is the mass anisotropy parameter. They hypothesized that physical quantities derived from the Gibbs free energy should obey the same scaling behavior. Hence, measurements of these properties should be scalable with the magnetic field in the same way through the reduced field $h$. These properties would include resistivity, critical current density, and flux-line-lattice melting temperature. They thus proposed the scaling $J_c(H,\theta) = J_c(h(\theta))$ for the critical current.

Blatter et al. [8, 11] expanded on and generalized this idea by noting that the conventional method for accommodating anisotropy into a phenomenological model of superconductivity is to introduce an anisotropic effective mass tensor into the Ginzburg–Landau or London equations and then proceed to repeat calculations made for the isotropic case. Instead, they proposed to rescale the anisotropic problem to a corresponding isotropic one at the level of the basic phenomenological equations. The scaling rules can then be applied to the known results of the isotropic model to obtain the anisotropic results with no great effort. This approach is schematically represented in Fig. 4.1.

The primary result is then a scaling rule as given in (4.1), where the desired quantity is $Q$ and the known isotropic result is $\tilde{Q}$.

$$Q(\theta, H, T, \xi, \lambda, \varepsilon, \gamma_B) = s_Q \tilde{Q}(\epsilon_\theta H, T/\varepsilon, \xi, \lambda, \gamma_B/\varepsilon) \tag{4.1}$$

Blatter et al. use different conventions from those now typical in critical current research, so in (4.1), $\varepsilon$ is the reciprocal of our anisotropy factor, i.e., $\gamma = 1/\varepsilon = 5 - 7$ for YBCO and $\gamma_B$ is not anisotropy but is a measure of the disorder. Blatter et al. derived these results in the context of weak collective pinning theory, where $\gamma_B$

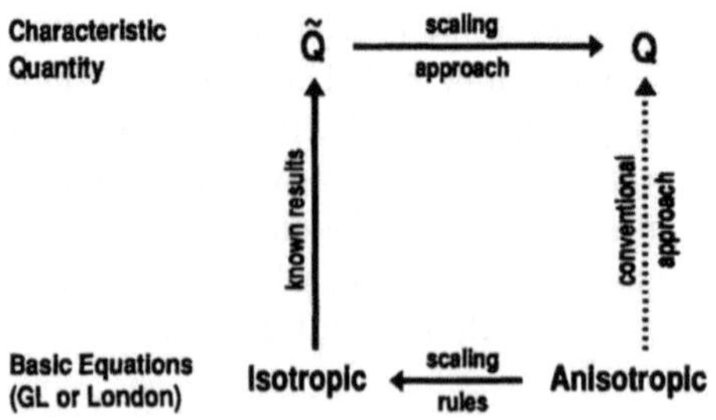


**Fig. 4.1** Scaling procedure to simplify the derivation of anisotropic models as proposed by Blatter et al. There are two paths to go from the basic equations of superconductivity to a model of the measured quantity Q, either via a direct calculation or via scaling rules, which transform the basic equations from anisotropic to isotropic. (Figure reproduced from [11] with permission)

describes short-range disorder in $T_c$. This scaling rule leads Blatter et al. to also predict for the planar critical currents in an anisotropic superconductor $J_c(H,\theta) = J_c(h(\theta))$, for large magnetic fields, consistent with Hao and Clem [12]. At low fields in the single vortex regime, a direct substitution of the scaling rules in the collective pinning theory gives a different result, $J_c(\theta) = \text{const}$, [8].

### 4.2.2 $J_c(h(\theta))$ Scaling of YBCO

The $J_c(H,\theta) = J_c(\epsilon_\theta H)$ scaling law was first applied to the critical currents in YBCO films by Xu et al. [13] and Kumar et al. [14]. For Xu et al., the scaling relationship worked reasonably well over the whole angular range, over a wide field range at high temperature. They plotted their data in the form of flux pinning force or 'Kramer scaling' with a driving force of $F_p^{eff} = J_c \epsilon_\theta H = F_0 h(\theta)^p (1 - h(\theta))^q$ and found a single scaling curve for $F_p^{eff}$ versus $h$ with no further rescaling of $F_p$. This is equivalent to finding a single curve for $J_c(\epsilon_\theta H)$ versus $h$ as in the experiments described over the next few paragraphs. Xu et al., who referenced Hao and Clem, did not argue that the scaling implied any particular type of pinning structure, rather they argued from the exponents of the Kramer scaling that planar pinning mechanisms were dominant with some point pinning. Kumar et al. referenced [8] and did not discuss pinning mechanisms; they found $\gamma = 3.4$ gave the best overlap of data.

The technique was not applied more commonly until Civale et al. used it to identify the effect of particular pinning structures [15, 16]. Civale et al. argued that the $J_c(\epsilon_\theta H)$ scaling would apply to *random defects* (uncorrelated disorder) only, and hence by identifying the component of critical current which followed the scaling rule, the components which corresponded to correlated defects could be separately identified. Figure 4.2 shows the process applied to make this assignment given a set of $J_c(\theta)$ curves at various values of $H$. Firstly, the $J_c(\theta)$ is replotted as $J_c(\epsilon_\theta H)$ with $\gamma$ as the only adjustable parameter. The procedure is repeated with every dataset for different $H$. If the correct value of $\gamma$ is chosen, then the data collapse onto a single curve, except for the regions where correlated pinning is contributing to $J_c$. A single curve can then be fitted to $J_c(\epsilon_\theta H)$ which covers the region of overlapping data. This fitting can then be mapped back to any single dataset $J_c(\theta)$ to identify the portion of the curve attributable to random pinning, and this contribution subtracted to quantitatively identify the correlated pinning. The result shown in Fig. 4.2 is typical for many YBCO films which have correlated pinning due to *ab*-plane pinning centers, either intrinsic pinning from the weakly superconducting spacing layers or stacking faults, and *c*-axis pinning due to either grain boundaries or twin plane boundaries.

This method can then be applied to investigate the temperature and field dependence of particular pinning contributions. This was done in detail by Gutierrez et al. [2] for a metal-organic deposited YBCO film with barium zirconate (BZO) inclusions.

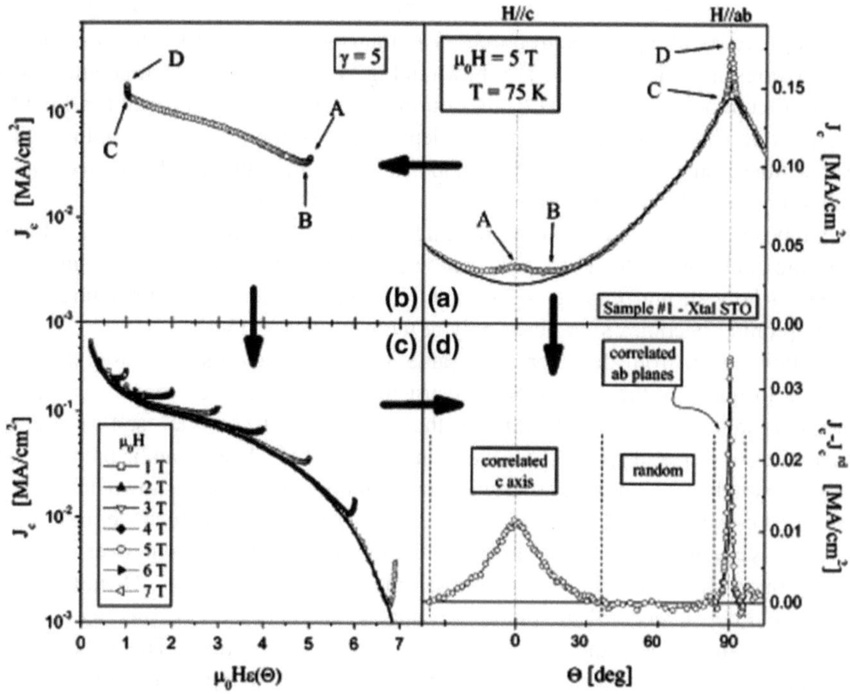


**Fig. 4.2** Application of the scaling method to separate isotropic and anisotropic components. **a** $J_c(\theta)$ for an YBCO film at B = 5 T, the solid line is the identified isotropic contribution to $J_c(\theta)$ from the scaling approach, **b** A scaled $J_c(h(\theta))$ dataset, where A, B, C, D are data points for $\theta = 0$, 15, 85, 90°, respectively. **c** scaled data for the full set of fields, the solid line is a fit to where the data overlap and is identified as the random isotropic contribution to $J_c$ **d** anisotropic contributions to $J_c(\theta)$ at 5 T obtained by subtracting the isotropic $J_c(\theta)$. (Figure reproduced from [15] with permission)

This paper reported the largest flux pinning forces in an HTS or LTS conductor to that date, a value of ~21 GN $m^{-3}$ at 77 K, and ~80 GN $m^{-3}$ at 65 K. This was an impressive 500% improvement over NbTi at 4 K.

The film incorporated a dense array of mostly randomly oriented BZO nanodots, i.e., embedded particles of ~15 – 30 nm diameter at a concentration of 10% mol. The film was also dense with further defects such as stacking faults (single Y-124 layers) and intergrowths (blocks of $YCuO_x$ or $BaCuO_x$). XRD analysis of the films showed them to have an internal strain of 0.56%, which was twice that of pure YBCO films prepared with the same technique.

The effect of these different pinning structures was resolved through the technique of Fig. 4.2, collapsing the $J_c(H,\theta)$ curves onto a single-scaled $J_c(\epsilon_\theta H)$. To achieve an overlap of the data required setting $\gamma_{eff} \sim 1.5$ compared to the usual $\gamma \sim 5 - 7$ for YBCO, that is, the effective anisotropy of this material is much lower than pure YBCO. The authors then went a step further by assuming critical currents due to weak collective pinning can be described by a temperature dependence $J_c^{wk}(T) = J_c^{wk}(0)\exp(-T/T_0)$ and those due to strong pinning can be described by $J_c^{str}(T) = J_c^{str}(0)\exp[-3(T/T^*)^2]$. The contributions to $J_c$ can thus be further decomposed by fitting the isotropic $J_c^{iso}(T)$ or anisotropic $J_c^{anis}(T)$ with a mixture of these expressions. Their results are shown in Fig. 4.3 for a range of temperatures and fields. Figure 4.3a shows for H//c the isotropic contribution is dominant at all fields. Figure 4.3b shows the full temperature dependence of the isotropic $J_c$ and that this can be fitted using the strong pinning model at high fields and a combination of weak and strong at 1 T. The final plot, Fig. 4.3c, shows how the complete $J_c(T)$ can be decomposed into weak and strong pinning components if we ignore the small anisotropic contribution for H//c. This can be generalized to a pinning phase diagram showing how strong pinning is dominant for higher fields and temperatures, and weak pinning becomes significant only at temperatures < 15 K.

Following on from this work, the same authors and others have used $J_c(\epsilon_\theta H)$ scaling to quantify the effects of nanostructuring on critical currents. Also using metal-organic deposition (MOD) as the fabrication technique, self-assembled nanowires of $Ce_{0.9}Gd_{0.1}O_{2-y}$ (CGO) or nanoislands of (La,Sr)$O_x$ can be prepared on single crystal substrates, [17]. Both surface modifications are shown to induce extended *c*-axis defects leading to an increase in critical currents for H//c. Using the $J_c(\epsilon_\theta H)$ scaling, the ratio of anisotropic pinning to total current over the (*H*, *T*) region is shown in Fig. 4.4 for the sample with LSO nanodots. Of note is the weak field dependence of the pinning. It appears the added *c*-axis pinning is quite effective to high fields, and is most effective at high temperatures. This is borne out by a direct comparison of the pinning engineered films with a pure YBCO reference film [17].

An important follow-up to this work was the investigation on the effects of strain in the MOD films [18]. It was found that nanocomposites of YBCO-$BaZrO_3$ and YBCO-$Ba_2YTaO_6$ enhanced the generation of strain—correlated with the interfacial area of the nanodots. The strain itself is of such a magnitude as to suppress pair formation and hence can be a source of vortex pinning. Through comparisons with

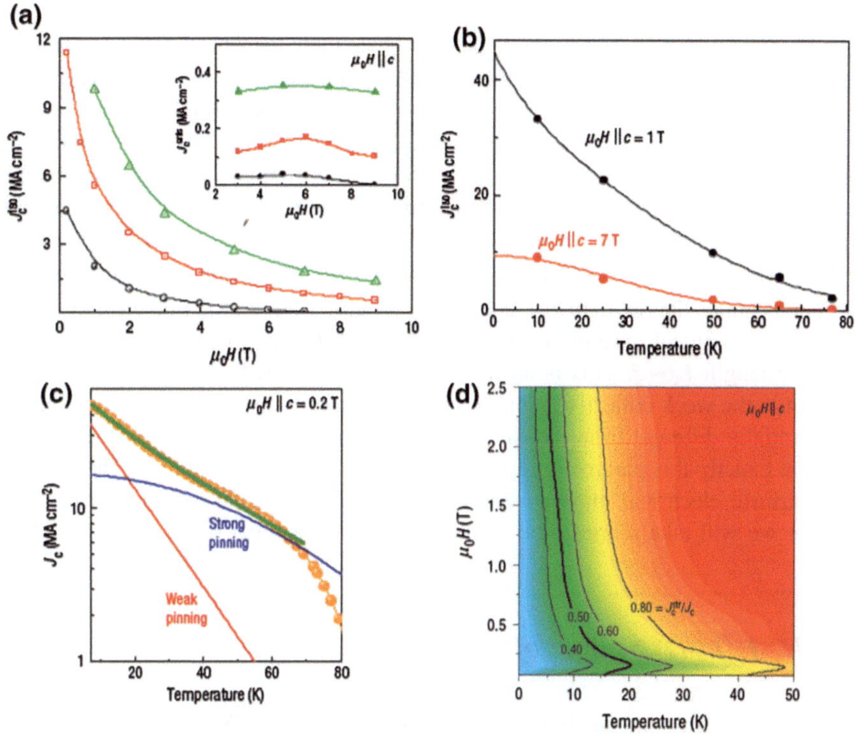


**Fig. 4.3** Separation of isotropic/anisotropic and weak/strong components of the critical current. **a** Field dependence $J_c^{iso}(B)$ and $J_c^{anis}(B)$ at different temperatures, **b** temperature dependence $J_c^{iso}(T)$ at two fields. 7 T data are fitted to the strong pinning expression, and 1 T data are fitted to a mixed weak-strong expression. **c** $J_c$ measured inductively (effectively the isotropic critical current only) and a fit of the weak and strong contributions. **d** Pinning phase diagram showing the contribution of strong pinning to the total critical current for H//c, as deduced from the fitting of **c**. (Reproduced from [2] with permission.)

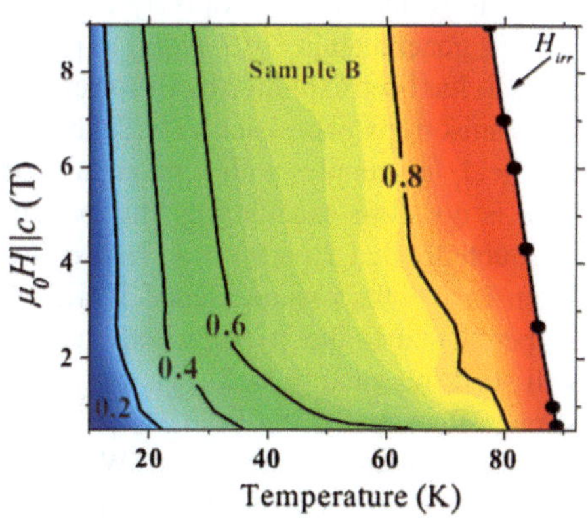


**Fig. 4.4** Vortex pinning phase diagram showing the ratio of anisotropic to total critical current $J_c^{anis}/J_c$ for *H*//*c* over a range of field and temperature for a YBCO sample prepared on a substrate with a surface preparation of LSO nanoislands. (Reprinted from Gutiérrez et al. [17] with permission)

different nanoparticle additions, they identify strain as a source of increased isotropic pinning force. These samples also have high densities of Y124 stacking faults. The result is samples with a low effective anisotropy of $\gamma_{eff} \sim 1.4$, compared to the confirmed mass anisotropy from $H_{c2}(\theta)$ of $\gamma \sim 5.9 - 6$ [18].

Despite the success of these approaches in assessing the effects of nanostructuring, we may harbor some reservations about the assignment of these components vis-a-vis weak/strong and isotropic/anisotropic pinning. The weak collective pinning theory on which Blatter et al. derived their formulation has an expression for critical current of $J_c \approx J_0[\xi/L_c]^2$ and only applies when the collective pinning length is much larger than the coherence length, $L_c \gg \xi$ [8]. For REBCO films which have critical currents on order of 10% of $J_0$, we would have a collective pinning length $L_c \sim \xi$. This is an inconsistency—high-performance REBCO films are not in the weak collective pinning limit. The use of anisotropy factors which are substantially different from the electronic mass anisotropy also invites questions as to what exactly this parameter represents. As confirmed by the authors themselves, the intrinsic electronic mass anisotropy has not changed. To explore these issues further we will take a brief look at similar scaling for other materials.

### 4.2.3 $J_c(h(\theta))$ Scaling of BSCCO

A scaling description can also be used on BSCCO (Bi, Pb)$_2$Sr$_2$Ca$_2$Cu$_3$O$_{11}$ wires as shown in Fig. 4.5 [19]. BSCCO is known to have a very complicated microstructure, so that a combination of *ab*-plane defects and point defects will exist. The material is also poorly textured compared with the YBCO thin films discussed thus far, and will have a FWHM of the rocking curve of order 12° [20]. In this plot, data for temperatures from 20–85 K have been combined by scaling the $I_c$ to the self-field value at each temperature. This data can be fitted with a (temperature-independent) anisotropy parameter, $\gamma \approx 8$, which is similar to results from YBCO, although this bears no relation to the real mass anisotropy of the material, which is conservatively estimated at $\gamma \approx 50$. [8, 21]

This general ability of $J_c(\epsilon_\theta H)$ to produce a convenient parameterization of large datasets has been exploited by Hilton et al. [22] and Pardo et al. [23] who have used the mass anisotropy expression to fit datasets of YBCO over wide ranges of temperature and field, for samples with $I_c$ maxima at perpendicular and/or parallel fields.

It is not clear exactly why this parameterization works so well. One reason is undoubtedly that $J_c(\epsilon_\theta H) \approx J_c(Hcos\theta)$ over much of the angular range, and this is insensitive to the $\gamma$ value. Hence, if critical current is proportional to the flux density perpendicular to the layers, where the pinning may be less strong, then this fitting will work reasonably well. It is only close to parallel field that the $\gamma$ value has a strong effect, but if the fitting is poor closer to parallel field, it is easy to invoke correlated pinning as the reason. We should therefore be sensitive to the possibility of misinterpretation using this technique.

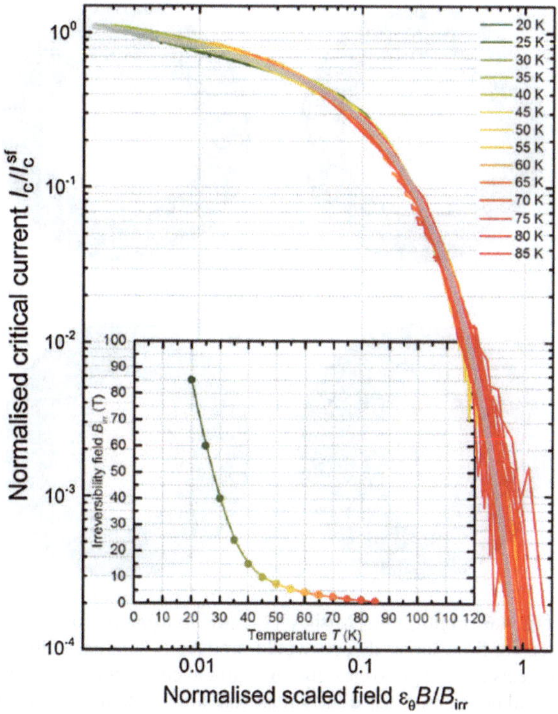


**Fig. 4.5** Scaling of BSCCO wire for temperatures of 20 – 85 K, and fields up to 8 T. The field was scaled using $\epsilon(\theta)B/B_{irr}$ with $\gamma = 8$. The $B_{irr}$ values used are shown in the inset. The critical currents are normalized to the self-field value at each temperature. Reprinted from [19] with permission

### 4.2.4 $F_p(h(\theta))$ Scaling of Ba-122

An interesting experimental exploration of the validity of scaling rules has recently been published by Mishev et al. [24]. They have studied $J_c(\theta)$ in $BaFe_2As_2$ (Ba-122)-based superconducting single crystals, with three samples covering weak, intermediate, and strong pinning regimes. Starting with a K-doped Ba-122 crystal with the lowest $J_c$, this sample was shown to obey $J_c(\epsilon_\theta H)$ scaling. For an example of strong pinning, they prepared a neutron-irradiated Co-doped Ba-122 sample, with isotropic defects which are larger than the coherence lengths. For this sample, $J_c(\epsilon_\theta H)$ scaling does not produce a collapsed curve, instead they found that an additional scaling factor is required $J_c(H,\theta) = \epsilon_\theta J_c(\epsilon_\theta H)$ which then produces a single curve for the data. As this is equivalent to scaling the flux pinning force, i.e., $J_c(H,\theta)\mu_0 H = \mu_0\epsilon_\theta H J_c(\epsilon_\theta H)$ or in the common Kramer form $F_p(H,\theta) = F_0 h(\theta)^m(1-h(\theta))^n$, we will refer to this as $F_p(\epsilon_\theta H)$ or $F_p(h(\theta))$ scaling. Mishev et al. motivate this scaling using an argument associated with Fig. 4.6 below.

Mishev et al. argue that the elementary pinning force, $f_p = E_p/d$, where $E_p$ is the pinning energy and $d$ is the relevant length scale, is for defects with a radius less than the coherence length, $r_d < \xi_{ab}$, $f_p \propto E_c r_d^3/\xi_{ab}\epsilon(\theta)$, with $E_c$ the condensation energy, which retains an angular dependence. This is illustrated in Fig. 4.6e, f. In contrast, for a large defect with $r_d \geq \xi_{ab}$, $f_p \propto E_c r_d \xi_{ab}\epsilon(\theta)/\epsilon(\theta) = \text{const}$. This is

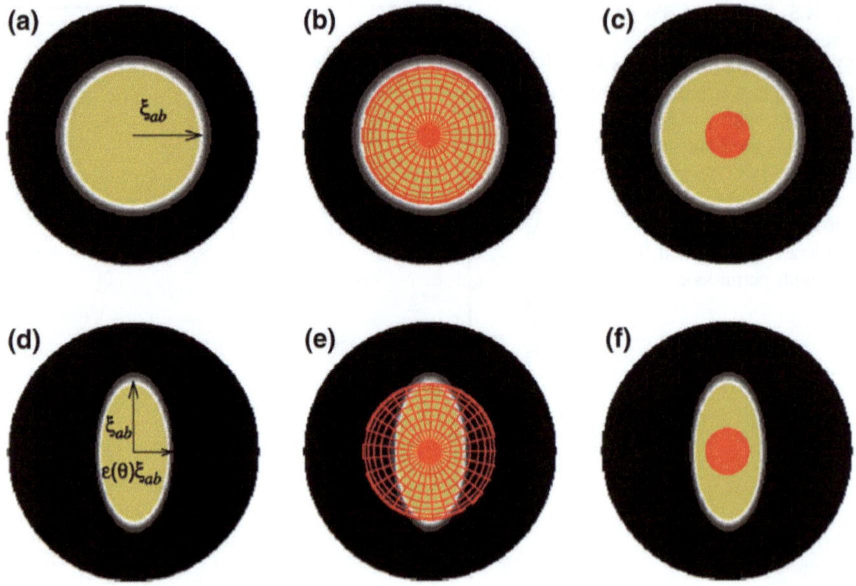


**Fig. 4.6** Figure from Mishev [24] giving a microscopic explanation as to why critical currents arising from larger defects would produce a different scaling rule from those arising from point defects **a** isotropic vortex core **b** isotropic core pinned by large defect, $r_d \geq \xi_{ab}$ **c** core pinning by small defect, $r_d < \xi_{ab}$, **d** the vortex core after a rotation by angle $\theta$ **e** pinning of the anisotropic core by a large defect, **f** pinning of the anisotropic core by a small defect. Reprinted from [24] with permission

illustrated in Fig. 4.6b, e. This fundamental difference in the elementary pinning force produces the difference in scaling of the macroscopic currents, that is, between $J_c(\epsilon_\theta H)$ scaling and $F_p(\epsilon_\theta H)$ scaling.

This argument introduces some additional puzzles, however. If for the case of $r_d \geq \xi_{ab}$, the elementary pinning is angle independent, and these defects are dominant, then how physically does the angular dependence of the critical currents arise? From our definition of the critical current, $\overrightarrow{J}_c \times \overrightarrow{\Phi}_0 + \vec{p}_{max} = 0$, the angular dependence in $J_c$ must arise from pinning interactions of some description.

This $F_p(\epsilon_\theta H)$ scaling has been proposed previous to the Mishev et al. paper to describe critical currents in an irradiated YBCO sample. Matsui et al. [25] prepared YBCO films with an MOD method and then applied low-energy Au irradiation to produce a large density of point like pins. Matsui et al. noticed that if they transform their $J_c$ data for the irradiated film into flux pinning force, the result is an interesting 'capping' of the magnitude of the flux pinning force, seemingly independent of field. This is shown in Fig. 4.8d. In an inset of Fig. 4.8c are calculated values of $F_p(H,\theta) = F_0 h(\theta)^m (1 - h(\theta))^n$, with $(m,\ n)$ = (0.5, 2), and the scaling used is $h(\theta) = \epsilon(\theta)H/H_0$ with $\mu_0 H_0$ = 8.3 T. This function is shown to fit the data in

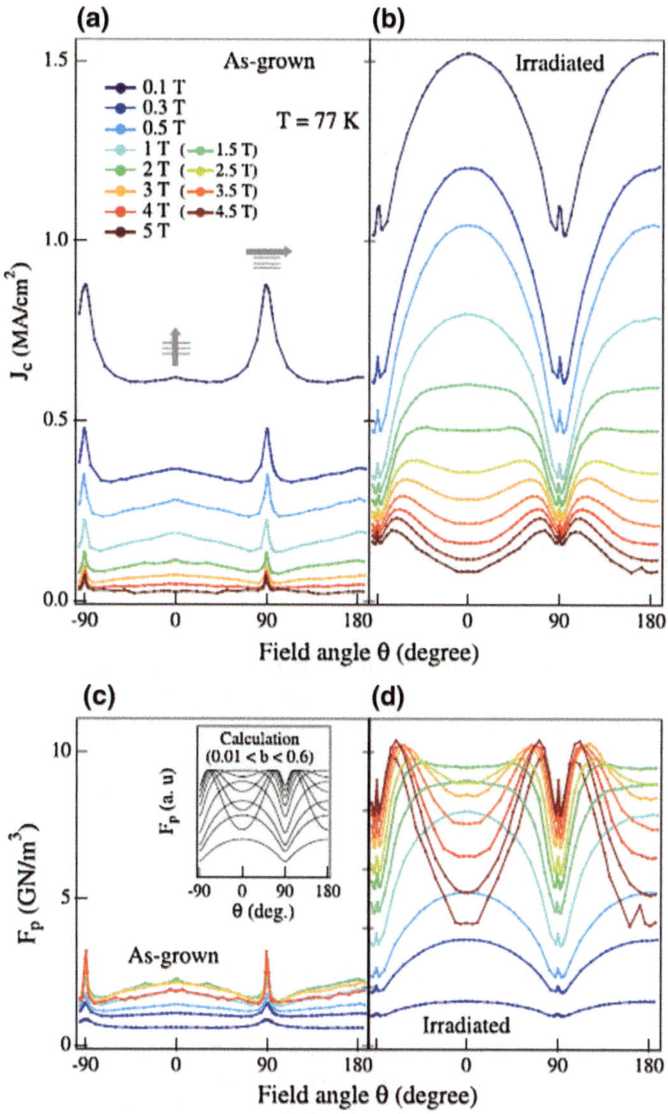


**Fig. 4.7** **a** Critical current for as-grown YBCO. **b** $J_c$ for irradiated YBCO, **c** flux pinning force for as-grown YBCO, and inset showing $F_p$ calculation as described in the text, **d** flux pinning force for the irradiated sample. Reprinted from [25] with permission

Fig. 4.7d reasonably well for the field range 2–5 T. This includes fitting the unusual shoulder features which lie at approximately 75°.

Matsui et al. offer the following explanation for their results. The $(m, n) = (0.5, 2)$ values coincide with the predictions of Kramer [26] for point pins, and given the large increase in critical current with the irradiation and hence introduction of point pins, they naturally ascribe the $J_c$ for B $\geq$ 2 T to strong point pinning. At lower fields, the fitting does not account for a large angle-independent contribution to $J_c$, and Matsui et al. modify the flux pinning expression with an offset and linear scaling. They justify this with a statistical argument around how pinning may be modified at lower vortex densities.

The $F_p(\epsilon_\theta H)$ scaling observed by Mishev et al. and Matsui et al. is indeed quite striking. We have offered an alternative explanation to the Matsui et al. results in a comment on their paper [27]. Our explanation centers on identifying the equation $f(b) = f_0 b^m (1-b)^n$ as a beta distribution and an outcome of averaging. In both papers, attempts are made to connect the scaling analysis to the microscopic physics. However, in both cases, this raises further questions. For example, in the Matsui et al. data, we may ask why point pinning should lead to the striking shoulder structures.

To make further progress in the next section, we examine direct models which connect the anisotropic critical currents to flux pinning.

## 4.3 Models of Pinning and Field Angle-Dependent Currents

In view of the open questions about scaling, it is of interest to describe some microscopic models of the critical current anisotropy. Rather than explaining the results through scaling of fundamental equations derived from the Ginsburg-Landau model or the London equations, these are more direct models which seek to link a particular defect structure with the form of $J_c(\theta)$ by considering the forces which are acting on flux lines. We describe the first model in some detail and the following two only briefly.

### *4.3.1 The Tachiki and Takahashi Model*

Tachiki and Takahashi [28] primarily focused on the pinning effect of the layer structure of the superconducting copper oxides. In these materials, the $CuO_2$ planes are strongly superconductive with a high superfluid density, and the regions between the $CuO_2$ layers have low superfluid density. The weakly superconducting layers therefore act as natural pinning centers, and when vortices are directed along these layers, they will be strongly pinned against perpendicular motion. Figure 4.8 shows the key idea of their papers. When the magnetic field is at an angle $\theta$, then the vortex forms a staircase structure in which it periodically lies along the *ab*-planes where it is pinned by the intrinsic layers, and intermittently, the vortex is deformed perpendicular to the layers where there is assumed to exist pinning parallel to the *c*-axis through twin planes or other correlated pinning. The twin plane spacing is assumed to be larger than the vortex lattice spacing and the intrinsic layer spacing.

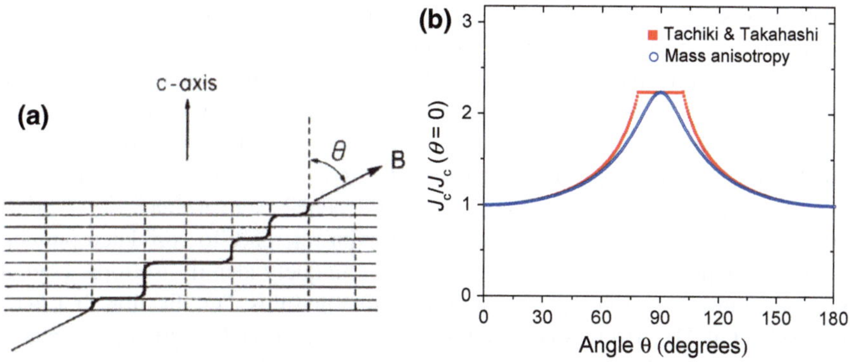


**Fig. 4.8** **a** Vortex at angle $\theta$, being pinned in the *ab*-plane by the intrinsic insulating planes, and pinned in the *c*-axis direction by correlated pinning such as twin planes. (Reproduced by permission from [28]) **b** The angular dependence of critical current according to (4.2), for a sample with $(J_{c\|}/J_{c\perp})^2 = 5$, and the mass anisotropy expression with $\gamma = 5$, and in both cases $J_c(B) \propto B^{-0.5}$

The authors then resolve the pinning force into Cartesian components, and derive expressions for $J_c$ when $J_c$ is a result of the Lorentz force exceeding the strength of the layered pinning and then a different expression for when the Lorentz force exceeds the strength of the *c*-axis pinning. Since the flux density from the parallel vortices is $B_x = Bsin\theta$, the critical current is $J_c = J_{c\|}(Bsin\theta)$ for the parallel vortices, that is, their pinning forces will be exceeded at this current, where $J_{c\|}$ is the critical current in the case of perfect alignment with the *ab*-planes. Tachiki and Takahashi take the parallel field $J_{c\|}(B)$ as field independent due to the high matching field for the intrinsic pinning, that is, $J_{c\|}(Bsin\theta) = J_{c\|}(0)$. The flux density for the *c*-axis oriented vortices is $B_x = Bcos\theta$, and the critical current density for these vortices is $J_c = J_{c\perp}(Bcos\theta)$ when the force parallel to the *c*-axis exceeds the pinning force in this direction. For this direction, they assume a functional dependence $J_{c\perp}(B) \propto B^{-\alpha}$. The overall critical current then depends on which component of pinning force is exceeded, thus

$$J_c(B;\theta) = \min\{J_{c\|}(0), J_{c\perp}(Bcos\theta)\} \tag{4.2}$$

Assuming the power law dependence $J_{c\perp}(B) \propto B^{-\alpha}$, we will have at a fixed field $J_c(\theta) \propto (cos\theta)^{-\alpha}$ for angles not too close to the parallel direction. Tachiki and Takahashi compared (4.2) with data available at the time, and particularly for angles away from parallel field, the fit was reasonably good. The main problem with (4.2) is that it contains an unphysical discontinuity and predicts an angle-independent $J_c$ at parallel field, depending on the ratio of $J_{c\|}/J_{c\perp}$, which is not consistent with experiments.

Despite this flaw, this model is instructive and rewards careful consideration. Firstly, it emphasizes that pinned vortices are immobilized in a plane. When they move, it may not be because all pinning forces available are exceeded, but only the forces necessary to prevent motion in a particular direction. Secondly, (4.2) again returns us to a scaling relation, even though the intention has been to consider pinning forces directly. It can be seen that this scaling relation is effectively the same as the mass anisotropy scaling for field angles near perpendicular. In Fig. 4.8b, we plot (4.2) and the $J_c(\epsilon_\theta B)$ expression assuming we have the power law dependence for the perpendicular field $J_{c\perp}(B) \propto B^{-0.5}$ and $\gamma = 5$. We therefore have reached similar outcomes for the angular dependence over much of the range, although this scaling has been derived solely with reference to correlated pinning.

The assumption that the $J_c(\epsilon_\theta H)$ scaling can uniquely identify isotropic random point pinning is therefore open to question. Note that this identification first made by Civale et al. [15, 16] was never a logical necessity; if random point pins create a scalable dataset, it does not *logically* follow that scaled data imply point pins must be the source.

### 4.3.2 Models with Anisotropic Vortices

There have been relatively few attempts to construct microscopic models for anisotropic superconductors and point pins. One detailed phenomenological treatment has been made by van der Beek et al. [29]. In their paper, they focus on the multiband iron-based superconductors so that the anisotropy of the coherence length and the penetration depth may be different. They also generalize the isotropic shape of the pins to include ellipsoidal defects. For the single vortex limit, they show that either relatively sharp *ab*-plane peaks or broad *c*-axis peaks can result from the interplay of the anisotropy and the defect size and shape.

The difficulty of constructing microscopic models of pinning was highlighted in a couple of short papers by Mikitik and Brandt [30]. They show how in anisotropic superconductors the force at which vortices move from a pin is not necessarily the same as the pinning force. This is an extension of the idea included in the Tachiki and Takahashi model, where at $J_c$ vortices become unpinned in the direction in the plane where the pinning force is lowest, not the direction of the Lorentz force. Mikitik and Brandt extend this analysis to three dimensions and take account of the fact that anisotropy will create an ellipsoidal pinning force profile in the plane perpendicular to the vortex. Their idea is illustrated in Fig. 4.9. For any pinned vortex, we construct a diagram showing the flux pinning force in any direction in the plane perpendicular to the vortex—$f_p(\Psi)$. If the vortex experiences a force in the direction $\Psi$, then for a critical force $f_c < f_p(\Psi)$, the vortex will already move in the direction $\Psi_1$ as the projection of this force at the angle $\Psi_1$ already exceeds $f_p(\Psi_1)$. We then have to define $J_c$ relative to the critical force $f_c$ not the pinning force $f_p$.

To construct a microscopic model which can make a credible prediction of the macroscopic $J_c(\theta)$ is hence a daunting task. Reflection on the Tachiki-Takahashi and Mikitik-Brandt models raises a further confounding factor. If we add a particular defect to our pinning landscape and see a significant increase in $J_c$, then it is

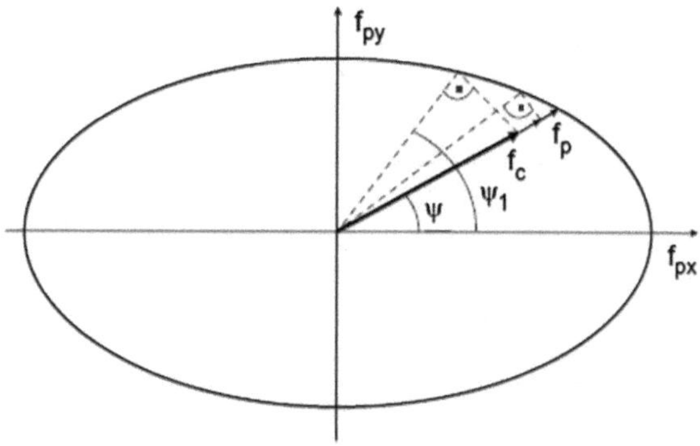


**Fig. 4.9** Diagram showing how the critical force $f_c$ at which a vortex will move in the direction $\Psi_1$ is below the pinning force $f_p(\Psi)$ in the direction of the applied force for point pins in an anisotropic superconductor. Reproduced by permission [30]

natural to suppose this new defect dominates the pinning. But Tachiki-Takahashi and Mikitik-Brandt show that this is not necessarily the case. The observed $J_c(\theta)$ can be dominated by ineffective pinning in certain directions. Adding pinning which prevents vortex motion in particular directions means that already existing pinning structures may become more effective and relevant in determining $J_c(\theta)$, not less relevant.

Experimentally, the complexity of pinning at the individual vortex level is now being explored through the novel use of scanning probe microscopes [31, 32]. These experiments confirm the complexity of vortex dynamics in response to forces even in quite idealized experimental situations. Another favorable direction for progress is to build computer models based on the Ginzburg–Landau equations which can give a detailed picture of vortex dynamics [33, 34]. At present, the complexity of these models and the computer resources required limit the physical size of the model system. Hence, questions concerning a complex pinning landscape are only just being addressed. A proper summary of these efforts is beyond the scope of this chapter.

In the next section, we will see how the Tachiki-Takahashi model can be extended in a direction which leads us into a different mode of thinking about critical current data, one that has the possibility of making sound inferences for the structure-property relationships.

## 4.4 The Vortex Path Model or Maximum Entropy Modeling

Rather than the vortex shown in Fig. 4.8a, imagine that the pinning landscape is disordered, and the pinned vortices are more like those of Fig. 4.10. The 'vortex path' through the material as depicted in Fig. 4.10 is a possible state of a pinned vortex which is compatible with an external field at the macroscopic angle $\theta$ of this path. The volume pinning strength, and hence $J_c$, is proportional to the density of such pinning paths through the sample at the angle $\theta$. A function for this density

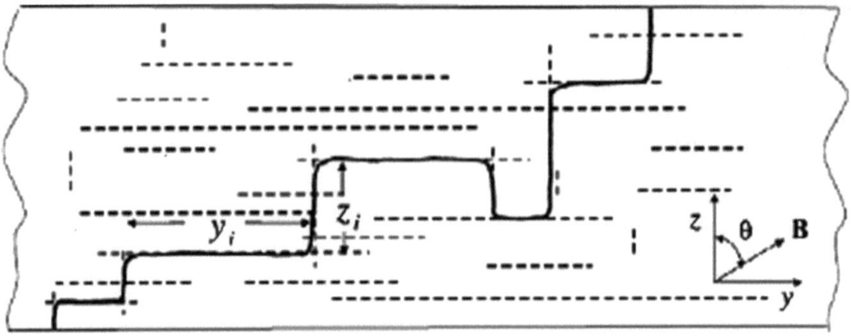


**Fig. 4.10** Representation of a vortex pinned both parallel and perpendicular to the *xy* plane and forming a macroscopic angle θ to the *z*-axis. The dotted lines represent pinning defects in the material. This vortex path and others existing in the material can be described as a directed random walk of *n* steps of length $y_i$ in the y-direction and $z_{-i}$ in the z-direction. An alternative description is to note the defects establish a mean value of *z*/*y* at which vortices will be pinned, and an associated variance to this mean. Reprinted from [35] with permission

with angle can then be constructed through probabilistic considerations. This leads to a model referred to as the 'vortex path model' [35].

A probability density for the pinning paths, $p(\theta)$, is derived by assuming simple rules for constructing such paths. Consider the vortex path as a directed random walk, where $y = n\lambda$ is the sum of *n* steps of average length λ in the *y*-direction. This will dictate the center of a peak (though not necessarily a maximum density) in the *y*-direction. Assuming there is no bias for steps in the *z*-direction and if the z steps are chosen from any distribution of finite variance, then for $z = \sum_{i=1}^{n} z_i$, the probability for finding a particular *z* value is $p(z) = (1/2\pi n\sigma^2)^{1/2}\exp(-z^2/2n\sigma^2)$, that is, the distribution converges to a Gaussian according to the central limit theorem. As $z/y = \tan\theta$, a transformation of random variables is required to go from $J_c(z/y) \propto p(z)$ to $J_c(\theta) \propto p(\theta)$, with the result

$$J_c(\theta) = \frac{J_0}{\sqrt{2\pi}\Gamma\sin^2\theta} \cdot \exp\left(-\frac{1}{2\Gamma^2\tan^2\theta}\right) \tag{4.5}$$

where $\Gamma = \sigma/\sqrt{n}\lambda$, and $J_0$ is a proportionality constant. This equation is referred to as an angular Gaussian. If instead of assuming the convergence to a Gaussian for $p(z)$, a heavy tailed Lorentzian distribution is chosen, $p(z) = (1/\pi)\gamma/(\gamma^2 + z^2)$, then the result is

$$J_c(\theta) = \frac{1}{\pi}\frac{J_0\Gamma}{\cos^2\theta + \Gamma^2\sin^2\theta} \tag{4.6}$$

where in this case $\Gamma = \gamma/\lambda$, and this is referred to as an angular Lorentzian.

It has been shown [35, 36] that these equations describe $J_c(\theta)$ for an incredibly wide range of samples, for isotropic and anisotropic superconductors. It may however seem that these equations have been derived using contrived rules. After all, are the paths formed by pinned vortices really such a random walk? Won't pinned vortices at large angles from the center angle of such a $J_c$ peak start to follow different rules, dictated by vortex tension, vortex-vortex interactions, or geometric effects of the sample?

Such questions become irrelevant if the distribution in the Cartesian coordinates is taken directly as $p(z/y) = (1/2\pi\,\Gamma^2)^{1/2}\exp(-(z/y)^2/2\Gamma^2)$, that is, a Gaussian in $z/y$. The model is now independent of the random walk description. A Gaussian is a maximum entropy distribution in which the mean and variance are specified [37]. That is, it is a maximization of the Shannon information entropy with constraints on the mean and variance of the distribution. This is now a maximum entropy model which says the pinning landscape is determining a mean value of $z/y$ at which vortices are pinned in the $zy$ plane (in this case, $<z/y> = 0$), and a variance to this orientation $<(z/y)^2> - <z/y>^2 = \Gamma^2$, but no further information is specified [38]. The angular Lorentzian can likewise be described as a maximum entropy distribution, where only the variance is specified. Any further effect of the physics is 'averaged out,' and we are only left with this information in the data. Maximum entropy can be loosely translated as 'maximum missing information.'

Philosophically, the shift from the vortex path model as describing actual vortex positions, to the maximum entropy description, is a shift from attempting to directly model the microscopic physics, to concentrating on the information content of the $J_c(\theta)$ data. Nonetheless, the vortex configurations described by the vortex path model are a useful physical picture if not taken too literally. The use of maximum entropy distributions to find structure-property relations is the methodology generally employed in spectroscopy, where Gaussian and Lorentzian functions are used, and many other forms of signal processing. Firstly, if the distributions describe the data, then the information content of the data is known with a high degree of confidence. Secondly, the information from the data fitting, in our case the parameters of the $J_c(\theta)$ fitting, can be correlated with information about the microstructure, and hence the structure-property relations of the system determined. We now give some examples from previously published work of applying this process.

Figure 4.11 shows results from Wimbush and Long [36], where a Nb thin film which has an array of vertical pores etched through the film at a spacing of 140 nm and a diameter of 75 nm. The addition of the nanopores increases $J_c$ by up to a factor of 50 over films without nanostructuring. It is known that Nb films will give a strong peak in $J_c(\theta)$ parallel to the plane of the film, so the authors conclude that the vertical pores are contributing to the overall $J_c(\theta)$ but are not creating a peak normal to the film surface. At the lower field, three maximum entropy components are present, a high uniform background, a narrow peak and a broader in-plane peak, both at ±90°. At the higher field, the narrow peak remains similar, but with an even broader peak at 90°, and a broad angular Gaussian at 0°.

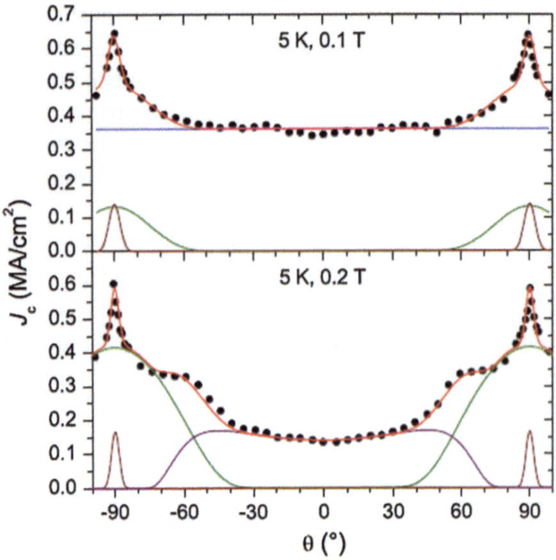


**Fig. 4.11** $J_c(\theta)$ for a Nb thin film which has been nanostructured with an array of vertical columnar pores: experiment (●), full fit (—), fit components (—,—,—,—). Reprinted by permission from [36]

The structure-property relations for the film are then elaborated as follows: The origin of the means at ±90° is the strong in-plane pinning. The mean at 0° at 0.2 T is due to the vertical pores. The origin of the variances is more speculative, but in this system, there are few sources of pinning. The in-plane peaks may be broadened slightly by surface roughness; this is posited as the source of the variance of the narrow peak. Larger broadening of the in-plane peak is achieved by interaction with the pores, that is, there is mixed pinning of the vortices by the in-plane pinning and the pores to broaden these peaks. The angular Gaussian at 0°, 0.2 T, is also a result of mixed pinning, this time with the pores as the dominant contribution and the in-plane pinning as the source of broadening. The constant background at 0.1 T must be due to a dominant contribution from the vertical pores in combination with the in-plane pinning. The field of 0.1 T is approximately the matching field for the pore density, and therefore, the effect of raising the field is to decrease the $J_c$ at 0° relative to the 90° value. The large region of constant $J_c(\theta)$ is easily accommodated in the maximum entropy analysis. In the vortex path model picture, we can think of the pores as fixing the density of possible pinning paths in a way which is angle independent.

The continual search for strategies to increase the critical current of coated conductors is leading to more complex pinning landscapes and resultant complex $J_c(\theta)$ data. The data shown in Fig. 4.12 are from a PLD YBCO sample with $Ba_2YNbO_6 + Gd_3TaO_7$ additions [36]. The resultant microstructure contains *c*-axis oriented $Ba_2Y(Nb,Ta)O_6$ segmented nanorods, *ab*-plane oriented $Y_2O_3$ platelets, and nanoparticles of $YBa_2Cu_4O_8$ superconducting phase.

The maximum entropy fitting in the range of 1–3 T finds there are four components with a mean of 0° (parallel to the *c*-axis). At 4 T, there is a crossover in

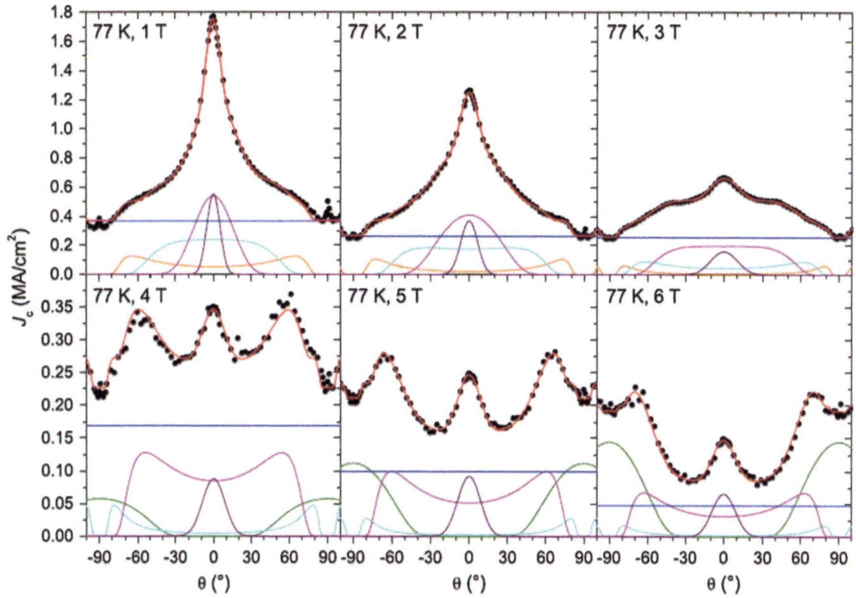


**Fig. 4.12** $J_c(\theta)$ for a YBCO film with $Ba_2YNbO_6 + Gd_3TaO_7$ additions. Reprinted by permission from [36]

behavior similar to the Nb sample of Fig. 4.11, and there is a ±90° Gaussian peak replacing one of the 0° components. The narrowest of these peaks remains relatively unchanged in width for all fields, in comparison with the other 0° peaks which become broader. This behavior indicates it has a very high matching field, also the unchanged width indicates the width may arise from the intrinsic angular spread of the defect species itself; hence, this component is identified with the segmented *c*-axis oriented nanorods. The next broadest of the components has a similar magnitude at all fields but becomes much broader. Hence, it is associated with the same defect species but additionally broadened with interactions with *ab*-planar pinning such as the $Y_2O_3$ platelets. A description of the origins of other components is given in [36].

The addition of BZO nanoparticles has been a common strategy to increase critical currents in YBCO films. As a final example of the use of vortex path model to find structure-property relations, we show in Fig. 4.13 data from Petrisor et al. [39] who prepared films of YBCO and YBCO + 10 mol% BZO using metal-organic deposition. The results from fitting $J_c(\theta)$ for the undoped and BZO-doped films are shown in Fig. 4.13.

In combination with an analysis of the microstructure, for the pure YBCO, Petrisor et al. assign the narrow *ab* Lorentzian to the intrinsic pinning broadened by lattice disorder including CuO precipitates. The *ab* Gaussian is attributed to the interaction between the intrinsic pinning and the orthogonal pinning by twin planes. The twin planes are also the source of the *c*-axis Gaussian peak. With the addition

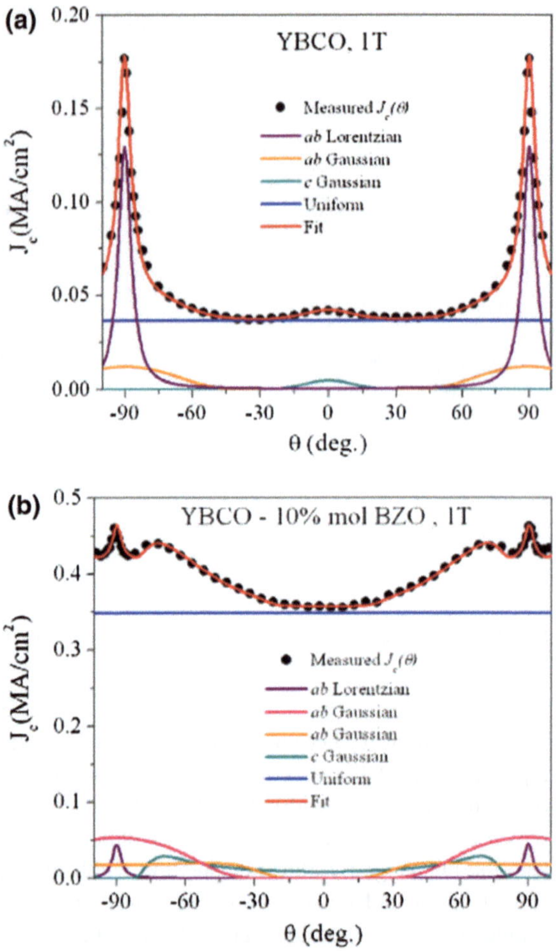


**Fig. 4.13** Angle dependence of critical currents for pure YBCO and YBCO +10 mol% BZO at 1 T. Reprinted with permission from [39]

of the BZO, an additional high-intensity *ab* Gaussian peak is observed which they attribute to the additional out of plane straining of the YBCO lattice due to the BZO inclusions which broadens a peak fixed by the intrinsic pinning. The *c*-axis peak under these conditions becomes extremely broadened by interaction with the intrinsic *ab*-plane pinning giving rise to the shoulders around ±70°. The additional strain of the BZO inclusions also gives rise to the very high isotropic component of the $J_c$, consistent with the results of Llordés et al. [18].

Other groups who have shown the usefulness of the vortex path model include the University of Turku group [40–42] who have used it to compare theoretical and measured values of the peak widths of YBCO films [40]. They also introduced using the pseudo-Voigt function which is a linear combination of (4.5) and (4.6) to get a better fit of peak shapes. In [41], they showed how the method can determine the contribution of *c*-axis aligned defects, even when a *c*-axis peak is not visible in the data. This group uses an explicit elliptical term in their fitting to account for the effect of mass anisotropy on point defect pinning. We do not believe this is necessary and would only consider adding this term if (4.5) and (4.6) could not account for the data without modification.

Other noteworthy publications showing the ability of the vortex path model to accurately fit data include Hänisch et al. [43] for iron arsenide superconductor films, Mikheenko et al. [44] for YBCO with BZO nanocolumns, and Talantsev et al. [45] used it to show a correlation between stacking fault density and *ab*-peak height, showing the dominance of this defect at low fields and high temperatures in MOD YBCO films.

## 4.5 Conclusions

Although a large literature exists of experimental data and a smaller literature of relevant models, a satisfactory consensus does not yet exist on understanding the structure-property relations of $J_c(\theta)$ data, and by extension critical currents generally. There is a consensus that samples in the regime of weak collective pinning from point pins should obey $J_c(\epsilon_\theta H)$ scaling. This behavior has been observed consistent with the mass anisotropy of the material. A $J_c(\epsilon_\theta H)$ scaling behavior has also been shown for BSCCO and YBCO samples, where scaling parameters are not consistent with the known electronic mass anisotropy. The origins of this scaling are not clear, but such behavior can arise through pinning from correlated defects, and it is not unreasonable to believe it can come from some kind of averaging over both correlated and uncorrelated pinning.

Intriguingly, $F_p(\epsilon_\theta H)$ scaling has also been observed for samples with added isotropic pinning centers. It has been proposed that this arises due to pinning from defects larger than the coherence lengths, although the microscopic explanation of this is not really convincing. It will be interesting to observe in the near future how common this form of scaling is for samples with high levels of defect engineering.

There are few microscopic models which aid our understanding of $J_c(\theta)$. This is perhaps not surprising as there are serious obstacles to constructing such models, particularly once mixed pinning is introduced. To accurately model pinning, vortices need to be treated as three-dimensional objects moving in three-dimensional space. We have presented the model of Tachiki and Takahashi and the key ideas of Mikitik and Brandt as giving worthwhile insight into the difficulties.

The vortex path model or maximum entropy modeling provides an alternative approach which avoids the difficulties of constructing or relying on the results from microscopic models. It is a statistical-based approach similar to conventional spectroscopy analysis which begins with determining the information content of the data. Combined with microstructural analysis, this approach can be used to build up knowledge of structure-property relations with a high degree of confidence.

## References


1. J.L. MacManus-Driscoll et al., Strongly enhanced current densities in superconducting coated conductors of $YBa_2Cu_3O_{7-x}$ + $BaZrO_3$. Nat. Mater. **3**, 439 (2004)
2. J. Gutiérrez et al., Strong isotropic flux pinning in solution-derived $YBa_2Cu_3O_{7-x}$ nanocomposite superconductor films. Nat. Mater. **6**, 367 (2007)
3. V. Selvamanickam et al., High critical currents in heavily doped (Gd, Y) $Ba_2Cu_3Ox$ superconductor tapes. Appl. Phys. Lett. **106**, 032601 (2015)
4. M. Miura et al., Mixed pinning landscape in nanoparticle-introduced $YGdBa_2Cu_3Oy$ films grown by metal organic deposition. Phys. Rev. B **83**, 184519 (2011)
5. S.R. Foltyn et al., Materials science challenges for high-temperature superconducting wire. Nat. Mater. **6**, 631 (2007)
6. A. Goyal et al., Irradiation free, columnar defects comprised of self-assmbled nanodots and nanorods resulting in strongly enhanced flux pinning in $YBa_2Cu_3O_7$-d films. Supercond. Sci. Technol. **18**, 1533 (2005)
7. M. Igarashi et al., Advanced development of IBAD/PLD coated conductors at FUJIKURA. Phys. Procedia **36**, 1412–1416 (2012)
8. G. Blatter, M. V. Feigel'man, V.B. Geshkenbein, A.I. Larkin, V.M. Vinokur, Vortices in high-temperature superconductors. Rev. Mod. Phys. **66**, 1125 (1994)
9. D.-X. Chen, J.J. Moreno, A. Hernando A. Sanchez, B.-Z. Li, Nature of the driving force on an Abrikosov vortex. Phys Rev. B. **57**, 5059 (1998)
10. J. Castro, A. Lopez, Comments on the force on pinned vortices in superconductors. J. Low Temp. Phys. **135**, 15 (2004)
11. G. Blatter, V.B. Geshkenbein, A.I. Larkin, From isotropic to anisotropic superconductors: a scaling approach. Phys. Rev. Lett. **68**, 875–878 (1992)
12. Z. Hao, J.R. Clem, Angular dependencies of the thermodynamic and electromagnetic properties of the high—Tc superconductors in the mixed state. Phys. Rev. B. **46**, 4833–5856 (1992)
13. X. Xu, J. Fang, X. Cao, K. Li, W. Yao, Z. Qi, Angular dependence of the critical current density for epitaxial $YBa_2Cu_3O_7$-d thin film. Solid State Commun. **92**, 501–504 (1994). X. Xu, J. Fang, X.W. Cao and K Li, A scaling formula of critical current density for anisotropic superconductors. J. Phys D: Appl. Phys. **29**, 2473 (1996)
14. G.R. Kumar, M.R. Koblischka, J.C. Martinez, R. Griessen, B. Dam, J. Rector, Angular scaling of critical current measurements on laser-ablated $YBa_2Cu_3O_7$−d thin films. Physica C **235–240**, 3053 (1994)
15. L. Civale et al., Understanding high critical currents in YBa2Cu3O7 thin films and coated conductors. J. Low. Temp. Phys. **135**, 87 (2004)
16. L. Civale et al., Angular-dependent vortex pinning mechanisms in $YBa_2Cu_3O_7$ coated conductors and thin films. Appl. Phys. Lett. **84**, 2121 (2004)
17. J. Gutiérrez et al., Anisotropic c-axis pinning in interfacial self-assembled nanostructured trifluoroacetate-$YBa_2Cu_3O_7$-x films. Appl. Phys. Lett. **94**, 172513 (2009)
18. A. Llordés et al., Nanoscale strain-induced pair suppression as a vortex-pinning mechanism in high-temperature superconductors. Nat. Mater. **11**, 329 (2012)
19. S.C. Wimbush, N.M. Strickland, N.J. Long, Low-Temperature scaling of the critical current in 1G HTS wires. IEEE Trans. Appl. Supercond. **25**, 6400105 (2015)
20. G. Grasso, Processing of high Tc conductors: the compound B,Pb(2223), in *Handbook of Superconducting Materials* v1, ed. by D.A. Cardwell, D.S. Ginley (IoP Publishing, Bristol and Philadelphia, 2003)
21. N. Clayton, N. Musolino, E. Giannini, V. Garnier, R. Flükiger, Growth and superconducting properties of $Bi_2Sr_2Ca_2Cu_3O_{10}$ single crystals. Supercond. Sci. Technol. **17**, S563–S567 (2004)
22. D.K. Hilton, A.V. Gavrilin, U.P. Trociewitz, Practical fit functions for transport critical current versus field magnitude and angle data from (RE)BCO coated conductors at fixed low temperatures and in high magnetic fields. Supercond. Sci. Technol. **28**, 074002 (2015)

23. E. Pardo, M. Vojenciak, F. Gomory, J. Souc, Low-magnetic-field dependence and anisotropy of the critical current density in coated conductors. Supercond. Sci. Technol. **24**, 065007 (2011)
24. V. Mishev et al., Interaction of vortices in anisotropic superconductors with isotropic defects. Supercond. Sci. Technol. **28**, 102001 (2015)
25. H. Matsui et al., Dimpling in critical current density versus magnetic field angle in $YBa_2Cu_3O_7$ films irradiated with 3-MeV gold ions. J. Appl. Phys. **114**, 233911 (2013)
26. E.J. Kramer, Scaling laws for flux pinning in hard superconductors. J. Appl. Phys. **44**, 1360 (1973)
27. N.J. Long, S.C. Wimbush, Comment on Dimpling in critical current density versus magnetic field angle in $YBa_2Cu_3O_7$ films irradiated with 3-MeV gold ions [J. Appl. Phys. **114**, 233911 (2013)]. J. Appl. Phys. **115**, 136101 (2014)
28. M. Tachiki, S. Takahashi, Anisotropy of critical current in layered oxide superconductors. Solid State Commun. **72**, 1083–6 (1989). M. Tachiki and S. Takahashi, "Intrinsic pinning in cuprate superconductors. Appl. Supercond. **2**, 305–313 (1994)
29. C.J. van der Beek, M. Konczykowski, R. Prozorov, Anisotropy of strong pinning in multi-band superconductors. Supercond. Sci. Technol. **25**, 084010 (2012)
30. G.P. Mikitik, E.H. Brandt, Flux-line pinning by point defects in anisotropic biaxial type-II superconductors. Phys. Rev. B **79**, 020506(R) (2009). Flux-line pinning by point defects in anisotropic type-II superconductors. Physica C **470**, S892–S893 (2010)
31. L. Embon et al., Probing dynamics and pinning of single vortices in superconductors at nanometer scales. Sci Rep. **5**, 7598 (2015)
32. O.M. Auslaender et al., Mechanics of individual isolated vortices in a cuprate superconductor. Nat. Phys. **5**, 35 (2009)
33. Q. Du, Numerical approximations of the Ginzburg-Landau models for superconductivity. J. Math. Phys. **46**, 095109 (2005)
34. I.A. Sadovskyy, A.E. Koshelev, C.L. Phillips, D.A. Karpeyev, A. Glatza, Stable large-scale solver for Ginzburg-Landau equations for superconductors. J. Comput. Phys. **294**, 639 (2015)
35. N.J. Long, Model for the angular dependence of critical currents in technical superconductors. Supercond. Sci. Technol. **21**, 025007 (2008)
36. S.C. Wimbush, N.J. Long, The interpretation of the field angle dependence of the critical current in defect-engineered superconductors. New J. Phys. **14**, 083017 (2012)
37. J.N. Kapur, *Maximum-Entropy Models in Science and Engineering*, 2nd edn. (New Age, New Delhi, India, 2009)
38. N.J. Long, Maximum entropy distributions describing critical currents in superconductors. Entropy **15**, 2585 (2013)
39. T. Petrisor et al., The vortex path model analysis of the field angle dependence of the critical current density in nanocomposite $YBa_2Cu3O_{7-x}$-$BaZrO_3$ films obtained by low fluorine chemical solution deposition. J Supercond. Nov. Magn. **27**, 2493–2500 (2014)
40. P. Paturi, The vortex path model and angular dependence of Jc in thin YBCO films deposited from undoped and $BaZrO_3$ doped targets. Supercond. Sci. Technol. **23**, 025030 (2010)
41. M. Malmivirta et al., Three ranges of the angular dependence of critical current of $BaZrO_3$ doped $YBa_2Cu_3O_{7-d}$ thin films grown at different temperatures. Thin Solid Films **562**, 554–560 (2014)
42. M. Malmivirta et al., The angular dependence of the critical current of baCeO3 doped $YBa_2Cu_3O_6$+x thin films. IEEE Trans. Appl. Supercond. **25**, 6603305 (2015)
43. J. Hänisch, K. Iida, F. Kurth, T. Thersleff, S. Trommler, E. Reich, R. Hühne, L. Schultz, B. Holzapfel, the effect of 45° grain boundaries and associated Fe particles on Jc and resistivity in Ba(Fe 0.9Co 0.1)2As2 thin films. Adv. Cryogenic Eng. **1574**, 260 (2014)
44. P. Mikheenko et al., Integrated nanotechnology of pinning centers in $YBa_2Cu_3$Ox films. Supercond. Sci. Technol. **23**, 125007 (2010)
45. E.F. Talantsev et al., Hole doping dependence of critical current density in $YBa_2Cu_3O_7$ conductors. Appl. Phys. Lett. **104**, 242601 (2014)

# Chapter 5
# Vortex Avalanches in Superconductors Visualized by Magneto-Optical Imaging

**Francesco Laviano**

## 5.1 Introduction

In this chapter, we deal with the general phenomena of the vortex avalanches (VAs). Soon after the discovery of Type II superconductivity, it was clear that the competition between the microstructural defects and the *Lorentz force* due to the supercurrent, in a superconductor exposed to a uniform magnetostatic field, corresponds to a (meta)stable magnetic field pattern, which was called the "critical state." In this state, vortices are effectively anchored to the defects until the *Lorentz force* surpasses the pinning force, which is coming from the energy gain due to occupation of the defects by the normal cores of vortices. Fluctuations, mainly due to thermal energy, contribute to de-pin vortices from defects, and thus the critical state is metastable. However, several experiments on superconductors demonstrated that the critical state is sometimes "not reached" or "escaped," during the measurement of magnetization or critical current, and "catastrophic" phenomena are observed. These electromagnetic instabilities cause dissipation, transition to the normal state, and mechanical damage of the specimen, in some cases. This class of phenomena was initially termed in different ways, such as "erratic magnetization" or "unpredictable normal state transition" or "flux jump." In fact, the abrupt redistribution of vortices is at the origin of the instability of the critical state, and, hereafter, we will generally refer to as "vortex avalanche."

It is evident that studying the conditions for generating—or better for avoiding—VAs is of chief importance for both fundamental understanding of the critical state and for practical applications of superconductors. Presence of vortex avalanche is indeed the most limiting factor for achieving the maximum critical current in power applications, such as wires and magnets.

F. Laviano (✉)
Department of Applied Science and Technology, Politecnico di Torino,
Corso Duca degli Abruzzi 24, Torino 10128, Italy
e-mail: francesco.laviano@polito.it



A. Crisan (ed.), *Vortices and Nanostructured Superconductors*, Springer Series in Materials Science 261, DOI 10.1007/978-3-319-59355-5_5

Since the bulk measurements of magnetization or critical current can display only the final and macroscopic effects of the VAs, the study of these phenomena is preferentially performed by magnetic imaging techniques. In particular, the magneto-optical imaging (MOI) allows one to visualize the magnetic field pattern, in real time, with microscopic resolution and on the macroscopic scale of the sample under investigation, and this technique is the principal tool for studying the VAs from the beginning to nowadays.

The scope of this chapter is twofold: reviewing the chief observations of flux instabilities by MOI[1] and trying to resume the state-of-the-art understanding of these phenomena in order to achieve a smart picture of the conditions that generate the instability and thus to assess the limits and the possible countermeasures for avoiding their occurrence.

In the first paragraph, I will describe the pioneering works, in which the VAs were detected for the first time, and I will give a brief description of the MOI technique. In the second paragraph, I will divide the collection of observations, made by MOI, mainly according to the experimental differences between unstable flux structure in bulk (3D) and in films (2D), and among different materials. It turns out the VAs are ubiquitous and the differences reside only in the conditions for triggering them depending on the geometry and on the material properties. In the third paragraph, I will present a qualitative overview of the current modeling of the instability phenomena, on the basis of theoretical models and of numerical simulations. Finally, a summary will be given in the last paragraph.

### *5.1.1 Flux Avalanche Phenomena in Superconductors: Historical Findings*

After the formulation of the "critical state model" by C.P. Bean in 1962 [1], the scientific community started to investigate the limits of Type II superconductors for what concerns critical currents and critical magnetic fields. From both kinds of experiments, namely electrical transport and magnetization measurements, unexpected features immediately came up. In particular, above a certain threshold when changing the applied magnetic field, the measured magnetization showed sudden collapses [2], eventually accompanied by heat release. One of the first magnetization curves, with evidence of flux jumps in a superconducting tube, is presented in Fig. 5.1 [3].

One striking feature of this finding was the experimental irreproducibility, namely that temperature, applied field, or current, for reproducing the experimental results, are different even in the same sample. However, some dominating factors

[1]This review will not cover the entire work on the study of vortex avalanches, because the number of published papers on the subject is impressive (more than 3000, in general, and more than 200 dealing with MOI, up to 2016).

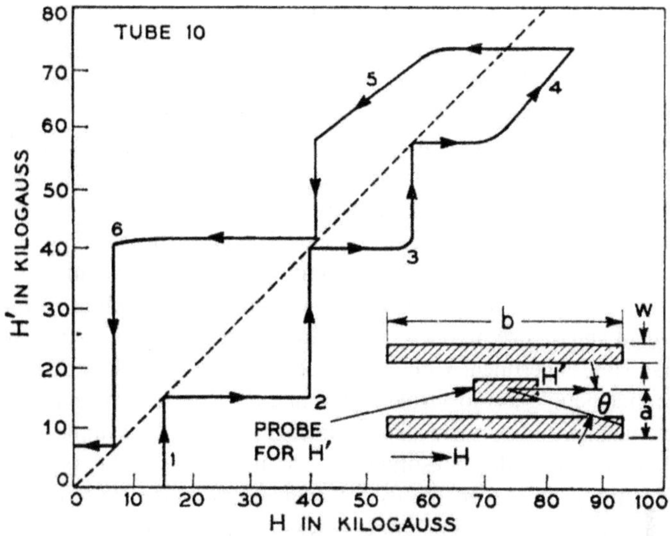


**Fig. 5.1** Magnetization curve of a superconducting tube ($Nb_3Sn$) that displays flux jumps. H' is the internal field measured in the tube, H is the applied field (graph taken from [3])

were identified, such as the most frequent occurrence of flux instability when the applied field is reversed and if the negative magnetization slope increases [4]. Another important aspect that was soon recognized is the feedback effect of the instability: the flux motion generates heat that increases the vortex velocity (a flux jump is indeed a thermo-magnetic instability). Therefore, the most catastrophic events correspond to a thermal runaway causing the transition to the normal state, eventual quenching of the superconductor and possible damage due to local melting of the specimen.

All these observations led to the formulation of two complementary models, in dependence on the material properties [5]: the "adiabatic critical state" [6] and the "isothermal instability" [7]. We will come back to these models and to their evolutions in the third section.

## *5.1.2 Magneto-Optical Imaging with an "Indicator"*

The MOI technique is a powerful tool for the visualization of the magnetic field distribution, and it is based on either *Kerr* (in reflection) or *Faraday* (in transmission) *effects*. By means of these effects, the local magnetization of the sample changes the phase of incident photons and, consequently, a linear dichroism comes out for reflected/refracted polarized light. Therefore, an optical microscope for polarization analysis, an electromagnet and an optical cryostat (if a low temperature characterization is necessary), constitute the basic experimental setup for MOI.[2] The MOI measurement is essentially a picture whose color/gray level is corresponding to the intensity of the local magnetic field (component perpendicular to the measurement plane for *Faraday effect*).

A direct observation of the magnetization of a sample, which can be a ferromagnet, a superconductor or a current-carrying conductor, needs that the material owns an appreciable magneto-optical activity (so called magneto-optical medium) and that its surface has treated to reach an optical quality. These two requirements are stringent, and very few materials meet them. In order to overcome this difficulty, magneto-optical active materials have been optimized for coupling with the sample magnetic field and, hence, to act as magneto-optical indicators by *Faraday effect*.

The first MOI experiment with an indicator was performed by Alers [8], which used a paramagnetic glass to visualize the intermediate state of a superconductor. Then, ferromagnetic films were deposited directly onto the sample under investigation [9], improving both the resolution due to the smaller thickness of the indicator and the sensitivity. However, the used ferromagnetic materials (europium chalcogenides) have a low Curie temperature (below 20 K), hampering the observation of the magnetic pattern in samples with higher critical temperature, $T_c$, such as the superconducting cuprates. A big improvement was achieved by Russians [10], with the development of optimized garnet films (whose parent compound is the Yttrium Iron Garnet), grown by liquid phase epitaxy on optical substrates and equipped with a thin mirror. The garnet materials with in-plane spontaneous magnetization represent the state of the art for what concerns MOI indicators [11], due to their giant Faraday rotation in the visible range (several tens of degrees per micron per T), defect-free lattice, and Curie temperature above 500 K.

The quantitative measurement of the magnetic field is achieved by means of suitable calibration procedures [12], and the real-time observation of the whole magnetic field distribution on the superconductor surface allows one to de-convolve the magnetic field map in order to reconstruct the supercurrent density distribution [13].

#### 5.1.2.1 "Regular" Critical State Observed by MOI

An example of a "regular" critical state measured by MOI in a superconducting sample is visible in Fig. 5.2. The sample is a superconducting film, $YBa_2Cu_3O_{7-x}$ (YBCO) deposited on YSZ substrate with a $CeO_2$ buffer layer ($T_c$ = 89 K), patterned by a standard optical lithography and wet etching. After zero field cooling, the magnetic field was applied in the direction perpendicular to the measurement plane. The superconductor initially expels the applied magnetic field because of the *Meissner effect* (dark area inside the sample that corresponds to zero induction field). When the local magnetic field surpasses the first critical field, the vortices

[2] It is customary to note that current big improvements in the MOI technique are consequence of the development and of the evolution of digital cameras and computers.

**Fig. 5.2** Quantitative MOI measurement of $B_z(x, y)$ (*top row*) and of $|J(x, y)|$ (*bottom row*) distribution in a superconducting YBCO film (thickness 215 nm), at T = 4.33 K. The applied magnetic field is specified in between the rows. The characteristic *dark lines* are called *discontinuity lines*. Bright spots and scratches are defects in the MOI indicator film

enter the superconductor and diffuse under the action of the *Lorentz force*. The lattice defects act as pinning centers, and an equilibrium is reached with the formation of a critical gradient in the vortex density (gray levels in the MOI measurements), characterized by a constant current density (*critical current density*, $J_c$), i.e., the critical state.

The critical state profile of the vortex density depends on the actual geometry and on the $J_c$, and it is well accounted for by existing models [14].

## 5.2 Observations of Magnetic Flux Avalanches

In this paragraph, we will concentrate on the experimental findings about VAs that were obtained by MOI measurements. Historically, VAs have attracted much attention in two different periods: in the sixties, in view of the application of "hard" Type II superconductors and in the last decade, after the ubiquitous appearance of the dendritic pattern in superconducting films.[3] We will distinguish experiments in

[3]It is worthy to report that recent attention to the topic was mainly stimulated by the work done in the Oslo group led by T.H. Johansen.

dependence on the dimensionality, namely 3D and 2D systems, and especially on the material type, because metallic superconductors show the instabilities easily, with respect to ceramic ones (i.e., cuprates).

### *5.2.1 Flux Jumps Observed in Bulk Superconductors*

The very first visualization of unstable flux structures was obtained by W. De Sorbo and Healy [15]. In that work, by means of MOI with a paramagnetic indicator, the authors presented several magnetic patterns of both Type I and Type II superconductors, exploring different magnetic histories, sample treatments for changing the microstructure, and, hence, the pinning strength, and they also checked the thermal stability of the observed patterns. It turned out that abrupt flux penetration occurs for "cold worked" samples, especially in lead and niobium.

Following this preliminary observations, Goodman and Wertheimer [16] developed a high-speed MOI apparatus for a systematic characterization of the avalanche phenomena. The results were spectacular at that time (clear images and frame-rate up to 12 kfps), as shown in Fig. 5.3 for a disk-shaped Nb sample [17].

The MOI indicator is a paramagnetic cerium phosphate thin glass, and the analyzer and the polarizer are near the crossed position, so that *dark contrast corresponds to zero or near zero magnetic induction, while bright contrast inside the specimen is corresponding to a nonzero vortex density* (as in most of the pictures in this chapter).

This sequence shows two separated flux jumps: in the first event, two avalanches are formed from the edges and then a third rounded avalanche is induced by further increasing the applied field. It is evident from these pictures that avalanches are influencing each other, the latter avoiding the former.

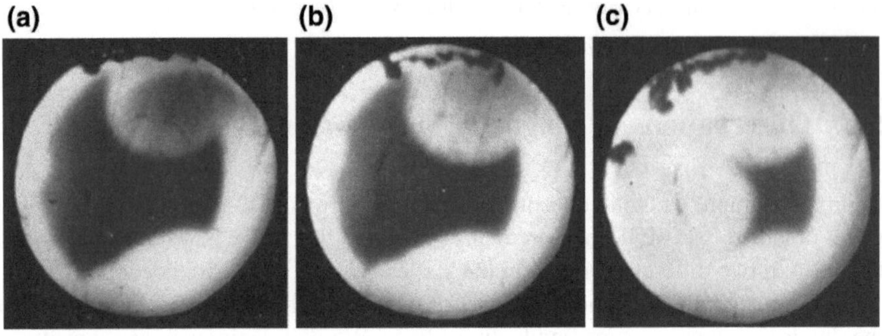


**Fig. 5.3** MOI of VAs in a Nb disk-shaped sample (diameter 13 mm, thickness 0.89 mm), at 1.8 K. **a** 1.5 ms after an applied field of 2130 Oe. **b** 2.5 ms with constant applied field. **c** 2 ms after increasing the field to 2810 Oe (from [17])

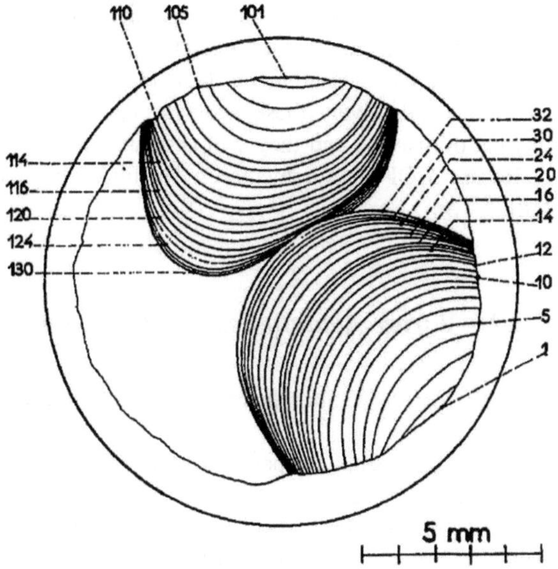


**Fig. 5.4** Contour lines depicted after MOI measurements of the flux front dynamics for two flux jumps in a Nb specimen (thickness 1.87 mm), T = 1.4 K, applied field of 800 Oe. The time interval between consecutive frames was 103 μs (the number of the frames is reported in the picture, sketch taken from [16])

A sequence of dynamic measurements of such flux instability is depicted in Fig. 5.4. The authors found that the flux front velocity, of the order of several m/s, is inversely proportional to the thickness of the sample and to $J_c$.

The authors also noticed that the induction of avalanches was disfavored by thinning the samples. Moreover, the rounded shape of VAs becomes more and

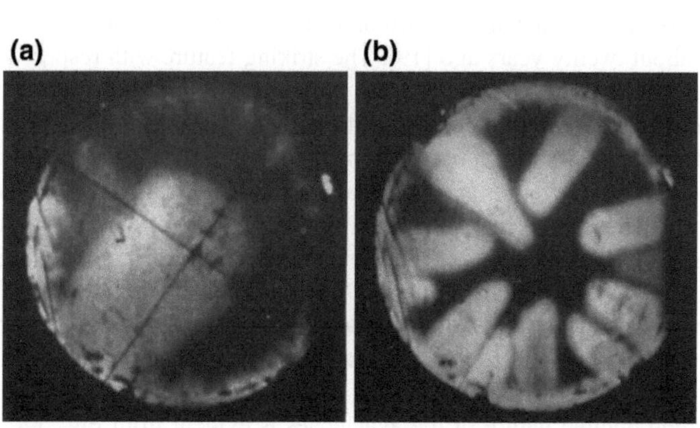


**Fig. 5.5** MOI of VAs in a Nb disk-shaped sample (diameter 13 mm, thickness 0.1 mm) **a** T = 4.2 K **b** T = 1.8 K (from [15], none information on the applied field is given). The cross is marked for achieving microscope focus on the indicator

more elongated with decreasing temperature, as visible in Fig. 5.5, while impurities in the specimen, i.e., lattice defects, generate branching of the pattern (on the contrary, in the measurements here reported, the avalanche flux profile is very smooth). Qualitatively these features depend on the ratio between magnetic and thermal diffusivity, and they will be clarified in the third section of this chapter.

Recent MOI measurements on Nb bulk samples confirmed all these findings [18].

## 5.2.2 *Vortex Avalanches in Superconducting Films*

### 5.2.2.1 Vortex Avalanches in Metallic Superconductors

Indeed, in thin metallic superconductors, VAs are straightforwardly arising, below a certain temperature threshold, $T_{th}$. This happens because of the thermo-magnetic nature of the instability, and even if slow applied field rate is used, the vortex pattern is becoming unstable for the heat release associated to the vortex diffusion. Moreover, current cryostats for MOI are working under vacuum, with cooling only through thermal conduction at the bottom of the sample in contact with the cold finger. Nevertheless, the magnetic diffusivity in metallic superconducting films is smaller with respect to cuprate superconductors; hence, the VAs do not generally cause a permanent damage in these systems. In the followings, I will separate the review of observations with respect to different superconducting elements and compounds.

#### Finger and Dendritic Flux Pattern in Nb Films

Fractal VAs in thin metallic superconductors, such as niobium, were clearly observed about twenty years ago [19]. The striking feature with respect to previous findings on bulk samples is indeed the pronounced branching of the avalanches (*dendritic pattern*) , with a precise evolution with temperature. This morphological evolution is reported in Fig. 5.6, from which three phases can be identified: above a first $T_{th}$, the flux penetration is smooth, and a stable critical state is formed; below this $T_{th}$, strongly branched avalanches are chaotically developing from sample edges; the VAs become less and less branched with further decreasing the temperature. The last flux pattern is quasi-1D, and it is called "finger-like" (Fig. 5.6 c).

The characteristic temperatures for avalanche formation and branching were mapped for Nb films [20], and the graph in Fig. 5.7 correlates them with the characteristic applied field for observing the first macroscopic flux jump.

The irreproducible nature of the avalanches is evident from Fig. 5.8, where the comparison of experimental reproducibility between the unstable flux patterns,

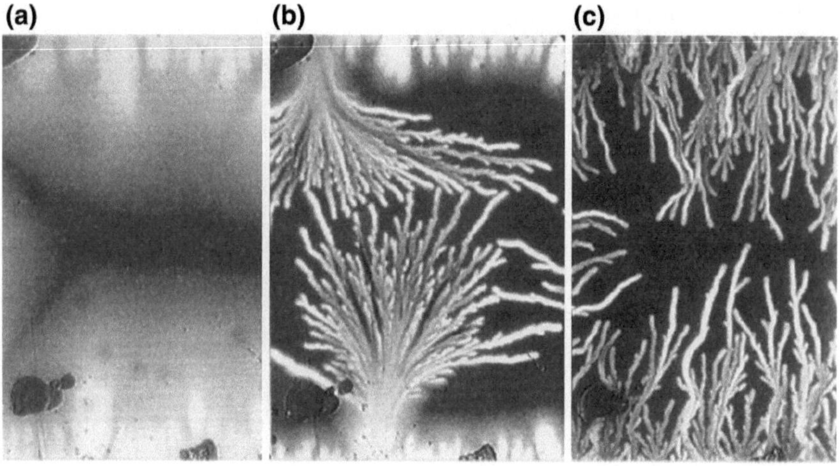


**Fig. 5.6** MOI measurements of magnetic field in a Nb film (thickness 500 nm, dimensions 3 mm × 8 mm, lateral field of view 2 mm). **a** T = 7.42 K. **b** T = 5.97 K. **c** T = 3.3 K (from [19])

**Fig. 5.7** Phase diagram of VAs in Nb films. The $B_{fj}$ value corresponds to the applied field when the first flux jump appears. Flux avalanche morphology is sketched in correspondence of the reduced temperature (actual temperature divided by $T_c$). HCA stems for huge compact avalanche (from [20])

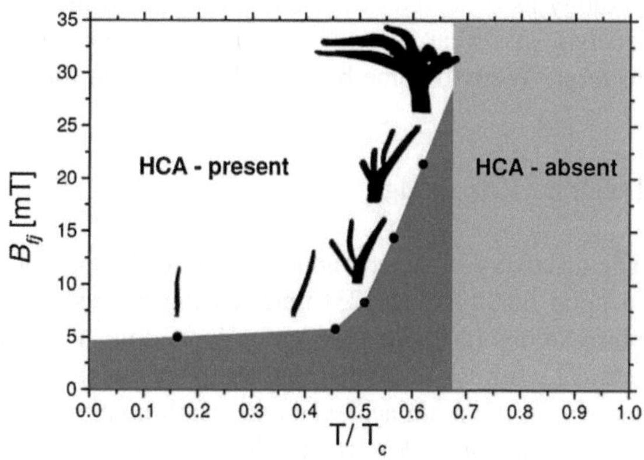


below $T_{th}$, and the critical state, above $T_{th}$, is done by superimposing several measurements that were performed in the same conditions (same applied field and temperature, always after zero field cooling).

It is worthy to note that starting locations of avalanches seem to be reproducible and that a small irreproducibility is present also on the boundary between the vortex penetrated part and the dark region corresponding to a flux-free Meissner state in Fig. 5.8b (colored parts). A further study accounts for the latter—apparently universal—behavior of vortex matter [21], because the metastable critical state seems

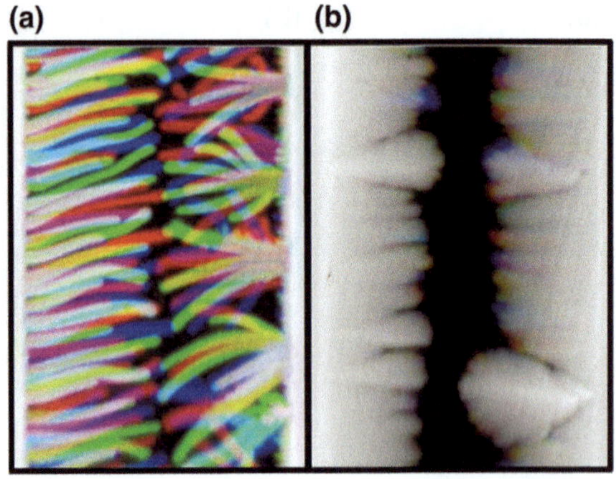


**Fig. 5.8** Several MOI frames were overlapped to show the irreproducibility of the VAs with respect to the critical state (applied field of 20 mT). **a** T = 4.2 K. **b** T = 6.7 K (from [20])

definitely composed of small size avalanches, whose size distributions follow power laws that have very similar critical exponents (in all the sample and over two decades of flux bundle size).

The disorder has also a big impact on the avalanche formation and its morphology as observed in nanostructured Nb films [22], in turns triggering the avalanches in "weak" points and reducing the irreproducibility of the observed patterns. We will come back on this aspect in the last paragraph.

## Dendritic VAs in $MgB_2$

The discovery of superconductivity in $MgB_2$ and the first MOI of magnetic pattern in good quality films [23] renewed the interest for the study of VAs. Since first observations [24], the Oslo group have done an impressive experimental and theoretical work for understating these phenomena and in this paragraph, I will report merely on the main achievements. High-quality $MgB_2$ films with same $T_c$ of bulk samples (about 39 K) display nice dendritic patterns, as shown in Fig. 5.9.

The thermo-magnetic nature of the instability was experimentally demonstrated in an elegant way, by covering with a metal foil half of the surface of the superconducting film [25]. The result is shown in Fig. 5.10, and it is evident that VAs are hampered by the better thermal stability induced by the superimposed metal foil.

Similarly, to the analysis performed on bulk samples, the quantitative measurements were devoted to find the characteristic "first-jump" applied field [26], the $T_{th}$, and the correlation with the sample geometry. The experimental results are summarized in Fig. 5.11 and in Fig. 5.12 [27].

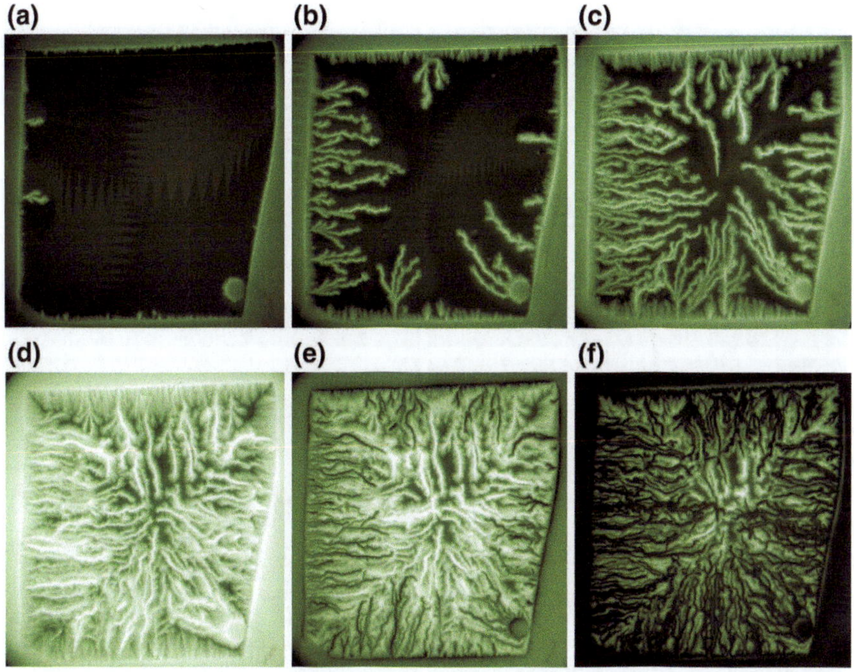


**Fig. 5.9** MOI of an $MgB_2$ film (400 nm thickness), at T = 3 K, after zero field cooling. From **a** to **f** applied fields perpendicular to the film were of 5, 10, 18, 35, 18, and 0 mT (from [23])

These results are indeed confirming the importance of two main factors that trigger VAs: the thermal diffusivity that decreases with temperature and the magnetic stray field that increases with lateral size.

Another chief parameter is the value of $J_c$. An impressive experiment performed on an $MgB_2$ film deposited on vicinal substrate [28], which induces anisotropy in the $J_c$ distribution, demonstrated that the $T_{th}$ is dependent on the $J_c$ strength (see Fig. 5.13). This achievement is consistent with the absence of thermo-magnetic instability in ultra-pure $MgB_2$ films, i.e., with low $J_c$ [29].

The distribution of current density in the presence of VAs reveals interesting features (Fig. 5.14): in the neighborhood of dendrites, the value of $J_c$ is comparable to the region where the critical state is established, while inside the dendrites, the value of the current density approaches zero [30]. This finding is suggesting that different vortex phases are coexisting in the sample, and that during the avalanche formation, a vortex liquid phase or even a filamentary normal phase could be generated. This aspect is still under debate.

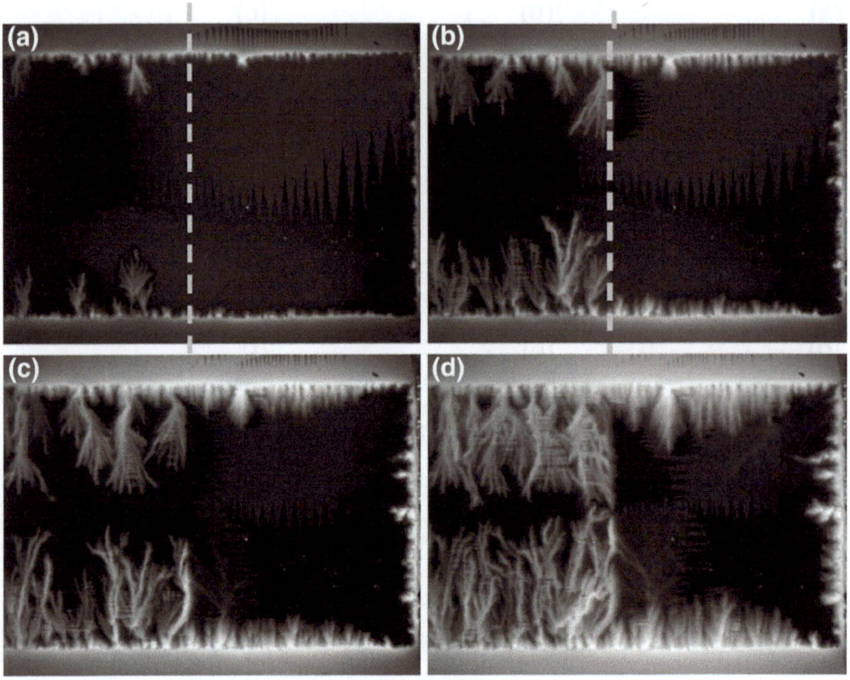


Fig. 5.10 MOI of an $MgB_2$ film (300 nm thickness, size $10 \times 3\ mm^2$), partly covered by a 10-μm thick Al foil (on the right part with respect to the dashed line). T = 3.5 K, from **a** to **d** applied fields were 3.4, 5, 8.5, and 12 mT (from [25])

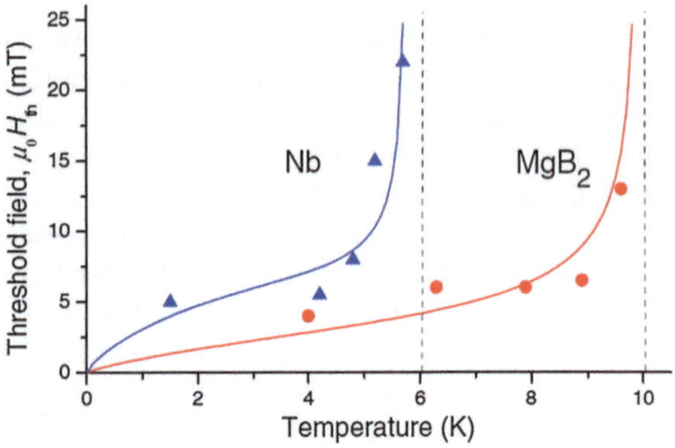


**Fig. 5.11** Temperature dependence of the applied magnetic field for the first flux jump. Experimental data are obtained for a 5-mm wide $MgB_2$ strip (●) and for a 1.8-mm wide Nb film (▲) [20]. The full lines are theoretical fits. The dashed lines show the $T_{th}$ for the onset of VAs (from [27])

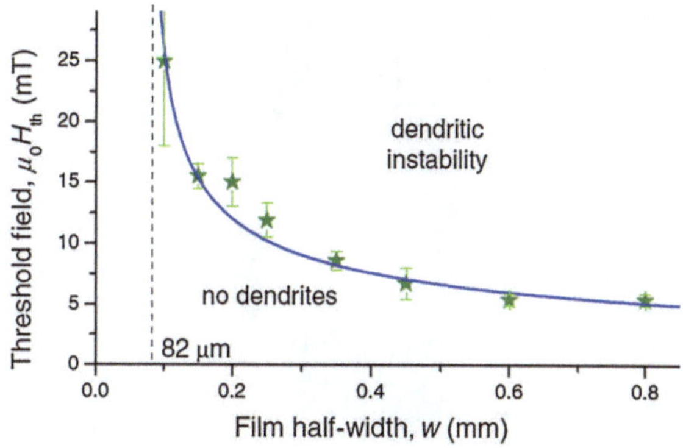


**Fig. 5.12** First-jump magnetic field in $MgB_2$ strips of different widths (symbols) fitted with a theoretical curve (*full line*) (from [27])

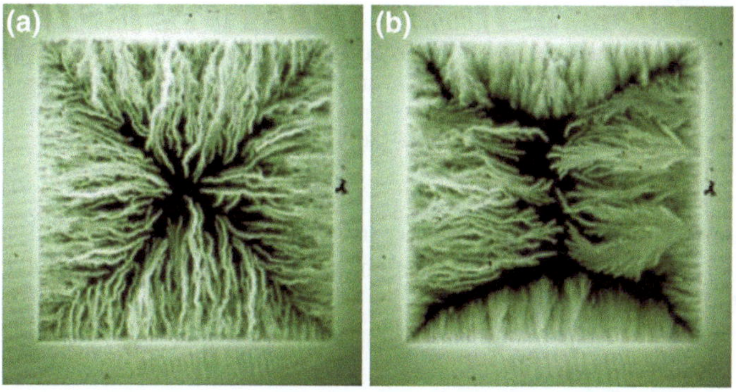


**Fig. 5.13** MOI of dentritic pattern in a $MgB_2$ film on vicinal substrate (thickness 200 nm, lateral size 5 × 5 $mm^2$), applied field of 16 mT. **a** T = 8 K. **b** T = 10 K (from [28])

## Avalanches Observed in Other Metallic Superconductors

Dendritic VAs were detected in other metallic superconducting films, following the attention prompted by their discovery in $MgB_2$ films. Noticeably the morphology of avalanches changes dramatically with engineered pinning centers, like with antidot arrays in Pb films [31]. These results are summarized in Fig. 5.15.

Metallic compounds such as NbN [32] and $Nb_3Sn$ [33] display dendritic patterns closely resembling those of $MgB_2$ films, as reported in Fig. 5.16. VAs have been observed also in borocarbide superconductors [34], and recently they have been visualized also in the $MoSi_x$ [35] and MoGe [36], low-$T_c$ superconductors.

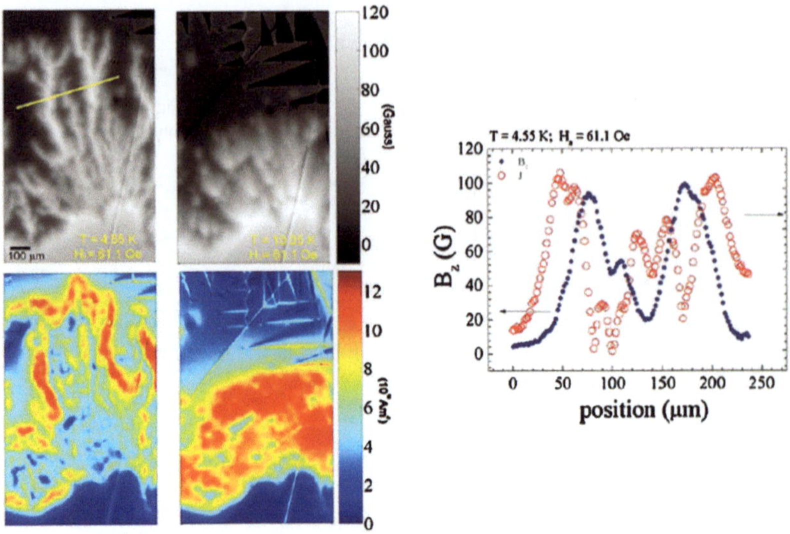


**Fig. 5.14** Quantitative measurement of magnetic field (*top row* on the *left*) and of current density (*bottom row* on the *left*) distribution in a part of an $MgB_2$ film (thickness 90 nm). Temperature and applied fields are indicated in the pictures. The profiles on the *right* are traced along the line in the picture on the first quadrant on the *left* (from [30])

The same general features, such as irreproducibility and abrupt dynamics, as well as the presence of a $T_{th}$ and of a characteristic $B_{fj}$ were confirmed in all the other metallic systems.[4]

#### 5.2.2.2 Vortex Avalanches in Cuprate Superconductors

Historically, the dendritic VAs in a cuprate superconductor were firstly visualized in the pioneering experiments of the P. Leiderer group [37]. The unstable vortex pattern was either spontaneously generated by a small defect on the sample edge or triggered by a short laser pulse, always at the sample border [38]. The observed pattern is very similar to that of metallic compounds, as presented in Fig. 5.17. The speed of propagation is much higher though. Since the authors equipped their MOI setup with a stroboscopic system with sub-ns resolution [39], the short dynamics of the VAs was carefully characterized.

Figure 5.18 shows measurements of both the sizes of the avalanche branches in dependence on time for several sample thicknesses and the speed of propagation in dependence on the sample thickness.

[4]Noticeably, at the moment of completing this chapter, none observation of vortex avalanches in iron-based superconductors has been reported.

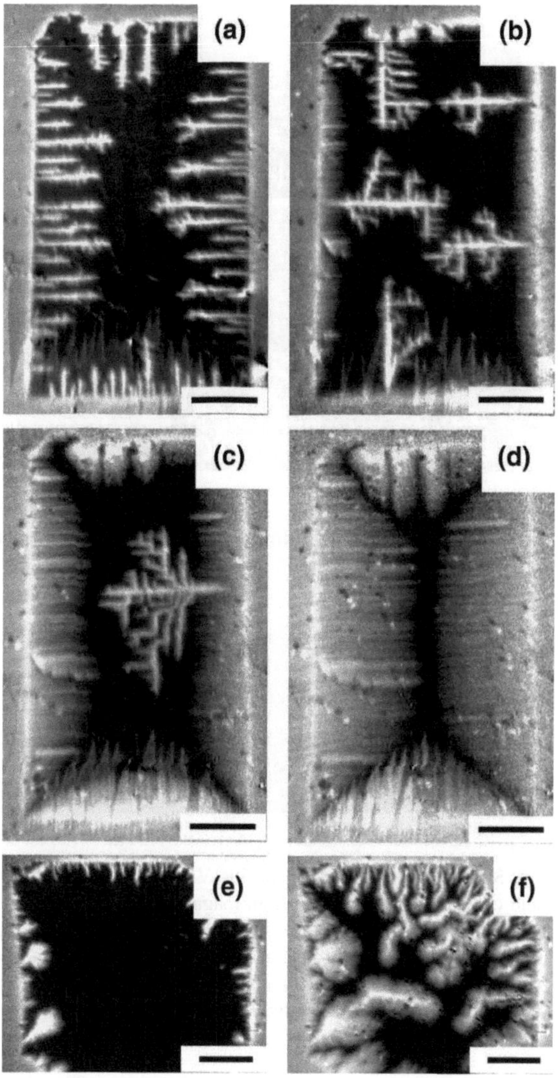


**Fig. 5.15** MOI measurements of Pb samples with (from (**a**) to (**d**)) and without antidotes (**e**), (f). **a** T = 4.5 K, applied field of 1.2 mT. **b** T = 5.5 K, applied field of 1.2 mT. **c** T = 6 K, applied field of 1.5 mT. **d** T = 6.5 K, applied field of 1.5 mT. **e** T = 2.5 K, applied field of 0.4 mT. **f** T = 2.5 K, applied field of 1.2 mT. Scale bar corresponds to 0.5 mm (from [31])

In the second graph, a comparison with the data from the study made on Nb bulk sample [17] reveals that the speed of propagation of vortex instabilities is always inversely proportional to the sample thickness. It is important to note that vortex moving at so high speed is accompanied by strong electric field/released heat and in most of the experiment the YBCO films were permanently damaged by the occurrence of VAs. The damage could be due to both heat and mechanical shock, since the measured speed of vortices well surpasses the sound velocity in the material; this conclusion is supported by the experiment presented in Fig. 5.19.

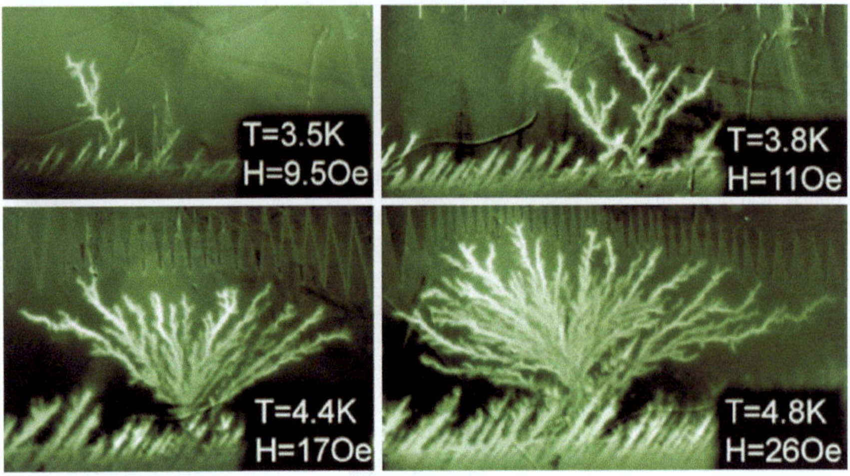


**Fig. 5.16** First dendrites formed in a zero-field cooled NbN film (thickness 290 nm), during increasing applied field. Dendrites formed at higher temperatures are characterized by a stronger branching (from [32])

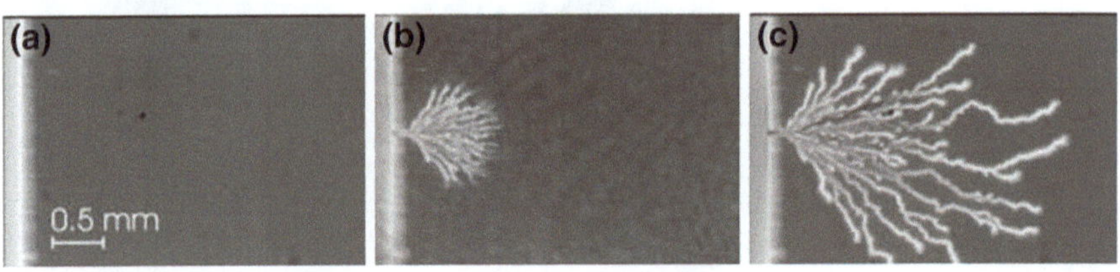


**Fig. 5.17** Propagation of a dendritic VA in an YBCO film at 10 K and in an external magnetic field of 20.6 mT. Regions of high magnetic field appear bright. The picture size is 4 × 2.5 mm. **a** The superconductor before the perturbation pulse is applied. One can clearly notice the enhancement of the field at the sample edge (*left-hand side* of the picture). **b** Double-exposure picture. The exposures were taken 37.6 and 43.3 ns after nucleation of the dendrite. **c** Final state of the dendrite (picture taken after 30 s) (from [39])

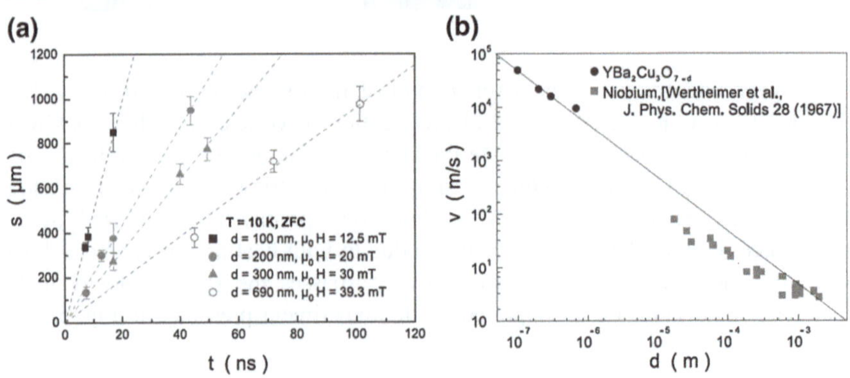


**Fig. 5.18** **a** Length s of the dendrite branches for several sample thicknesses. **b** Dendrite velocity as derived from (**a**) (*circles*), and data from [16] (*squares*) (from [39])

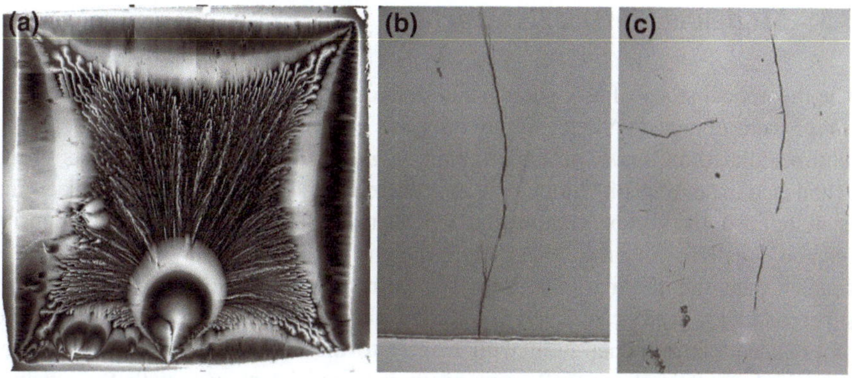


**Fig. 5.19** **a** MOI of remanent state of an YBCO film (thickness 500 nm, lateral size 8 mm), at T = 4.5 K after the application of a 200-mT applied field. **b** Crack that was formed on the YBCO surface after the VA. **c** Same crack was reproduced in the indicator film, demonstrating that a thermo-mechanical shock wave accompanied the VA

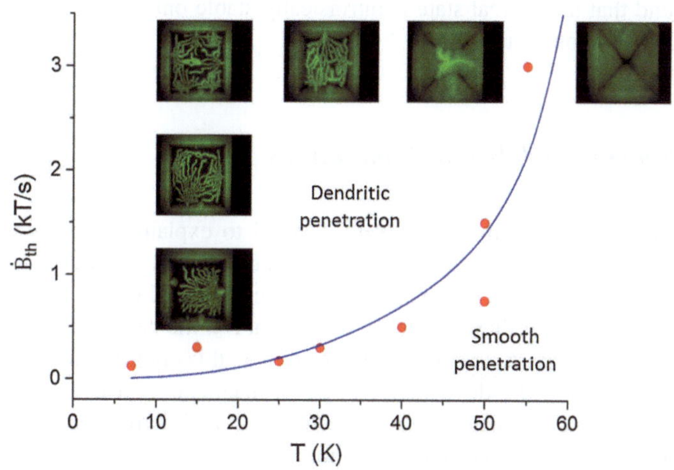


**Fig. 5.20** Measured applied field rate, $dB_{th}/dt$, versus temperature (circles are experimental data); each point represents the minimum applied field rate that is required to trigger the instability, at a specific temperature. The solid line is a fit [41]. The inset images show the evolution of the dendritic morphology (from [41])

Further studies [40] have shown that a critical parameter for inducing the VA in cuprate superconductors is the applied field rate. In the graph of Fig. 5.20, a correlation between applied field rate and the temperature is reported [41].

## 5.3 Modeling Vortex Avalanches

The theoretical study of VA phenomena started as soon as first observations of flux jumps were obtained. A relevant analysis was performed by Swartz and Bean [6], who postulated an *adiabatic critical state model* for VAs. This is the basis of the actual understanding of "limited" VAs, where the positive feedback between the heat induced by vortex motion and the decrease of the critical current density remains localized. In turns, this picture is valid if the magnetic diffusivity, $D_\mu$, is smaller than the thermal one, $D_\theta$. The corresponding field profiles inside the superconductor were calculated for a slab geometry, which resulted in a more pronounced vortex penetration with respect to the classical "isothermal" critical state, and an estimation of $B_{fj}$ was given in a consistent way with the existing experiments on VAs.

A more dramatic scenario was given by the introduction of the "isothermal" model [7], where the isothermal runaway of vortices can better explain the transition to the normal state of the whole superconductor. This is the case when a thermal wave is propagating faster than the vortices. Anyhow, a stability criterion is significant for the formation of VAs: Wipf was the first who recognized the importance of the critical current density dependence on (local) temperature [42] and he found that the critical state is intrinsically stable only if the pinning force is increasing with temperature.

### 5.3.1 *Recent Models and Simulations*

In recent models, the main efforts were devoted to explain the branching of the VAs. Since this is a specific characteristic of superconducting films, it was clear that the nonlocal dynamics of a confined superconductor should play a major role [43]. The main ingredients of these models are the local $D_\theta$, the heat transfer through the coolant bath or substrate, the Joule heating due to flux motion, and the Maxwell equations. Neglecting the dependence of $J_c$ on magnetic field and considering an exponential relation between the local electric field and current density (i.e., nonlinearity of vortex diffusion) , several authors found impressive agreement with the observed dendritic pattern. An example of this kind of calculation is presented in Fig. 5.21.

In these calculations, the presence of edge defects that trigger the instabilities was crucial and this fact implies a nonuniversal behavior, in contradiction with previous statements [44].

Another model that extends the results of the adiabatic approach for a superconducting slab [45] to superconducting films was developed by the Oslo group [46]. It quantitatively accounts for the observed $B_{fj}$ in superconducting films, as reported in the fit of Figs. 5.11 and 5.12. The corresponding formula for $B_{fj}$ is:

**Fig. 5.21** Simulated flux penetration in superconducting films with different widths and $J_c$. In **a**, the $J_c$ is ¼ with respect to **b** (same number of edge defects). **c** Larger film width, with higher field sweep rate with respect to the previous images. The arrow points a defect-free region where a larger VA developed. In **d** a larger number of edge defect is present (from [43])

$$B_{\text{fj}} = \sqrt{2\mu_0 C T^*}\sqrt{\frac{d}{\omega\pi}}, \quad T^* \equiv jc\left|\frac{\partial j_c}{\partial T}\right|^{-1}$$

where $C$ is the heat capacity, $d$ is the thickness, and $w$ is the width of the superconducting stripe line.

In this framework, the flux penetration is always producing a local thermal jump, which can be microscopic and, in this sense, it is a dynamic understanding of the isothermal critical state formation. Moreover, a critical electric field value for the formation of VAs was found and it turned to be proportional to the sample thickness.

A further development was obtained by introducing the above exposed ideas with the equation of motion for the current density, taking into account the thin film geometry [47]. In this approach [48], the nonlinear diffusion of vortices is modeled by the standard equation that links the local electric field with the current density:

$$\mathbf{E} = \frac{\rho \mathbf{J}}{d}, \quad \rho \equiv \begin{cases} \rho_0 & J > J_c(T), \\ \rho_0 (J/J_c)^{n-1} & \text{otherwise.} \end{cases}$$

where $\rho_O$ is the normal state resistivity and $n$ is the flux creep exponent.

The current density is dependent on temperature in the following way:

$$J_c = J_{c0}(1 - T/T_c), \quad n = n_1(T_c/T) - n_0.$$

Then, the current density is expressed through a local magnetization function *g (x, y)*:

$$\partial g/\partial y = J_x, \quad \partial g/\partial x = -J_y,$$

By means of the Biot-Savart law, the time-derivative of g in the Fourier space can be found in the following way:

$$\frac{\partial g}{\partial t} = \mathcal{F}^{-1}\left[(2/k)\mathcal{F}\left(\frac{1}{\mu_0}\frac{\partial B_z}{\partial t} - \frac{\mathrm{d}H_a}{\mathrm{d}t}\right)\right].$$

Here, the direct and inverse Fourier transformations are used, $k$ is the in-plane wave vector module, and $H_a$ is the applied field. This approach well describes the electrodynamics of superconducting films as demonstrated in several papers.

The unstable vortex patterns come out when coupling the previous equation with the equation for local heat diffusion:

$$c\frac{\partial T}{\partial t} = \nabla \cdot (\kappa \nabla T) - (h/d)(T - T_0) + \mathbf{J} \cdot \mathbf{E}/d.$$

where $c$ is the heat capacity, $\kappa$ the thermal conductivity, and $h$ the heat exchange with the cooling bath.

The resulting simulations are fully consistent with experiments and shed light on the flux diffusion in entire superconducting thin samples, with arbitrary geometries [49]. Figure 5.22 contains an example of the simulated dendritic VA with this approach.

**Fig. 5.22** Simulated dendritic pattern in a superconducting rectangular sample (T = 0.22 $T_c$, applied field is 0.14 times $J_c$) (from [49])

## 5.4 Summary and Perspectives

The most relevant features of VAs are the abrupt magnetic flux penetration and the consequent local heating of the material. The conditions for observing the VAs are a temperature below a certain threshold, the application of a magnetic field above a threshold value or better surpassing a certain field rate, and the sample geometry, for what concerns external factors. These variables are all dependent on the intrinsic material parameters, such as the thermal and the magnetic diffusivity, which are influenced by disorder in addition to the chemical composition of the superconductor.

Anyway, it is clear after this panorama that VAs are ubiquitous in superconductors. Therefore, much attention has to be paid for controlling and for avoiding these phenomena, apart studying them for fundamental purposes. A suggestive interpretation of VAs rises from the analogy of "lighting in superconductors" [48], because VAs are viewed as the consequence of nonlinear electric field response to the magnetic field excitation of superconductors. The confirmation of Snell's laws for VAs' propagation across zones with different thermal diffusions (for instance observed in a NbN film that was partially covered with a Cu layer [50] and reproduced by simulations [51]) gives a strong support to this picture. So, the VAs can be thought as electromagnetic shock waves [52] propagating through the superconducting medium, with a corresponding heat generation that causes dissipation and in dependence on the thermal diffusion can generate permanent damage to the superconducting device.

Moreover, VAs are surely a complex phenomenon: they are responsible of noisy magnetization if limited to a part of the sample [53] and the statistics of flux jumps follow a power–law relation with respect to their size and to the size of the sample (the SOC scenario seems to be not satisfactory for explaining the vortex diffusion phenomena though).

The general dependence on sample geometry still deserves further study as revealed by nonconnected geometries [54]: There, the generation of dendrites with opposite signs of the magnetic flux and different fractal structures gives clues of a stronger dependence on the demagnetization properties in dependence on the particular sample shape (Fig. 5.23).

In fact, VAs are nucleated at particular "hot spots" [55] at sample edges, where the electric field reaches maximum values and, hence, these locations correspond to the maxima of thermal heating too.

In addition, these "hot spots" are inducing a screening current bending that in turns determine a radial distribution of the Lorentz forces and thus the propagating avalanche is prone to develop 2D structure of flux diffusion [30].

All these observations have triggered complementary studies for finding solutions to avoid VAs.

A decisive result has been achieved by patterning inside the samples holes with different geometries [56]. It turns out that circular holes are the most efficient way for stopping the propagation of dendritic VAs in superconducting films [57]. This finding is clearly demonstrated by the measurements presented in Fig. 5.24.

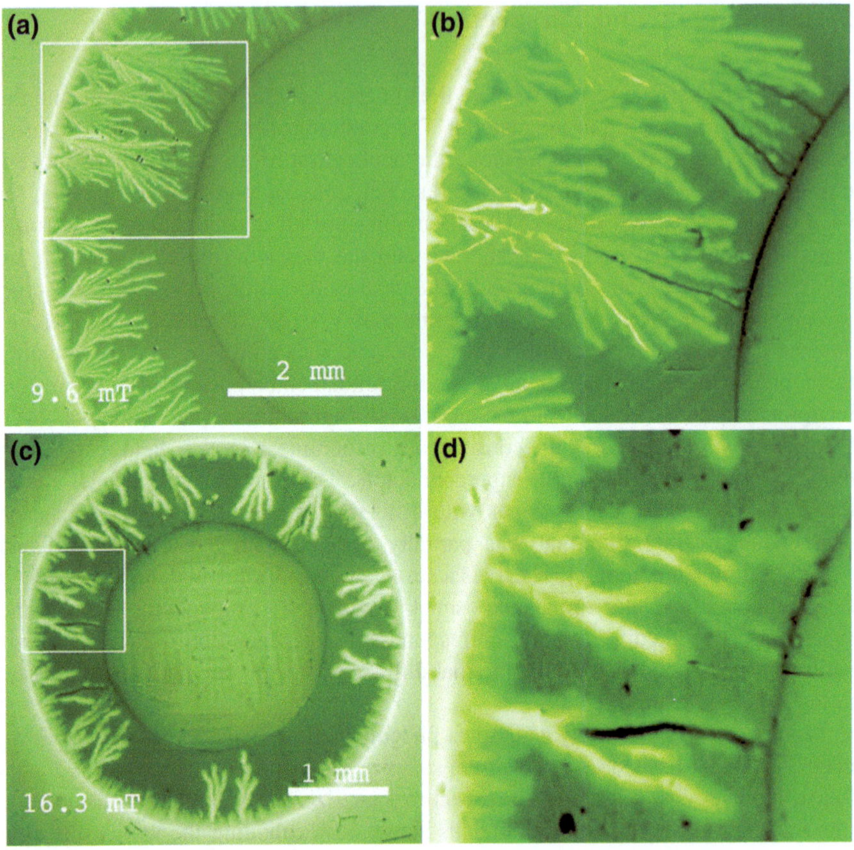


**Fig. 5.23** MOI of VAs in an $MgB_2$ superconducting ring (thickness 500 nm). **a**, **b** larger ring (outer radius 5 mm, inner radius 3 mm). Applied field of 9.6 mT. **c**, **d** smaller ring (outer radius 2 mm, inner radius 1.2 mm). Applied field 16.3 mT (taken from [54])

**Fig. 5.24** MOI of magnetic field distribution in a Nb film with holes of different geometries. From *left* to *right*, the applied field is 2.25, 3.15, and 4.20 mT (from [56])

Further studies on the control of VAs by means of micro- and nano-patterning are in progress [58].

Another important factor that can be optimized in view of limiting VAs is the thermal stability of the superconductor. It is clear that improving the thermal conduction or introducing a thermal diffuser by depositing a metal layer is effective for reducing the occurrence of VAs. In addition, recent experimental and theoretical studies demonstrate that the metal layer deposited directly or in proximity with the superconductor surface is also efficient for braking the electric field shock wave [59] and protect the superconducting film from eventual damages [60]. A precursor temperature oscillatory mode [61] was also predicted and this aspect surely deserves deeper investigation.

In summary, much work has been done for studying VAs, from the understanding of their development and propagation, of the causes of nucleation, of the conditions for favoring their presence, to various recipes for limiting or avoiding them.

An important outcome is their ubiquitous presence, in all the studied materials, in dependence on the external conditions, which is an indication that VAs are a basic ingredient of the vortex matter behavior and thus of both the critical state formation and its stability. The recent results indeed point to a review of the critical state picture in which the VAs on a small scale represent the basic element of the vortex diffusion process [62]. The final pattern depends on the sample properties, in particular on the balance between disorder and demagnetization.

Anyhow it is already sure that VAs are a key factor for mastering the superconducting devices under magnetic field, in any power application and for superconducting electronics [63], from dc to high frequencies [64], so that their study is really of chief importance and, in this respect, the MOI technique constitute an invaluable experimental tool.

**Acknowledgements** Support from EU Cost Action MP1201 is acknowledged. The author thanks Dr. Laura Gozzelino for reading the manuscript.

## References

1. C.P. Bean, Phys. Rev. Letters **8**, 250 (1962)
2. F. Rothwarf et al., Bull. Am. Phys. Sot. **7**, 189 (1962); P.S. Swartz and C.H. Rosner, J. Appl. Phys. **33**, 2292 (1962); Y. B. Kim, C. F. Hempstead and A. R. Strnad, Phys. Rev. **129**, 528 (1963); W. De Sorbo. Phys. Rev. **130**, 2177 (1963)
3. Y.B. Kim, C.F. Hempstead, A.R. Strnad, Phys. Rev. **129**, 528 (1963)
4. M.A.R. LeBlanc Phys. Lett. **6**, 140 (1963)
5. B.B. Goodman, Rev. Mod. Phys. **36**, 12 (1964)
6. P.S. Swartz, C.P. Bean, J. Appl. Phys. **39**, 4991 (1968)
7. K. Yamafuji, M. Takeo, J. Chikaba, N. Yano, F. Irie, J. Phys. Soc. Jap. **26**, 315 (1969)
8. P.B. Alers, Phys. Rev. **105**, 104 (1957)
9. H. Kirchner, Phys. Lett. A **26**, 651 (1968); H.-U.Habermeier, H. Kronmüller, Appl. Phys. **12**, 297 (1977); R. P. Huebener, *Magnetic Flux Structures in Superconductors* (Springer, 1979)

10. A.A. Polyanskii, V.K. Vlasko-Vlasov, M.V. Indenbom, V.I. Nikitenko, Sov. Tech. Phys. Lett. **15**, 872 (1989)
11. L.A. Dorosinskii, M.V. Indenbom, V.I. Nikitenko, Yu.A. Ossip'yan, A.A. Polyanskii, V.K Vlasko-Vlasov, Physica C **203**, 149 (1992)
12. C. Jooss, J. Albrecht, H.Kuhn, S. Leonhardt, H. Kronmüller, Rep. Prog. Phys. **65**, 651 (2002)
13. F. Laviano, D. Botta, A. Chiodoni, R. Gerbaldo, G. Ghigo, L. Gozzelino, S. Zannella, E. Mezzetti, Supercond. Sci. Technol. **16**, 71 (2003)
14. E.H. Brandt, Rep. Prog. Phys. **58**, 1465 (1995)
15. W. De Sorbo, W.A. Healy, Cryogenics **4**, 257 (1964)
16. B.B. Goodman, M. Wertheimer, Phys. Lett. **18**, 236 (1965)
17. M.R. Wertheimer, J. de G. Gilchrist, J. Phys. Chem. Solids, **28**, 2509 (1967)
18. R. Prozorov, D.V. Shantsev, R.G. Mints, Phys. Rev. B **74**, 220511 (2006)
19. C. A. Durán, P. L. Gammel, R. E. Miller, D. J. Bishop. Phys. Rev. B **52**, 75 (1995)
20. M.S. Welling, R.J. Westerwaal, W. Lohstroh, R.J. Wijngaarden, Physica C **411**, 11 (2004)
21. E. Altshuler, T. H. Johansen, Y. Paltiel, Peng Jin, K. E. Bassler, O. Ramos, Q. Y. Chen, G. F. Reiter, E. Zeldov, C. W. Chu, Phys. Rev. B **70**, 140505(R) (2004)
22. V. Vlasko-Vlasov, U. Welp, V. Metlushko, G.W. Crabtree, Physica C **341–348**, 1281 (2000)
23. T.H. Johansen, M. Baziljevich, D.V. Shantsev, P.E. Goa, Y.M. Galperin, W.N. Kang, H. J. Kim, E.M. Choi, M.-S. Kim, S.I. Lee, Supercond. Sci. Technol. **14**, 726 (2001)
24. T.H. Johansen, M. Baziljevich, D.V. Shantsev, P.E. Goa, Y.M. Galperin, W.N. Kang, H. J. Kim, E.M. Choi, M.-S. Kim, S.I. Lee, Europhys. Lett. **59**, 599 (2002)
25. M. Baziljevich, A.V. Bobyl, D.V. Shantsev, E. Altshuler, T.H. Johansen, S.I. Lee, Physica C **369**, 93 (2002)
26. F.L. Barkov, D.V. Shantsev, T.H. Johansen, P.E. Goa, W.N. Kang, H.J. Kim, E.M. Choi, S.I. Lee, Phys. Rev. B **67**, 064513 (2003)
27. D.V. Denisov, D.V. Shantsev, Y. M. Galperin, Eun-Mi Choi, Hyun-Sook Lee, Sung-Ik Lee, A.V. Boby, P. E. Goa, A. A. F. Olsen, T. H. Johansen, Phys. Rev. Let. **97**, 077002 (2006)
28. J. Albrecht, A.T. Matveev, J. Strempfer, H.-U. Habermeier, D.V. Shantsev, Y.M. Galperin, T. H. Johansen, Phys. Rev. Lett. **98**, 117001 (2007)
29. Z.X. Ye, Q. Li, Y.F. Hu, A.V. Pogrebnyakov, Y. Cui, X.X. Xi, J.M. Redwing, Q. Li, Appl. Phys. Lett. **85**, 5284 (2004)
30. F. Laviano, D. Botta, C. Ferdeghini, V. Ferrando, L. Gozzelino, E. Mezzetti—Thermo-magnetic instability as limiting mechanism for electrical current density in MgB2 thin films, in "Magneto-Optical Imaging", ed by T.H. Johansen, D.V. Shantsev, Kluwer Academic Publishers (2004), p. 237
31. M. Menghini, R.J. Wijngaarden, A.V. Silhanek, S. Raedts, V.V. Moshchalkov, Phys. Rev. B **71**, 104506 (2005)
32. I.A. Rudnev, D.V. Shantsev, T.H. Johansen, A.E. Primenko, Appl. Phys. Lett. **87**, 042502 (2005)
33. I.A. Rudnev, S.V. Antonenko, D.V. Shantsev, T.H. Johansen, A.E. Primenko, Cryogenics **43**, 663 (2003)
34. S.C. Wimbush, B. Holzapfel, Ch. Jooss, J. Appl. Phys. **96**, 3589 (2004)
35. F. Colauto, M. Motta, A. Palau, M.G. Blamire, T.H. Johansen, W.A. Ortiz, IEEE Trans. Appl. Supercond. **25**, 7500704 (2015)
36. M. Motta, F. Colauto, J.I. Vestgården, J. Fritzsche, M. Timmermans, J. Cuppens, C. Attanasio, C. Cirillo, V.V. Moshchalkov, J. Van de Vondel, T.H. Johansen, W.A. Ortiz, A.V. Silhanek, Phys. Rev. B **89**, 134508 (2014)
37. P. Brull, D. Kirchgassner, P. Leiderer, P. Berberich, H. Kinder, Ann. Phys. (Leipzig) **1**, 243 (1992)
38. P. Leiderer, J. Boneberg, P. Brull, V. Bujok, S. Herminghaus, Phys. Rev. Lett. **71**, 2646 (1993)
39. U. Bolz, B. Biehler, D. Schmidt, B.-U. Runge, P. Leiderer, Europhys. Lett. **64**, 517 (2003)
40. M. Baziljevich, E. Baruch-El, T.H. Johansen, Y. Yeshurun, Appl. Phys. Lett. **105**(2014), 012602 (2014)

41. E. Baruch-El, M. Baziljevich, B.Ya. Shapiro, T.H. Johansen, A. Shaulov, Y. Yeshurun, Phys. Rev. B **94**, 054509 (2016)
42. S.L. Wipf, Phys. Rev. **161**, 404 (1967)
43. I. S. Aranson, A. Gurevich, M. S. Welling, R. J. Wijngaarden, V. K. Vlasko-Vlasov, V. M. Vinokur, U. Welp, Phys. Rev. Lett. **94**, 037002 (2005)
44. E. Altshuler, T.H. Johansen, Rev. Mod. Phys. **76**, 471 (2004)
45. R.G. Mints, A.L. Rakhmanov, Rev. Mod. Phys. **53**, 551 (1981)
46. D. V. Shantsev, A. V. Bobyl, Y. M. Galperin, T. H. Johansen, S. I. Lee, Phys. Rev. B **72**, 024541 (2005)
47. E.H. Brandt, Phys. Rev. B **52**, 15442 (1995)
48. J.I. Vestgarden, D.V. Shantsev, Y.M. Galperin, T.H. Johansen, Scientific Reports **2**, 886 (2012)
49. J.I. Vestgården, Y.M. Galperin, T.H. Johansen, Physica C **479**, 92 (2012)
50. P. Mikheenko, T.H. Johansen, S. Chaudhuri, I.J. Maasilta, Y.M. Galperin, Phys. Rev. B **91**, 060507(R) (2015)
51. J.I. Vestgården, Y.M. Galperin, T.H. Johansen, J. Low Temp. Phys. **173**, 303 (2013)
52. I.L. Maksimov, Physica C **301**, 235–240 (1994)
53. F. Colauto, E.M. Choi, J.Y. Lee, S.I. Lee, V.V. Yurchenko, T.H. Johansen, W.A. Ortiz, Supercond. Sci. Technol. **20**, L48 (2007)
54. Å.A. Falnes Olsen, T.H. Johansen, D. Shantsev, E. Choi, H.-S. Lee, H. Jung Kim, and S.-I. Lee. Phys. Rev. B **74**, 064506 (2006)
55. T.H. Johansen, D.V. Shantsev, A.A.F. Olsen, M. Roussel, A.V. Pan, S.X. Dou, Surf. Sci. **601**, 5712 (2007)
56. F. Colauto, J.I. Vestgarden, A.M.H. de Andrade, A.A.M. Oliveira, W.A. Ortiz, T.H. Johansen, Appl. Phys. Lett. **103**, 032604 (2013)
57. J.I. Vestgarden, F. Colauto, A.M.H. de Andrade, A.A.M. Oliveira, W.A. Ortiz, T.H. Johansen, Phys Rev B **92**, 144510 (2015)
58. M. Motta, F. Colauto, J.I. Vestgarden, J. Fritzsche, M. Timmermans, J. Cuppens, C. Attanasio, C. Cirillo, V.V. Moshchalkov, J. Van de Vondel, T.H. Johansen, W.A. Ortiz, A.V. Silhanek, Phys. Rev. B **89**, 134508 (2014)
59. J.I. Vestgården, P. Mikheenko, Y.M. Galperin, T.H. Johansen, Supercond. Sci. Technol. **27**, 055014 (2014)
60. P. Mikheenko, J.I. Vestgården, S. Chaudhuri, I.J. Maasilta, Y.M. Galperin, T.H. Johansen, AIP Adv. **6**, 035304 (2016)
61. J.I. Vestgarden, Y.M. Galperin, T.H. Johansen, Phys Rev B **93**, 174511 (2016)
62. A. J. Qviller, V.V. Yurchenko, Y. M. Galperin, J. I. Vestgarden, P. B. Mozhaev, J. B. Hansen, T. H. Johansen, Phys. Rev. X **2**, 011007 (2012)
63. Y. Tsuchiya, Y. Mawatari, J. Ibuka, S. Tada, S. Pyon, S. Nagasawa, M. Hidaka, M. Maezawa, T. Tamegai, Supercond. Sci. Technol. **26**, 095004 (2013)
64. G. Ghigo, F. Laviano, L. Gozzelino, R. Gerbaldo, E. Mezzetti, J. Appl. Phys. **102**, 113901 (2007)

# Chapter 6
# Behavior of the Second Magnetization Peak in Self-nanostructured $La_{2-x}Sr_xCuO_4$ Single Crystals

**Lucica Miu, Alina M. Ionescu, Dana Miu, Ion Ivan and Adrian Crisan**

## 6.1 Introduction

The peak effect on the dc magnetic hysteresis curves of superconducting single crystals with relevant, randomly distributed quenched disorder (vortex pinning centers) has attracted a continuous interest. The onset of the spectacular increase of the effective critical current density $J_c$ with increasing external magnetic field $H$ associated to this peak can appear just below the upper critical magnetic field $H_{c2}$, as reported early for weakly pinned low-temperature $T$ superconductors [1–4]. This "anomalous" peak effect (PE) was first attributed to the softening of the vortex lattice and to the easy compliance of the latter to the pinning structure, enhancing the pinning force density $F_p$. According to the collective pinning theory [3], short-range order in the vortex system persists in a correlation volume $V_c$, and $F_p \propto V_c^{-1/2}$. As $H$ approaches $H_{c2}$, the shear modulus and the nonlocal tilt modulus soften rapidly [4], $V_c$ decreases, and a sudden $F_p(H)$ increase is expected. However,

L. Miu (✉) · A.M. Ionescu · I. Ivan · A. Crisan
National Institute of Materials Physics, Bucharest, Magurele, Romania
e-mail: elmiu@infim.ro

A.M. Ionescu
e-mail: alina.ionescu@infim.ro

I. Ivan
e-mail: ion.ivan@infim.ro

A. Crisan
e-mail: adrian.crisan@infim.ro

A.M. Ionescu
Faculty of Physics, University of Bucharest, Bucharest, Romania

D. Miu
National Institute of Laser, Plasma, and Radiation Physics, Bucharest, Magurele, Romania
e-mail: dana.miu@inflpr.ro



A. Crisan (ed.), *Vortices and Nanostructured Superconductors*, Springer Series in Materials Science 261, DOI 10.1007/978-3-319-59355-5_6

the $J_c(H)$ maximum not far from $H_{c2}$ was evidenced for superconductors with a low Ginzburg-Landau parameter, as well, where the decrease in the elastic moduli is not appreciable [5]. Subsequent observations of vortex arrangements revealed a strong connection between the PE and the disordering of the vortex system with dynamic history effects [6–10]. A somehow similar PE occurs in the case of high-temperature superconductors with weak random pinning, like that in untwined $YBa_2Cu_3O_7$ (YBCO) single crystals, for example [11–13], close to the dc irreversibility line (IL), in the domain of the first-order vortex-lattice melting transition which would be expected for "clean" specimens [14, 15]. This was considered to be a "pre-melting" peak, where a significant contribution to vortex system disordering comes from thermally induced vortex fluctuations.

The onset field for the $J_c(H)$ increase can be located far below the IL, generating a fishtail shape of the dc magnetic hysteresis curves [16]. In highly anisotropic specimens, such as $Bi_2Sr_2CaCu_2O_8$ (Bi-2212) single crystals, the fishtail effect [17, 18] is influenced by the crossover field toward two-dimensional vortex fluctuations $B_{2D} \sim \Phi_0\varepsilon^2/s^2$, where $\Phi_0$ is the magnetic flux quantum, $\varepsilon$—the anisotropy parameter ($\varepsilon^2$ representing the ration between the effective masses of quasiparticles moving perpendicular to the $c$-axis and along it [19]), and $s$ is the distance between the superconducting $CuO_2$ layers (of the order of 1 nm [20]). Consequently, the fishtail effect in highly anisotropic superconductors has a small shift with temperature and disappears in the low-$T$ region [21, 22]. For superconductors of moderate anisotropy, like $La_{2-x}Sr_xCuO_4$ single crystals (LSCO), the peak field is usually well below $B_{2D}$. Here, we discuss the fishtail effect in LSCO, with the corresponding second peak on the dc magnetization curves (in increasing $H$) denoted as the second magnetization peak (SMP).

There is no consensus yet about the origin of the fishtail effect, and the problem is now revisited, since the SMP is a common feature on the dc magnetization curves of iron-based superconducting single crystals, as well [23–27]. Many interesting SMP models and mechanisms have been proposed so far, by considering, for example, the influence of surface barriers [17, 28], a crossover in the collective pinning [29], formation of a percolation-like network of reversible regimes [30], the competition between the field dependences of current density and effective pinning energy scales [31], vortex entanglement [32], or the occurrence of thermo-magnetic instabilities [33]. Alternatively, an order-disorder transition in the vortex system (caused by pinning [34] or by thermal and quenched disorder-induced vortex fluctuations [35]) was taken into account. Moreover, the rhombic-to-square vortex-lattice transition [36, 37] has been proposed as the cause of the SMP. It was also suggested that the emergence of an SMP requires the simultaneous presence of strong and weak pinning centers [38, 39].

At present, the crossover in the collective pinning regime [29] (where the SMP is simply due to a faster relaxation at low $H$) is reconsidered [40, 41]. On the other hand, the pinning-induced disordering of the quasi-ordered vortex solid at low $H$ (the Bragg vortex glass, BVG, stable against dislocation formation [42, 43]) as the source of the SMP is encouraged by the often reported crossover elastic (collective) vortex creep–plastic creep across the SMP [23, 44–51]. However, it has to be noted that the change of magnetic relaxation is an effect, not the cause for an SMP. In the pinning-induced disordering scenario, $J_c$ increases owing to a better accommodation of vortices to the pinning centers in the disordered vortex phase (where dislocations proliferate).

In this context, the behavior of the SMP in specimens where some constraints for the vortex system naturally appear could offer useful information about its actual nature. For this purpose, the self-nanostructured (striped) superconductors seem to represent good candidates. As known, charge inhomogeneity at nano- and mesoscales is a characteristic aspect of superconductors resulting upon doping of an antiferromagnetic Mott insulator. Besides the inhomogeneity caused by nonuniformly distributed heterovalent substituents or extra oxygen atoms, the competition between magnetism and the kinetic energy of mobile carriers can lead to stripe formation [52–63]. These stripes are simply viewed as charged rivers (charge stripes) in the superconducting layers, separating domains of oppositely phased antiferromagnetism (spin stripes) [57]. The superstructure is most stable for $x \sim 1/8$ and manifests itself through an anomalous minimum in the critical temperature $T_c(x)$, known as the "1/8 anomaly" [58, 59]. The weakening of superconductivity in LSCO is present for $x$ roughly between 0.095 and 0.135 [60, 61], and the static stripe order (SSO) is reached due to the existence of some local crystal lattice distortions [62] and/or to the presence of impurities in the $CuO_2$ layers [63].

Here, we analyze the behavior of the SMP in LSCO single crystals of various doping, starting with an overdoped specimen (Sect. 6.3), where a pronounced SMP appears up to close to $T_c$. The main aspect observed by decreasing $x$ is the complete disappearance of the fishtail effect when $H$ is parallel to the $c$-axis upon entering the doping range of static stripe structures ($x \sim 1/8$), while it reappears at $x \leq 0.10$ (Sect. 6.4). This corresponds to the BVG instability detected for this field orientation through small-angle neutron scattering (SANS) experiments [64]. The relaxation results support the scenario in which the SMP is generated by the pinning-induced disordering of the low-$H$ BVG. The ac magnetic response and the vortex phase diagram are discussed in Sect. 6.5 and Sect. 6.6, respectively.

## 6.2 Samples and Experiments

The LSCO specimens considered here ($0.076 \leq x \leq 0.196$) were grown by the traveling-solvent floating zone method, as reported in [61]. The sample length was 2–3 mm, whereas the width and the thickness were around 1 mm, leading to a demagnetization factor $D \sim 0.4$. A special attention has been paid to the actual Sr content $x$, which was determined with an accuracy of $\pm 0.005$ by inductively coupled plasma optical emission spectrometry. $T_c$ was taken at the onset of the diamagnetic signal in a dc measurement ($H = 10$ Oe). The sample notation is S$x$, with $x$ in %, and the results discussed below are for an overdoped specimen (S19.6, with $T_c \sim 30.5$ K), two samples close to the borders of the 1/8 anomaly (S13 and S10, having $T_c \sim 34$ K and 28 K, respectively), a specimen inside the doping range of the 1/8 anomaly (S12, $T_c \sim 27.5$ K), and for two strongly underdoped single crystals (S8.5, $T_c \sim 23.5$ K, and S7.6, $T_c \sim 18$ K).

Standard (zero-field cooling, ZFC) dc magnetic hysteresis curves and the magnetization relaxation were registered with a commercial Quantum Design Magnetic Property Measurement System (MPMS). Well below $T_c$ and above the field for the first full vortex penetration (in increasing $H$), or at any $H$ in decreasing field, the irreversible magnetic moment was identified with the measured moment $m$. Close to the IL, the well-known procedure for the extraction of the irreversible component was applied. The volume magnetization $M$ was considered for a rough comparison of the pinning strength. The self-field critical current density determined with the Bean model [65, 66] at $T \sim 0.4T_c$ decreases from $\sim 1.2 \times 10^5$ A/cm$^2$ for S19.6 to $\sim 6.4 \times 10^3$ A/cm$^2$ in the case of S7.6. Since the full penetration maximum in $|m(H)|$ is pronounced at low $T$, the search for a fishtail effect in strongly underdoped specimens has been done in decreasing $H$. The magnetic relaxation was registered over a time window $t_w \sim 1500$–$2500$ s. For samples with relevant pinning, $\ln(|m|)$ versus $\ln(t)$ is linear for a moderate $t_w$ (when $T$ is not too close to the IL), and one determines a normalized magnetization relaxation rate $S = -\Delta\ln(|m|)/\Delta\ln(t)$ averaged over $t_w$ [67].

The ac magnetic measurements at usual frequencies and amplitudes have been performed with the ac and dc fields perpendicular to the $(a, b)$ planes. For comparison, the ac magnetic response of a pnictide single crystal with $D$ close to unity was also registered.

## 6.3 Analysis of the Second Magnetization Peak in Overdoped $La_{2-x}Sr_xCuO_4$ Single Crystals

Overdoped LSCO single crystals exhibit a well-developed SMP for both field orientations (see Figs. 6.1 and 6.2), and the temperature variation of the onset field $H_{on}$ and the peak field $H_p$ can be determined in a wide $T$ range. Due to the large $H_p$ values at low $T$ and the limitation of $H$ accessible in experiments, the $H_{on}(T)$ dependence is more often discussed.

The SMP can still be observed at high temperatures, as shown in Fig. 6.2. The relatively strong random pinning up to close to $T_c$ in S19.6 ($H$ parallel to the $c$-axis) is mainly due to the random Sr distribution and the charge mean-free path $l$ fluctuations introduced by doping [68]. It is not clear if a microscopic charge separation occurs in optimally doped and overdoped LSCO (expected to generate pinning effective, hole-rich "normal" particles [69]).

The first notable aspect in Fig. 6.1 is the increase of $H_{on}$ with decreasing temperature even in the low-$T$ range. This made questionable the SMP model based on the pinning-induced disordering of the BVG across the SMP. For static conditions (no macroscopic currents in the sample), the disordering line in the $(H, T)$ plane was derived from the equality between the pinning energy $E_p$ and the elastic energy $E_{el}$

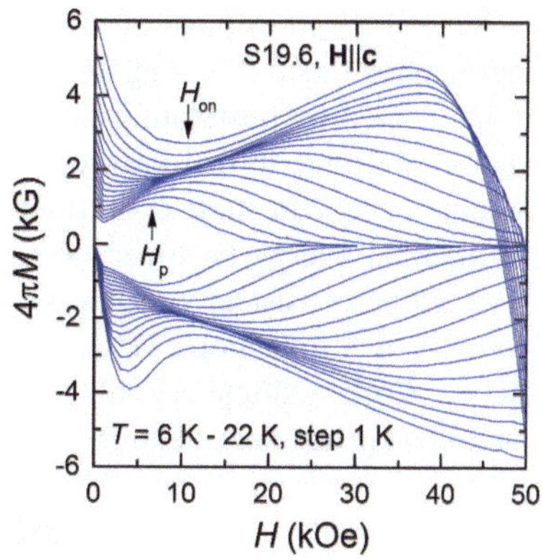


**Fig. 6.1** dc magnetization curves $M(H)$ of the overdoped LSCO sample S19.6 with the external magnetic field $H$ parallel to the $c$-axis. The SMP onset field $H_{on}$ and the peak field $H_p$ (indicated by *arrows* for $T = 6$ K and 22 K, respectively) increase at low temperatures

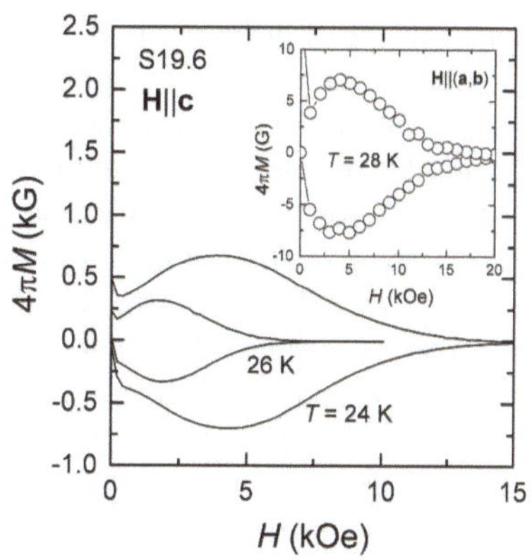


**Fig. 6.2** In the case of optimally doped and overdoped specimens, a well-developed SMP appears even at high temperatures, as shown in the main panel for S19.6 at $T = 24$ K and $T = 26$ K ($H$ parallel to the $c$-axis), and at $T = 28$ K ($H$ perpendicular to the $c$-axis) in the inset

in the vortex system, directly related to the superconductor parameters, such as the ($a$, $b$) plane magnetic penetration depth $\lambda$, the in-plane correlation length $\xi$, $\varepsilon$, and the pinning parameter $\gamma$. Following [35], for example, near the order-disorder transition (for $H$ along the $c$-axis and $T$ well below the IL)$E_{el} = \varepsilon\varepsilon_0 c_L^2 a_0$ and $E_p = U_{dp}(L_0/L_c)^{1/5}$, where $\varepsilon_0 = (\Phi_0/4\pi\lambda)^2$ is the vortex line tension, $c_L = 0.1$–$0.3$ is the Lindemann number, $a_0 \sim (\Phi_0/B)^{1/2}$ is the intervortex spacing (with $B$—the magnetic induction), $U_{dp} = (\gamma\varepsilon^2\varepsilon_0\xi^4)^{1/3}$ is the single-vortex depinning energy, $L_0 \approx 2\varepsilon a_0$ is the characteristic length for the longitudinal fluctuations, and $L_c = (\varepsilon^4\varepsilon_0^2\xi^2/\gamma)^{1/3}$ is the size of the coherently pinned vortex segment. In the case of $\delta T_c$ pinning (caused by local $T_c$ variations), $\gamma \propto \lambda^{-4}$, and the equality between $E_p$ and $E_{el}$ leads to a transition field $H_t(T) \propto [\xi(0)/\xi(T)]^3$, independent of $\lambda$. For $\delta l$ pinning, resulting from local $l$ variations, the transition field (often associated with $H_{on}$) increases with temperature in the high-$T$ domain, but in both cases the order-disorder line is temperature independent at low $T$ even if the pinning-caused and thermal vortex fluctuations are considered [70]. This would be at odds with the increase of $H_{on}$ and $H_p$ by decreasing temperature (see Fig. 6.1).

Several attempts have been made to explain the upward curvature in $H_{on}(T)$ at low temperatures. A strong $H_{on}(T)$ decrease with increasing $T$ in the entire range was obtained by postulating that both thermal and disorder-induced vortex fluctuations contribute to the destruction of the BVG [36]. However, compared with the pinning energy in static conditions (no macroscopic current in the specimen) [36], the thermal energy becomes negligible when $T \rightarrow 0$. The increase of $H_p$ at low $T$ would be consistent with the mechanism which considers the rhombic-to-square vortex-lattice transition [71] as the origin of the SMP [37, 38]. In LSCO [37] and $BaFe_{2-x}Co_xAs_2$ single crystals [38] with $H$ along the $c$-axis, the thermally induced rhombic-square vortex-lattice transition gives a transition field

$$H_t \propto (T_0 - T)T^{-\nu}C^{1-\nu}, \tag{6.1}$$

where $T_0 = 0.92$–$0.95T_c$, $C \propto \lambda^2$, and $\nu = 0.9$–$0.95$, which make $H_t$ practically independent of the superfluid density $n_s \propto 1/\lambda^2$.

Alternatively, it has been proposed [72] that when the density $J$ of the macroscopic currents induced in the sample is finite ($J \propto |m|$) the pinning energy in the energy balance relation should be substituted by an effective value, proportional to the actual vortex-creep activation energy $U[J(t), T, H]$. According to the general vortex-creep relation, for a relaxation time $t$ larger than a macroscopic time scale for creep $t_0$, $U = T\ln(t/t_0)$ [19]. As known, in the low-$T$ range the pinning potential is weakly temperature dependent (i.e., the intrinsic variation of $U$ with $T$ can be neglected), and the main role of the thermal energy is to change the $J$ interval probed over $t_w$ [73]. When $T \rightarrow 0$, the probed $J(t)$ shifts toward the ideal (creep free) critical current density $J_{c0}$, reducing the effective pinning. Obviously, this should cause significant changes in the behavior of the vortex system relative to that for static conditions ($J = 0$). In the approximation of a factorized $U$ [74], one can take $U(J) \propto T$ and $U(H) \propto H^{\alpha}$, with $\alpha > 0$ [44], because around $H_{on}$ the vortex creep is elastic (collective), where $U$ increases with $H$. When the activation energy $U$ for elastically creeping vortex bundles overcomes the energy for the plastic vortex deformation, which for $H$ parallel to the $c$-axis is proportional to $\varepsilon\lambda^{-2}H^{-1/2}$ [19] (independent of $J$), dislocations in the vortex system start to occur (at the vortex bundle limits), and the BVG disordering sets in. This is supported by the large vortex bundle creep regime detected at $H_{on}$ (Sect. 6.4).

The pinning-induced disordering of the BVG appears to be a continuous process (which extends up to $H_p$), following a grain boundary scenario [75]. In this dynamic approach,

$$H_{on}(T) \propto (\varepsilon\lambda^{-2}/T)^{\beta} \propto (\varepsilon x/T)^{\beta}, \tag{6.2}$$

where $\beta = 1/(\alpha + 1/2)$. In the case of $FeSe_{0.4}Te_{0.6}$ and $PrFeAsO_{0.60}F_{0.12}$ samples [50, 51], $\alpha \sim 0.4$, which leads to $\beta$ of the order unity. For strongly underdoped LSCO, where $H_{on}$ decreases toward $\Phi_0/\lambda^2$, $\beta \rightarrow 2$ [76].

One notes the relatively strong $H_{on}(n_s)$ dependence in (6.2). The determination of the exponent $\alpha$ will be addressed later.

Another aspect in Fig. 6.1 is the agglomeration of the $M(H)$ curves for $T$ between ~9 and ~ 18 K, which is in agreement with the $M(T)$ variation registered in increasing temperature from the main panel of Fig. 6.3 ($H$ = 10 kOe, applied in ZFC conditions). This behavior is due to the $H_{on}(T)$ dependence plotted in the inset of Fig. 6.3, which exhibits a more rapid decrease with increasing temperature in the above interval.

As pointed out in [72], (6.2) suggests that the $H_{on}(T)$ variation from the inset of Fig. 6.3 may be related to the presence of two-band superconductivity. An inflection-like point in $n_s(T)$ at low fields for this $T$ range has been observed in [77]. The peculiar $H_{on}(T)$ evolution from the inset of Fig. 6.3 has been detected for other multiple-band superconductors [50, 51]. It is worthy to note that the temperature dependence of the induction $B_{on} = H_{on} + 4\pi M(H_{on})(1 - D)$ (where $D \sim 0.43$ for S19.6 in $H$ along the $c$-axis) is similar to $H_{on}(T)$.

The pinning-induced BVG disordering in the dynamic conditions of standard dc magnetic measurements can then explain the increase of $H_{on}$ at low $T$ through relation (6.2), where the relatively strong variation of $H_{on}$ with $\lambda$ is confirmed by the particular $H_{on}(T)$ dependence from the inset of Fig. 6.3. The often reported decrease of $H_{on}$ and $H_p$ with time becomes straightforward. As illustrated in Fig. 6.4, at high relaxation levels the SMP shifts to lower magnetic fields. This is because for a longer relaxation time $J$ is lower, $U$ increases, and it will balance the plastic vortex deformation energy (independent of $J$ and decreasing with $H$) at lower fields.

The existence of an ordered (disordered) vortex phase is easily evidenced by analyzing the magnetization relaxation. Figure 6.5 shows, as an example, ln($m$) versus ln($t$) at several temperatures for S19.6 in $H$ = 10 kOe applied along the $c$-axis. At a given $T$, the constant field was reached by decreasing $H$ from the

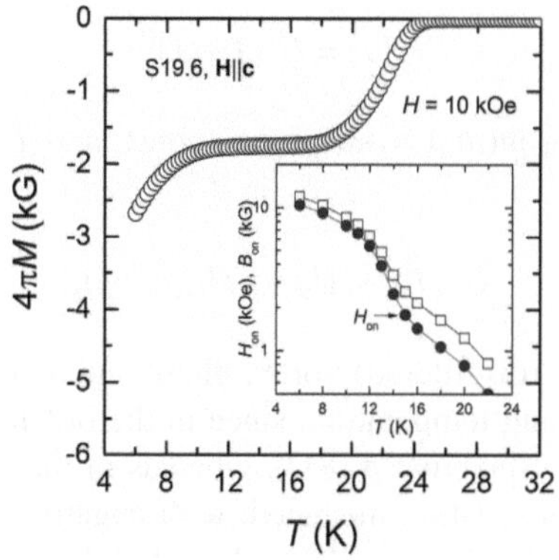


**Fig. 6.3** Main panel: The $M(T)$ variation registered for S19.6 on warming with the field $H$ = 10 kOe applied along the $c$-axis at low temperatures in ZFC conditions. There is a plateau in $M(T)$ for $T$ between ~9 and ~18 K, where some agglomeration of the $M(H)$ curves in Fig. 6.1 appears. Inset: Temperature dependence of the onset field $H_{on}(T)$ plotted in semilogarithmic scales, exhibiting an inflection-like point in the above temperature interval. The induction at the onset field $B_{on}$ included in the inset has a similar temperature variation

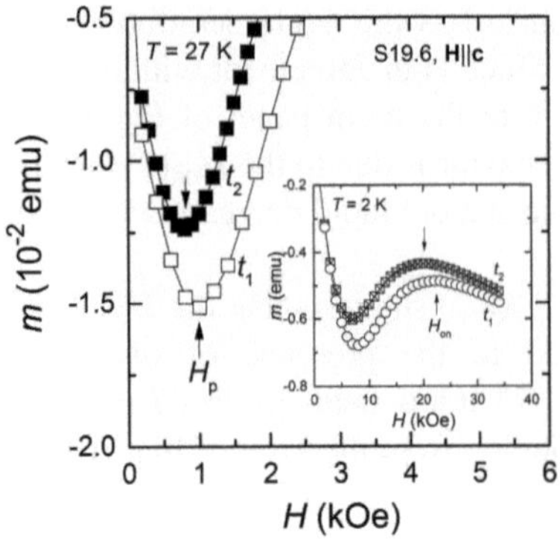


**Fig. 6.4** Main panel: Evolution of the peak field $H_p$ for S19.6 at $T = 27$ K with the relaxation time $t$, where the magnetic moment $m$ was registered at $t = t_1 = 28$ s and at $t_2 = 1200$ s after $H$ was applied parallel to the $c$-axis in zero-field cooling conditions. $H_p$ increases when the induced current density $J \propto |m(t)|$ is larger. The onset field $H_{on}$ behaves similarly (as illustrated in the inset for $T = 2$ K)

high-field region. The absolute value of the slope of the $m(t)$ curves in the log-log representation is the averaged normalized relaxation rate $S = -\ \Delta\ln(|m|)/\Delta\ln(t)$. For the actual purpose, it is better to consider the normalized vortex-creep activation energy $U^*(J) = -\ Td\ln(t)/d\ln(|m|)$. This can be derived using the general creep relation and the parameterization of the actual vortex-creep activation energy [78] $U(J) = (U_c/p)[(J_{c0}/J)^p - 1]$, where $U_c$ is a characteristic pinning energy (independent of $J$) and $p$ is the vortex-creep exponent (−1 in the Kim-Anderson model [79]). With $T$ or $J$ as the explicit variable, for a small relaxation over $t_w$ [$p(J)$, $t_0(J)$ = const.] one obtains [80]

$$U^*(J) = U_c(J_{c0}/J)^p, \tag{6.3}$$

or, in the approximation $\ln(t/t_0) = \ln(t_w/t_0)$ = const. and $T$ significantly below the IL,

$$U^*(T) \sim U_c + pT\ln(t_w/t_0). \tag{6.4}$$

Thus, for an ordered (disordered) vortex phase and a constant $t_w$, $U^*$ increases (decreases) with increasing temperature, since in the ordered phase one has elastic (collective) vortex creep (positive $p$ [81]), whereas in the disordered vortex phase one has plastic creep (dislocation mediated, with negative $p$) [44].

As argued in [82], the $U^*(T)$ dependence determined for a moderate, constant $t_w$ is more sensitive to changes in the vortex-creep process than $U(J)$ extracted with the widely used Maley technique [73]. The $U^*(T)$ variation [where $U^*(T) = T/S(T)$] resulting for S19.6 ($H = 10$ kOe parallel to the $c$-axis) is shown in Fig. 6.6 and reveals the crossover elastic vortex creep–plastic creep at a crossover temperature $T_{cr} =$ 15.9 K, where $U^*(T)$ is maximum. According to (6.3) and (6.4), $U^*(T_{cr}) = U_c$, since

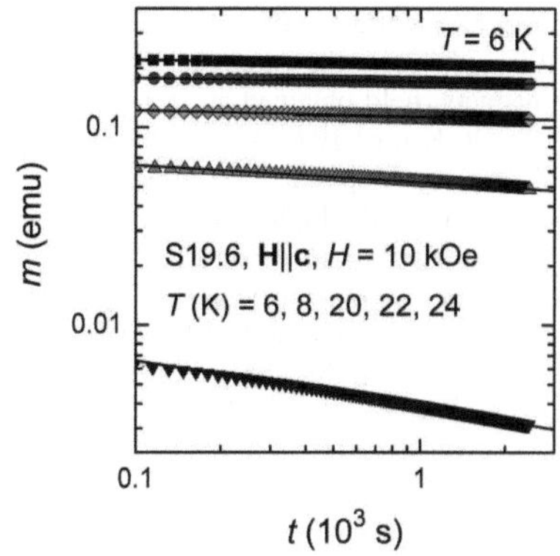


**Fig. 6.5** Magnetic relaxation curves *m(t)* in a log-log plot for S19.6 in $H = 10$ kOe (oriented along the *c*-axis) reached at various temperatures by decreasing *H* from the high-field region. The absolute value of the slope in a linear fit (*the continuous line*) is the normalized magnetization relaxation rate *S* averaged over $t_w$

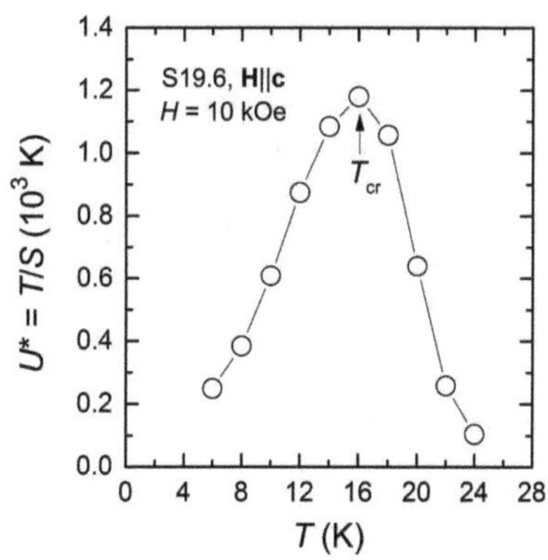


**Fig. 6.6** Resulting temperature variation of the normalized vortex-creep activation energy $U^* = T/S$ (S19.6 in $H = 10$ kOe along the *c*-axis). According to (6.4), the nonmonotonous $U^*(T)$ reflects the crossover elastic-plastic creep at the crossover temperature $T_{cr}$ indicated by an *arrow*

*p* vanishes at $T_{cr}$. The magnetic hysteresis curve at $T = 16$ K (Fig. 6.1) indicates that the field $H = 10$ kOe from Fig. 6.6 approximately represents the midpoint between $H_{on}$ and $H_p$ for $T = T_{cr}$, suggesting a continuous disordering of the BVG between $H_{on}$ and $H_p$.

The creep crossover is confirmed by the $U^*(H)$ dependence, as illustrated in Fig. 6.7. The crossover field $H_{cr} \sim 0.8$ kOe (corresponding to the $U^*$ maximum) approximately represents the midpoint between $H_{on}$ and $H_p$ at $T = 26$ K (see the main panel of Fig. 6.2).

Going back to the hysteresis curves from Fig. 6.1, with a well-developed SMP, one can discuss the temperature dependence of *m* (at constant *H*). Figure 6.8 (main panel) illustrates the *m(T)* variation registered for S19.6 on warming, after the external field $H = 30$ kOe was applied parallel to the *c*-axis in ZFC conditions at $T = 5$ K (ZFCW), or above $T_c$ (FCW). The chosen *H* value is significantly above the field range where the *M(H)* curves overlap (Fig. 6.1). The FCW *m* for $H = 30$ kOe is small (around $-1.01 \times 10^{-3}$ emu) and appears practically constant between 5 K and 20 K (with a variation of ~1%). There is no peak in the ZFCW $|m(T)|$, but the decrease of $|m|$ slows down around $T \sim 13.5$ K, where the external field

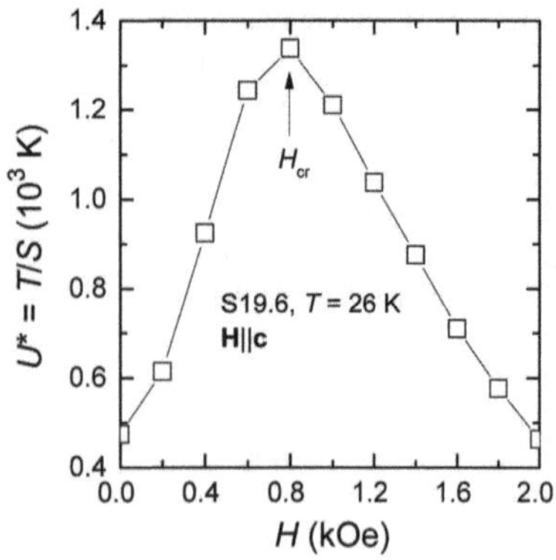


**Fig. 6.7** Variation of $U^* = T/S$ with the external magnetic field obtained at $T = 26$ K for S19.6 in decreasing $H$ (parallel to the $c$-axis) from the high-field region. The crossover elastic creep (at low $H$)–plastic creep (at high $H$) across the SMP manifests itself through a maximum in $U^*(H)$ at the crossover field $H_{cr} \sim 0.8$ kOe

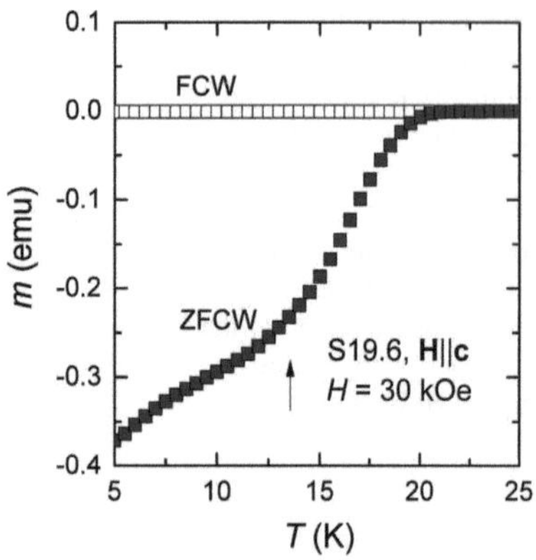


**Fig. 6.8** Temperature dependence of the magnetic moment $m$ of S19.6 registered on warming after the field $H = 30$ kOe oriented along the $c$-axis was applied above $T_c$ (FCW), or in zero-field cooling conditions at $T = 5$ K (ZFCW). The *arrow* points the $T$ value around which the temperature variation of the ZFCW $m$ slows down (~13.5 K)

$H = 30$ kOe approaches the peak field $H_p$ (see Fig. 6.1). This is because the increase of the effective pinning accompanying the vortex accommodation on the pinning structure is smeared out due to relaxation and by the continuous, strong decrease of $|m|$ with increasing temperature. From $U = (U_c/p)[(J_{c0}/J(t_1))^p - 1] = T\ln(t_1/t_0)$ and $J \propto |m|$, where $t_1$ is the measuring time at a certain $T$, one obtains $m(t_1, T) = m(t_0, T)[1 + pT\ln(t_1/t_0)/U_c]^{-1/p}$ ($p \neq 0$). For $p = 0$, $U^* = U_c$ [see (6.3)], equivalent with $U(J) = U_c(J_{c0}/J)$. In this situation, $m(t_1, T) \propto J(t_1, T) = J_{c0}(t_1/t_0)^{-T/U_c}$, neglecting the intrinsic temperature variation of $J_{c0}$ and $U_c$, as well as the $t_0(J, T)$ dependence. However, for $H = 12$–$15$ kOe, where the $H_{on}(T)$ anomaly is maximum (Fig. 6.3, inset) and the $M(H)$ curve overlaps (Fig. 6.1), a nonmonotonous $M(t_1, T)$ variation was observed (Fig. 1 of [72]). Without the anomalous $H_{on}(T)$ behavior, a peak in the ZFCW $|m(T)|$ is unlikely to appear.

## 6.4 Disappearance of the Second Magnetization Peak in the Presence of Static Stripe Order

By decreasing $x$, a pronounced SMP can still be observed at $x = 0.147$ [69]. However, in the case of S12, in the doping range of static stripe order (SSO), the SMP disappears at all temperatures if $H$ is parallel to the $c$-axis [83], as shown in Fig. 6.9. The specimen with $x = 0.125$ investigated in [69] behaves similarly. For S13, just below the upper $x$ border of the 1/8 anomaly, the SMP can clearly be seen at high temperatures, but not below ~18 K, as illustrated in the inset of Fig. 6.9.

Keeping $H$ along the $c$-axis, the SMP reappears (at low $H$) in strongly underdoped specimens [76] ($x \leq 0.10$, see Fig. 6.10), where the stripe structure is more diffuse [64]. Just above the lower $x$ border of the 1/8 anomaly, it was found that no SMP is present below ~8 K ($x = 10$) [84]. As discussed in [83], there is a good correlation between the SMP suppression from Fig. 6.9 and the development of SSO in LSCO detected in neutron scattering measurements [85]. The SMP is absent at all temperatures in S12, since at this doping the temperature $T_m$ for the melting of the static stripe phase into mobile stripes is slightly above $T_c$. Moreover, in Fig. 4 of [85] one has $T_m(x = 0.13) \sim 20$ K. By taking into account possible, small differences in the actual doping, this is roughly the temperature value below which the SMP in S13 smears out and disappears (see the inset of Fig. 6.9).

At the same time, the SANS experiments performed on LSCO single crystals cooled to ~3 K in $H$ parallel to the $c$-axis [64] revealed the instability of the Bragg vortex glass in the presence of SSO (see Fig. 1 of [64]). For $x \sim 1/8$, where the onset of static incommensurate spin density wave (SDW) order at $T_{SDW}$ (equivalent to $T_m$ in [85]) is above $T_c$, the BVG is unstable in the whole ($H$, $T$) domain, and the SMP does not appear for S12 (see the main panel of Fig. 6.9).

By decreasing doping below 1/8, the SANS signal indicates a vortex lattice (BVG) and the SMP reappears in strongly underdoped LSCO (see Fig. 6.10), meaning that the static BVG is the necessary ingredient for an SMP. One notes that while the SMP is present at low doping, the structural vortex-lattice transition is missing in Fig. 1 of [64].

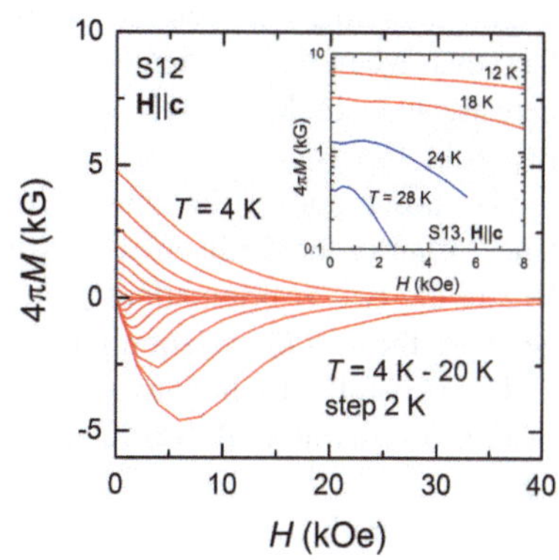


**Fig. 6.9** Main panel: dc $M$ ($H$) curves registered for S12 (in the doping range of the 1/8 anomaly) for $H$ oriented along the $c$-axis. No SMP was observed up to $T_c \sim 27.5$ K. The inset shows that for S13 the SMP can be seen above ~18 K, but it disappears at lower temperatures

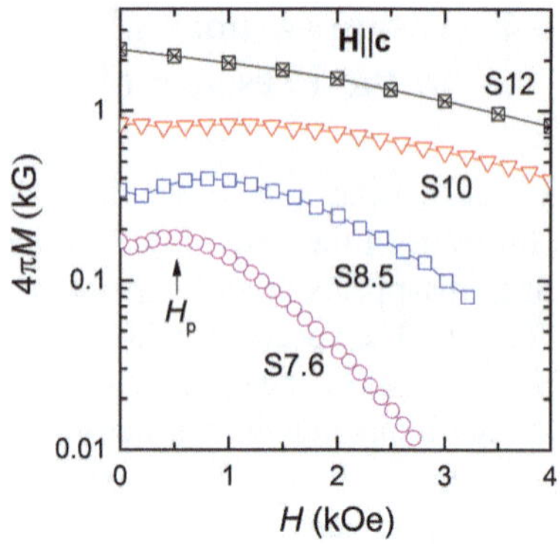


**Fig. 6.10** While the SMP is absent in S12, strongly underdoped LSCO single crystals (S10, S8.5, S7.6) exhibit an SMP, with the peak field $H_p$ detected on the descending $M(H)$ branch ($H$ parallel to the $c$-axis, $T \sim 0.4T_c$)

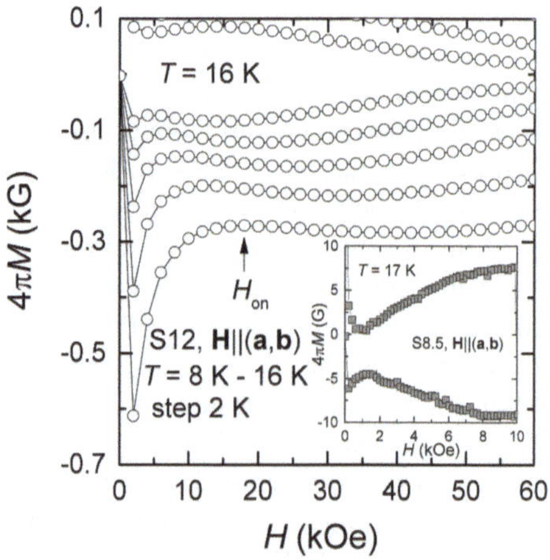


**Fig. 6.11** When $H$ is parallel to the $(a, b)$ planes, the SMP is present for all considered samples. This is shown in the main panel for S12, with the onset field $H_{on}$ (indicated by an *arrow*) increasing with decreasing temperature in the low-$T$ range, and for S8.5 at $T = 17$ K in the inset

In qualitative agreement with relation (6.2), where $\varepsilon$ decreases at low doping ($\varepsilon \sim 1/60$ in S7.6 and $\sim$1/15 for S19.6 [20, 86]) and $1/\lambda^2 \propto x$ [87], the SMP in Fig. 6.10 develops at lower $H$ values, far below the field for complete suppression of pair tunneling between the $CuO_2$ layers in LSCO [88]. This is confirmed by the fact that when $H$ is oriented along the $(a, b)$ planes the SMP appears for all investigated single crystals (including those with static stripes, as illustrated in Fig. 6.11 for S12).

As shown above, the analysis of magnetization relaxation can signal the existence of an ordered (disordered) vortex phase. The relaxation curves registered for S12 in the two $H$ orientations (as exemplified in Fig. 6.12 for $T = 8$ K and 12 K) generate the $U^*(T)$ variation plotted in Fig. 6.13.

For $H$ parallel to the $c$-axis of S12 (no SMP, see the main panel of Fig. 6.9), $U^*$ in Fig. 6.13 is around 100 K at $H = 10$ kOe and decreases with increasing temperature, (6.4) indicating a disordered vortex phase in the investigated $T$ range, in agreement with the SANS results [64]. When $H$ is along the $(a, b)$ planes of the

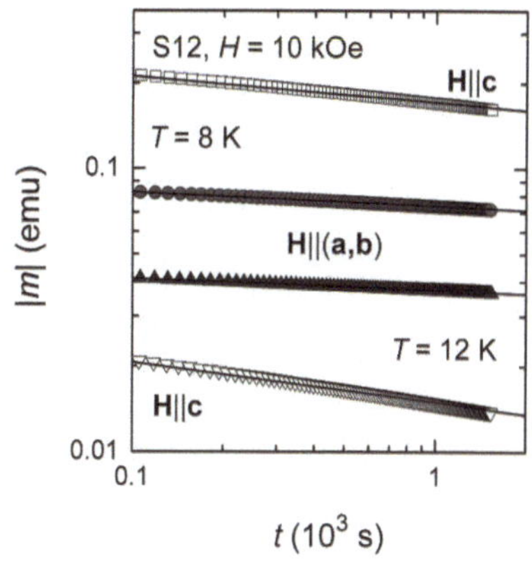


**Fig. 6.12** Relaxation of the irreversible magnetic moment, $|m|$ versus $t$ (log-log scales), for S12 in $H = 10$ kOe parallel to the $c$-axis (open symbol), or along the ($a$, $b$) planes, at $T = 8$ K and 12 K. The absolute value of the slope in a linear fit (the *continuous line*) is the normalized relaxation rate $S$ averaged over $t_w$. One can see that the relaxation is faster for $H$ perpendicular to the stripe structure

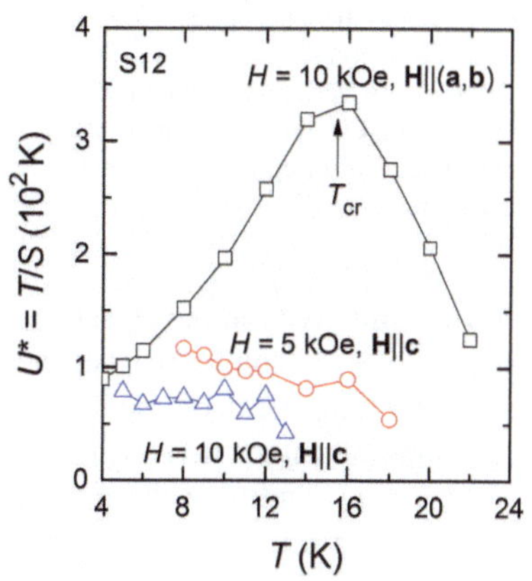


**Fig. 6.13** Resulting $U^*$ ($T$) for $H = 10$ kOe oriented along the ($a$, $b$) planes, and $H = 5$ kOe and 10 kOe applied parallel to the $c$-axis. When $H$ is perpendicular to the static stripe structure, the vortex system is disordered, and no SMP is present. With $H$ along the ($a$, $b$) planes, the SMP develops (see Fig. 6.12), and the vortex dynamics changes at the creep crossover temperature $T_{cr} \sim 15.5$ K ($H = 10$ kOe)

same specimen (the SMP develops, Fig. 6.11), the crossover elastic creep (ordered vortex phase) plastic creep (disordered phase) is present at $T_{cr} = 15.5$ K ($H = 10$ kOe). The magnetic hysteresis curve registered at $T = 16$ K from Fig. 6.11 supports the above observation that the constant $H$ in the $U^*(T)$ determination approximately represents the midpoint between $H_{on}$ and $H_p$ at $T = T_{cr}$.

The characteristic pinning energy $U_c$ for the elastic creep regime below $T_{cr}$ is small. A linear extrapolation in Fig. 6.13 ($H = 10$ kOe, along the $CuO_2$ planes) for $T \leq 8$ K leads to $U_c \sim 20$ K. In this case, according to (6.4), at relatively high temperatures (below $T_{cr}$), $U^*(H) \propto U(H) \propto H^{\alpha}$.

The $U^*(H)$ variation at $T = 12$ K is plotted in the main panel of Fig. 6.14, and the derived exponent $\alpha \sim 0.2$ is in good agreement with the exponent $\beta \sim 1.4$ appearing in the experimentally determined $H_{on}(T)$ variation at low temperatures shown in the inset of Fig. 6.14.

The weaker vortex pinning for $H$ perpendicular to the stripe plane [83] (see Fig. 6.13) is supported by the pinning enhancement through strain-induced destabilization of the static stripe structure reported in [89]. The spatial landscape for

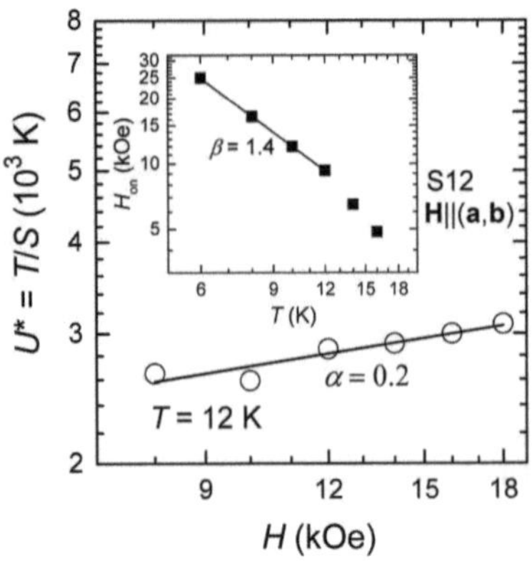


**Fig. 6.14** Main panel: Magnetic field dependence of $U^*$ ($\propto U$, see text), as determined for S12 in $H$ parallel to the $CuO_2$ planes from the magnetic relaxation curves at $T = 12$ K with $H$ close to $H_{on}$ (elastically creeping vortices). The linear fit in the log-log plot leads to $U^*(H) \propto H^{0.2}$. Inset: Temperature variation of the onset field $H_{on}$ (log-log scales) extracted from $M(H)$ curves such as those plotted in the main panel of Fig. 6.11, taking the form $H_{on}(T) \propto T^{-1.4}$. The magnetic field exponent $\alpha$ is in agreement with the temperature exponent [$\beta$ in relation (6.2)]. The *continuos lines* represent a linear fit

superconductivity in the presence of SDW order is complex [90] and can lead to the appearance of (coreless) Josephson vortices and/or staggered flux. In such a situation, the pinning decrease would be associated with a diminished contribution of core pinning interactions to the pinning force density.

One may think that the absence of SMP in the main panel of Fig. 6.9 (S12 in $H$ parallel to the $c$-axis) is due to poor pinning. However, in this case, one should have an ordered vortex system, exhibiting a first-order melting transition or a PE, like in clean single crystals, which is not the case. Moreover, compared to that in S12, pinning is even weaker in strongly underdoped specimens, but the SMP develops (Fig. 6.10).

The BVG instability in static conditions ($J = 0$) at the 1/8 anomaly for $H$ parallel to the $c$-axis (leading to the complete disappearance of the SMP) was discussed in [64] as due to the real-space competition between the magnetic correlation length (of several tens of nanometer around the 1/8 anomaly [85]) and the intervortex spacing. The essential point is that this is not related to pinning. In general terms, the existence of the Bragg vortex glass implies the rearrangement of the vortex system (at scales larger than the in-plane vortex correlation length [19]) to lower its energy in the vortex pinning potential [91]. The presence of a well-developed static stripe structure seems to impede the vortex rearrangement when $H$ is perpendicular to the stripe plane. With no correlation between the static stripes in adjacent $CuO_2$ layers, an impediment would be the vortex location in the weakly superconducting regions outside the charge stripes, if one considers the results from [92]. Actually, the stripe structure in neighboring layers is rotated by $\pi/2$, as idealized in Fig. 6.15 [93], and the vortex rearrangement for a stable BVG may be affected by the in-plane anisotropy of $\lambda$ induced by the presence of the SSO [94].

The above constraints are less effective for a diffuse SSO, the BVG reappears with decreasing doping below $x = 1/8$ [64], and the SMP sets in (Fig. 6.10). On the other side, when $H$ is parallel to the $(a, b)$ planes, the static stripe order produces inhomogeneities in the interlayer coupling, the pinning source for Josephson vortices [19]. In this field orientation, vortices can slide along the $CuO_2$ planes, the rearrangement for a stable three-dimensional BVG becomes possible, and the SMP is present even for S12 (see Fig. 6.10). Vortex sliding along the static spin stripes is frustrated for the stripe orientation from Fig. 6.15.

As known, a disordered vortex phase can dynamically order with increasing $J$. During standard magnetic measurements at low temperatures, the induced $J$ approaches $J_{c0}$, the effective pinning decreases, the intervortex interactions prevail, and elastic creep is expected [80]. In specimens with weak bare pinning, this dynamic ordering appears in the low-$T$ range [95]. Generated by $J(T)$, the vortex-creep crossover associated to the dynamic ordering of the vortex system is not accompanied by an SMP. Thus, the necessary condition for the SMP appearance is the existence of BVG (with long-range order or a polycrystalline BVG [8]) in static conditions.

In principle, the existence of the (low-field) BVG does not contradict the SMP model based on a more rapid relaxation at low $H$ [29, 40, 41], in the single-vortex collective pinning domain (where the vortex correlation length $L_c$ becomes lower than the intervortex spacing $a_0$ [29, 96]). Indeed, with (6.4), it results

$$S(T) = T/U^*(T) \sim T/[U_c + pT\ln(t_w/t_0)], \tag{6.5}$$

i.e., a larger $S$ at low $H$, in the single-vortex collective creep regime, where the creep exponent $p = 1/7$ [81]. According to this SMP model, a low $p$ should be present around $H_{on}$, and, by increasing $H$, a creep crossover toward bundle (three-dimensional) creep should be detected, where $p = 1.5$ (small vortex bundles) or 7/9 (a large bundle volume) [81].

In this context, for S12 in $H$ parallel to the static stripe structure, for example, where the SMP is present, it is instructive to plot $U^*$ versus $1/J$ instead of the $U^*(T)$ variation from Fig. 6.13, to extract $p$ with (6.3). The $U^*(1/J)$ dependence (where $J$ represents an average over $t_w$) is shown in Fig. 6.16, in log-log scales. With the applied field ($H$ = 10 kOe) around $H_{on}(T)$ (see Fig. 6.11, main panel), the exponent

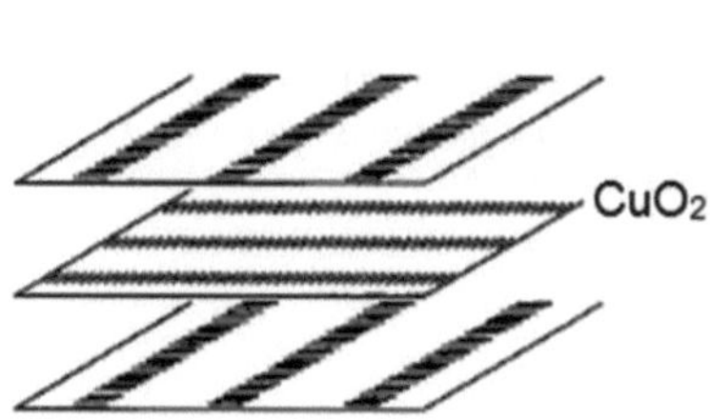


**Fig. 6.15** Sketch showing the relative orientation of static spin stripes (*black*) and charge stripes (*white regions* in between) for neighboring $CuO_2$ planes [94]

(determined for $T$ between 8 K and 12 K) is $p = 0.74 \sim 7/9$, meaning a large vortex bundle creep regime. Moreover, in the case of $FeSe_{0.5}Te_{0.5}$ single crystals investigated in [96], the crossover single-vortex collective creep–bundle creep at 4.2 K appears at $H \sim 0.5$ kOe, whereas $H_{on}(T = 4.2$ K) determined for similar specimens is above 20 kOe [41]. This casts some doubts on the interpretation of the SMP in terms of collective pinning only [29, 40, 41].

It is worthy to note that the appropriate choice of the $T$ range and of the relaxation time interval to diminish the influence of the temperature variation of the vortex pinning potential and of thermo-magnetic instabilities at low $T$ is essential for getting the real vortex-creep exponent $p$, revealing the dynamic state of the vortex system.

The first condition is roughly fulfilled at $T \leq T_c/2$. The decrease of $p$ in the low-$T$ range (like in Fig. 6.16) has been explained through the influence of microflux jumps (MFJ) [50]. The change of $U^*$ due to the presence of MFJ can be different, even for the same specimen and measurement system. As discussed in [50], in the case of moderate pinning (i.e., when $T$ and $H$ are not too low), the MFJ are expected to occur mainly at early relaxation stages (where the electric field is higher). When $J$ decreases due to such MFJ at short $t$, the effect can be assimilated with the introduction of a "supplemental relaxation time" $t_j$ (increasing with decreasing $T$ and/or $H$), and over the $t_w$ considered above one determines a higher $U^*$. Using the general vortex-creep relation and the definition formula for $U^*$, one obtains that the resulting $U^*$ will be enhanced by a factor $\sim(1 + t_j/t_w)$, leading to a smaller $p$ (Fig. 6.16). On the other hand, in the case of dynamic relaxation measurements [40, 41], where $t_w$ is shifted to shorter $t$, as well as for the dc relaxation at low $T$ and $H$, the microflux jumps will invade the relaxation time window, and a faster apparent relaxation occurs. This means a smaller apparent $U^*$ at low temperatures, and, consequently, a larger $p$.

Thus, the disappearance of the fishtail effect at the 1/8 anomaly illustrated in Fig. 6.9 is directly related to the absence of the low-$H$ Bragg vortex glass, which is unstable in the presence of static stripes when vortices are perpendicular to the stripe plane [64]. The change of the sign of the creep exponent between $H_{on}$ and $H_p$

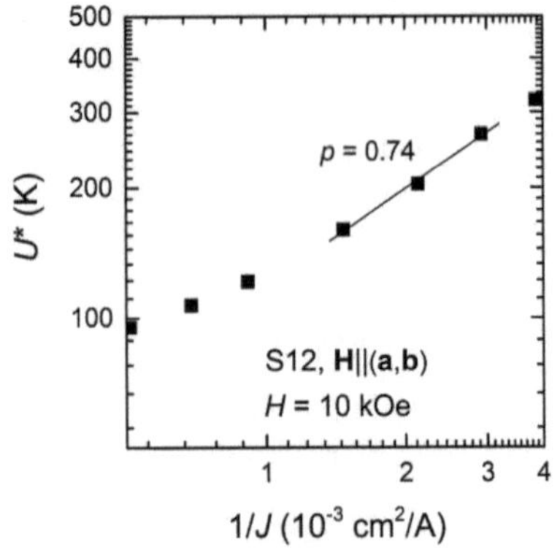


**Fig. 6.16** $U^*$ versus $1/J$ in log-log scales for S12 in $H = 10$ kOe oriented along the ($a$, $b$) planes. When this field value is around $H_{on}$ (see the main panel of Fig. 6.11), the creep exponent determined with (6.4) *(as the slope of thecontinuous line)* is $p = 0.74 \sim 7/9$

shown above supports strongly the pinning-induced BVG disordering as the actual scenario for the SMP. With the static BVG as the necessary ingredient for an SMP, one can explain the puzzling behavior of some samples, showing no SMP (while the vortex pinning strength is in the range). For example, in specimens with superconducting domains of limited extension there is no BVG [91], and, consequently, no SMP [97]. The multi-domain structure of FeSe crystals [98, 99] may have a similar effect.

## 6.5 The Ac Magnetic Response of Specimens Exhibiting a Second Magnetization Peak

The ac magnetic response at usual frequencies $f$ and amplitudes $h_{ac}$ [100] is often considered for the investigation of the dynamic vortex phase diagram and the pinning potential of superconductors [12, 13, 101–109]. However, while a strong influence of the PE (close to the IL) on the ac signal was reported (see [103], for example), the conditions in which the presence of an SMP modifies significantly the ac response are not clear. Figure 6.17 shows the ac signal of the striped specimen (S12) in $H$ parallel the $c$-axis, when no SMP is present, whereas Fig. 6.18 illustrates the ac response of S19.6 (in the same field orientation), with a well-developed SMP (Fig. 6.1). The response signal was registered on warming, with the ac field ($h_{ac} = 3.9$ Oe) along the $c$-axis, after cooling the sample from above $T_c$ to $T = 10$ K in constant $H$ (FCW). The FCW runs are usually preferred [103], since on cooling the vortex system nucleates at high temperatures, and may remain trapped in metastable, strongly pinned configurations.

As known [110, 111], the in-phase component of the ac magnetic moment ($m'$) is directly related to the presence of the screening current density $J$, whereas the out-of-phase component ($m''$) is a measure of dissipation. The peak in $m''(T)$ at $T_p$ (see Figs. 6.17 and 6.18) is interpreted in terms of the critical state model [111, 112], where the maximum dissipation corresponds to the first full penetration of the

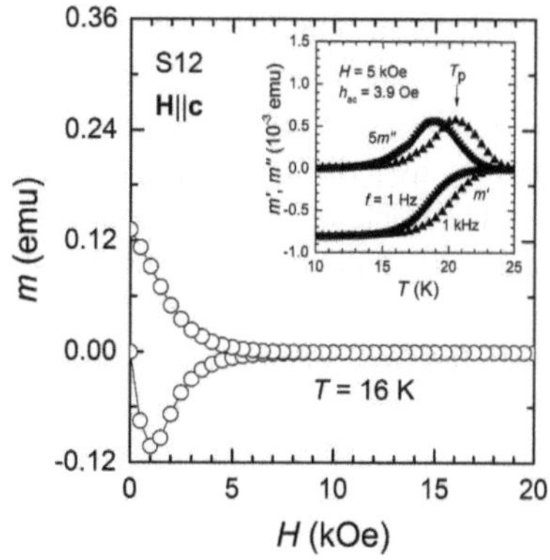


**Fig. 6.17** Main panel: The hysteresis $m(H)$ curve at $T = 16$ K for the striped specimen S12 with $H$ along the $c$-axis (no SMP). Inset: The ac signal ($h_{ac} = 3.9$ Oe, $f = 1$ Hz and 1 kHz) registered on warming, in $H = 5$ kOe applied in field-cooling conditions (FCW) along the $c$-axis. The peak temperature $T_p$ (indicated by an *arrow*) shifts to lower values by decreasing $f$

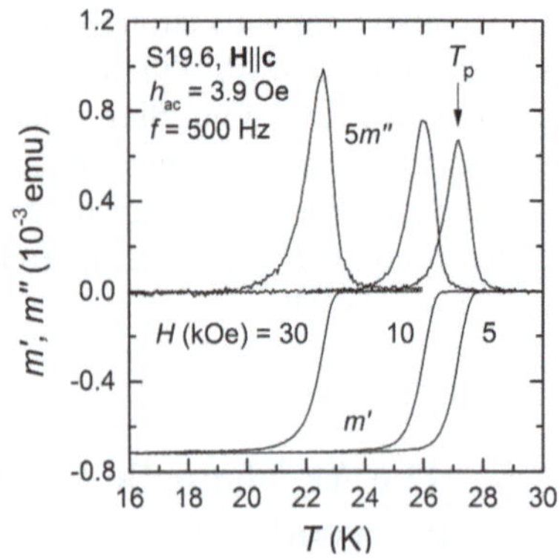


**Fig. 6.18** FCW ac magnetic response of S19.6 in $H = 5$, 10, and 30 kOe parallel to the $c$-axis ($h_{ac} = 3.9$ Oe, $f = 500$ Hz). While this sample exhibits a well-developed SMP (Fig. 6.1), the ac response is similar to that from Fig. 6.17, in the absence of an SMP. The hysteretic (dissipative) regime appears close to the IL

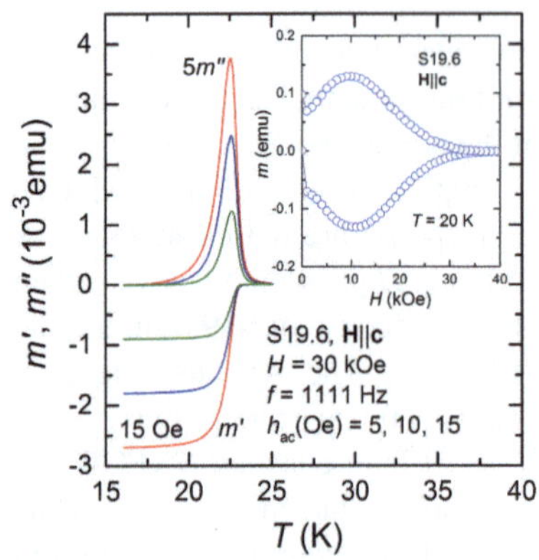


**Fig. 6.19** Main panel: The FCW ac magnetic response of S19.6 in $H = 30$ kOe parallel to the $c$-axis at several ac field amplitudes $h_{ac}$ and the frequency $f = 1111$ Hz. The hysteretic regime appears only above ~20 K, in the vicinity of the IL, as indicated by the $m(H)$ curve registered at $T = 20$ K, plotted in the inset. No sign for a significant influence of the SMP on the ac signal is present

critical state (generated by the ac field) starting from low temperatures, and a thermally activated vortex hopping process. For our LSCO specimens with small demagnetization effects, a relevant critical state related ac magnetic response at the used $h_{ac}$ develops only close to the IL. The undistorted $m''(T)$ maximum means the absence of a PE, in agreement with the dc magnetic hysteresis curves. In the case of S19.6, with relatively strong pinning and exhibiting an SMP, the picture does not modify even at higher $h_{ac}$, as shown in the main panel of Fig. 6.19. The hysteretic (dissipative) ac regime for $H = 30$ kOe appears above ~20 K, i.e., in the vicinity of the IL, as indicated by the dc $m(H)$ curve at $T = 20$ K from the inset.

The nonhysteretic, Meissner-like signal present below the IL in Figs. 6.17, 6.18, and 6.19 indicates no critical state penetration, in qualitative agreement with the occurrence of a skin effect [113]. In the low-$T$ region, $m'$ is frequency independent (see the inset of Fig. 6.17), and the dissipation ($m''$) is very small. The real part of the complex magnetic susceptibility ($m'/h_{ac}$) is practically independent of $h_{ac}$. This linear ac response (the Campbell regime [114]) is determined by (reversible) intravalley vortex oscillations, with the ac field penetrating the sample on a distance $\lambda_C$ (the Campbell penetration depth) larger than $\lambda$. For S19.6, the crossover from the linear regime to nonlinearity (when $J$ is below $h_{ac}/\lambda_C$) [101] appears in the vicinity of the IL even at $h_{ac}$ = 15 Oe (Fig. 6.19), owing to the relatively large values of the critical current density (Sect. 6.2).

It is instructive to see the ZFC ac magnetic signal obtained by varying $H$ at constant $T$, in the conditions where a pronounced SMP develops (Fig. 6.1). The ZFC ac response of S19.6 in $H$ parallel to the $c$-axis at $T$ = 26 K is plotted in Fig. 6.20. It can be seen that there are no appreciable changes of the magnetic signal around $H_p \sim 2$ kOe (see Fig. 6.2). The response is $f$ independent across the SMP, with a small $m''$, indicating a linear regime. The slight decrease of $|m'|$ with increasing $H$ in Fig. 6.20 is due to the increase of $\lambda_C$ with $B$ [114]. While the dependence of $\lambda_C$ on the vortex state under strong pinning conditions has been recently predicted [106], the existence of an SMP on the dc $m(H)$ curves remains practically irrelevant for the results of noninvasive ac measurements (in the Campbell regime) from Figs. 6.18, 6.19, and 6.20. We can understand now the ac results obtained for 2H-$NbSe_2$ single crystals in $H$ parallel to the ($a$, $b$) planes ($D \sim 0$) from [115], where it was claimed the identification of a crossover from weak collective pinning to a strong pinning regime which is not associated with a peak in the ac signal.

The situation radically changes if the crossover from the linear regime to nonlinearity is attained well below the IL, toward the domain of $H_{on}(T)$. At high enough $h_{ac}$, the displacements of vortices overcome the width of the pinning potential well, and the intervalley motion can dominate. The hysteretic penetration of the ac critical state (generating the nonlinear response) is described by the characteristic Bean length ($\sim h_{ac}/J$) [74]. To reach the nonlinear response well below the IL for S19.6 by increasing $h_{ac}$ is difficult. Another way is to use thin specimens with the large side perpendicular to the external magnetic field. For specimens with $D$ close to unity, strong demagnetization effects are present, and $h_{ac}$ at the sample edge will be enhanced by a factor $\sim 1/(1-D)$.

Figure 6.21 shows the magnetic hysteresis curve $m(H)$ and the field dependence of the ZFC ac magnetic response components ($m'$ and $m''$) for a thin $BaFe_2(As_{0.68}P_{0.32})_2$ single crystal (with weaker pinning) at $T$ = 26 K in $H$ oriented

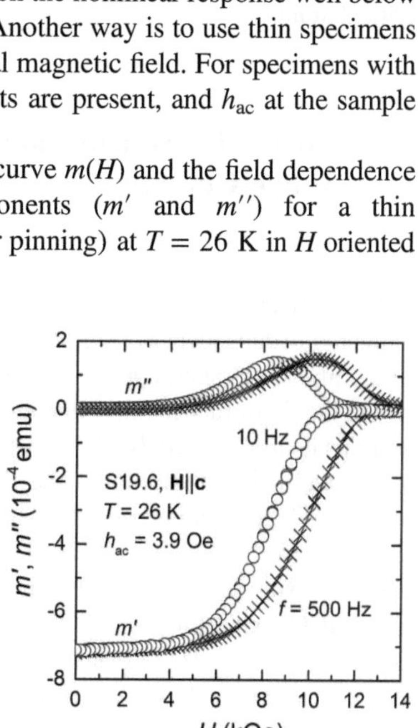


**Fig. 6.20** ZFC ac magnetic response of the overdoped specimen S19.6 ($D$ = 0.43), registered as a function of $H$ (oriented along the $c$-axis) at $T$ = 26 K ($h_{ac}$ = 3.9 Oe, $f$ = 1 Hz and 1 kHz). The presence of the SMP ($H_p \sim 2$ kOe, see Fig. 6.2) does not induce significant changes of the ac magnetic signal at usual $f$ and $h_{ac}$

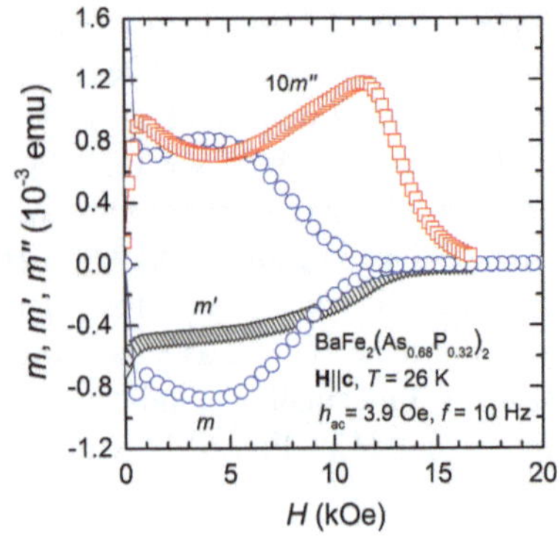


**Fig. 6.21** Magnetic field $H$ dependence of the ZFC dc magnetic moment $m$ and of the ZFC ac magnetic response components ($m'$ and $m''$) at $T = 26$ K for a thin $BaFe_2(As_{0.68}P_{0.32})_2$ single crystal with $T_c \sim 27.5$ K. $H$ was oriented along the $c$-axis (perpendicular to the largest side), when $D$ is close to unity

along the $c$-axis (i.e., perpendicular to the largest side). This sample is similar to that investigated in [116], having $D$ above 0.9. The penetration of the ac critical state is appreciable even far below the IL ($m$ and $m'$ are within the same order of magnitude). The modification of the ac magnetic response induced by the well-developed SMP with $H_{on} \sim 1$ kOe and $H_p \sim 4$ kOe becomes obvious. The critical state penetration (with $m''$ as its measure) has a local maximum at $H_{on}$, and a local minimum at $H_p$.

The behavior of $m''(H)$ in Fig. 6.21 is in agreement with the enhancement of the effective critical current density between $H_{on}$ and $H_p$. However, $|m'|$ does not exhibit a maximum at $H_p$, since the increase of the current density induced by the ac field around the peak field is compensated by a smaller penetration depth of the ac critical state.

## 6.6 Vortex Phase Diagram of $La_{2-x}Sr_xCuO_4$ Single Crystals

It is well established that in superconducting single crystals with weak quenched disorder (untwined YBCO, as an example) the SMP line exhibits a nonmonotonous temperature dependence [117, 118], approaching the vortex-lattice melting line [14, 15] in the high-$T$ domain, approximately at the so-called upper critical point [119, 120], where the first-order melting line terminates. When vortex pinning increases, the end point tends to $T_c$, while $H_p(T)$ becomes a monotonically decreasing function [44, 45].

A careful investigation of the dc magnetic data of LSCO [83] shows that no features which could be related to a PE or to a first-order melting are present. In this situation, in the considered ($H$, $T$) domain, the dynamic vortex phase diagram of LSCO is simple, as illustrated in Fig. 6.22 (S19.6 in $H$ parallel to the $c$-axis). For LSCO exhibiting an SMP, the BVG is present in the low-$T$ domain up to $H_{on}(T)$. Between $H_{on}(T)$ and $H_p(T)$ there is a partially disordered vortex solid (PDVS), whereas above $H_p$ one has a disordered vortex system (DVS), which melts continuously on crossing the IL.

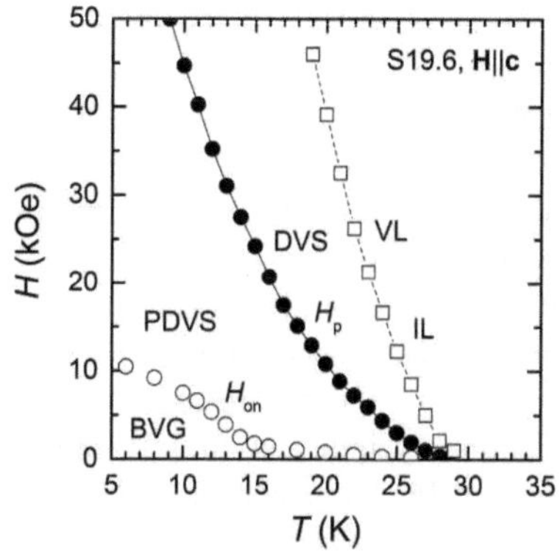


**Fig. 6.22** Vortex phase diagram of S19.6 in $H$ parallel to the $c$-axis in the dynamic conditions of dc magnetic measurements. The long-range quasi-ordered vortex solid (BVG) is present in the low-$H$ domain up to $H_{on}(T)$. Between $H_{on}$ and $H_p$, one has a partially disordered vortex system (PDVS). Above $H_p(T)$, the completely disordered vortex system (DVS) is expected to melt continuously across the dc irreversibility line IL extracted from the $m(H)$ curves from Fig. 6.1. The vortex liquid (VL) extends up to close to the $H_{c2}(T)$ line (not shown)

In principle, the dynamic vortex phase diagram of the LSCO samples considered here and in [83] in $H$ oriented along the $(a, b)$ planes should be similar to that from Fig. 6.22 (since for this $H$ orientation the SMP occurs), but shifted to higher fields, owing to anisotropy. The LSCO specimens with static stripes (S12, for example) in $H$ parallel to the $c$-axis have the simplest diagram, since there is no Bragg vortex glass, SMP, or PE.

Finally, it is worth mentioning that the experimentally determined vortex phase diagram is sometimes complicated by the simultaneous presence of an SMP and a PE, as reported using dc and ac measurements [121–123], which is not the case of the single crystals with dense, randomly distributed pinning centers investigated here. The existence of both SMP and PE would be in conflict with the disordering of the BVG across the SMP. According to this scenario, as presented in Sect. 6.3, above $H_p$ the pinning structure already accommodated vortices, and the PE should not appear. This aspect was discussed in [124, 125], and it was argued [125] that in global magnetic measurements the possible macroscopic inhomogeneities in the distribution of the pinning centers can lead to multiple peak structures (which may even be accompanied by a first-order vortex-lattice melting).

## 6.7 Conclusions

In summary, the investigation of the dc magnetic hysteresis curves of $La_{2-x}Sr_xCuO_4$ single crystals in a wide doping range reveals the complete disappearance of the SMP in the domain of well-developed static stripe order ($x \sim 1/8$) when the external magnetic field $H$ is perpendicular to the stripe planes. The evolution of the SMP with decreasing doping follows the instability of the Bragg vortex glass in static conditions at the 1/8 anomaly (detected using SANS experiments), where the well-developed static stripe order obstructs the vortex system rearrangement for a stable, three-dimensional BVG. This is in agreement with the temperature variation of the normalized vortex-creep activation energy.

The SMP reappears in strongly underdoped LSCO specimens ($x \leq 0.10$), where the static stripe structure is diffuse and the BVG was detected in SANS measurements at low $H$ (with no sign of the square-to-rhombic vortex-lattice transition). The observed behavior definitely indicates the quasi-ordered vortex solid at low $H$ as the necessary ingredient for the occurrence of a second magnetization peak. When the BVG instability is not caused by pinning, the SMP is missing. At the same time, when $H$ is oriented along the $(a, b)$ planes, the SMP occurs in the whole considered doping range ($0.076 \leq x \leq 0.196$). The presented analysis supports the scenario in which the SMP is generated by the pinning-induced disordering of the low-field Bragg vortex glass, the enhancement of the effective critical current density resulting from a better accommodation of vortices to the pinning structure in the disordered vortex phase. This is confirmed by the nonlinear ac magnetic response registered well below the dc irreversibility line. The increase of the characteristic fields for the SMP in the low-$T$ domain can be explained by taking into account the dynamic conditions of dc magnetic measurements.

No features which could be related to a first-order vortex-lattice melting or to a peak effect close to the irreversibility line have been observed for the investigated single crystals with dense, randomly distributed pinning centers in both dc and ac measurements. This leads to a simple dynamic vortex phase diagram of LSCO, especially for specimens with static stripes in $H$ perpendicular to the stripe plane.

**Acknowledgements** This work has been supported by the POC Project P_37_697 (28/01.09.2016). The kind assistance of the Alexander von Humboldt Foundation is gratefully acknowledged. Thanks are due to T. Adachi, K. Omori, Y. Koike, and J. Mosqueira, providing us with the single crystal specimens.

## References

1. W. De Sorbo, Rev. Mod. Phys. **36**, 90 (1964)
2. A.B. Pippard, Philos. Mag. **19**, 217 (1969)
3. A.I. Larkin, YuN Ovchinnikov, J. Low Temp. Phys. **93**, 409 (1979)
4. S. Bhattacharya, M.J. Higgins, Phys. Rev. Lett. **70**, 2617 (1993)
5. T. Matsushita, *Flux Pinning in Superconductors* (Springer, Heidelberg, 2007), p. 325
6. Y. Paltiel et al., Phys. Rev. Lett. **85**, 3712 (2000)
7. A.M. Troyanovski, M. van Hecke, N. Saha, J. Aarts, P.H. Kes, Phys. Rev. Lett. **89**, 147006 (2002)
8. Y. Fasano, M. Menghini, F. de la Cruz, Y. Paltiel, Y. Myasoedov, E. Zeldov, M.J. Higgins, S. Bhattacharya, Phys. Rev. B **66**, 020512(R) (2002)
9. I.K. Dimitrov, N.D. Daniilidis, C. Elbaum, J.W. Lynn, X.S. Ling, Phys. Rev. Lett. **99**, 047001 (2007)

10. M.M. Bermúdez, E.R. Louden, M.R. Eskildsen, C.D. Dewhurst, V. Bekeris, G. Pasquini, arXiv:1701.08798v1 [cond-mat.supr-con] 30 Jan 2017
11. W.K. Kwok, J.A. Fendrich, C.J. van der Beek, G.W. Crabtree, Phys. Rev. Lett. **73**, 2614 (1994)
12. J. Giapintzakis, R.L. Neiman, D.M. Ginsberg, M.A. Kirk, Phys. Rev. B **50**, 16001 (1994)
13. T. Ishida, K. Okuda, H. Asaoka, Phys. Rev. B **56**, 5128 (1997)
14. A. Houghton, R.A. Pelcovits, A. Sudbø, Phy. Rev. B **40**, 6763 (1989)
15. E. Zeldov, D. Majer, M. Konczykowski, V.B. Geshkenbein, V.M. Vinokur, H. Shtrikman, Nature (London) **375**, 373 (1995)
16. M. Däumling, J.M. Seuntjens, D.C. Larbalestier, Nature (London) **346**, 332 (1990)
17. V.N. Kopylov, A.E. Koshelev, I.F. Schegolev, T.G. Togonidze, Physica C **170**, 291 (1990)
18. B. Khaykovich, E. Zeldov, D. Majer, T.W. Li, P.H. Kes, M. Konczykowski, Phys. Rev. Lett. **76**, 2555 (1996)
19. G. Blatter, V.M. Feigel'man, V.B. Geshkenbein, A.I. Larkin, V.M. Vinokur, Rev. Mod. Phys. **66**, 1125 (1994), and references therein
20. T. Sasagawa, Y. Togawa, J. Shimoyama, A. Kapitulnik, K. Kitazawa, K. Kishio, Phys. Rev. B **61**, 1610 (2000)
21. S.L. Li, H.-H. Wen, Phys. Rev. B **65**, 214515 (2002)
22. L. Miu, Physica C **405**, 260 (2004)
23. R. Prozorov, N. Ni, M.A. Tanatar, V.G. Kogan, R.T. Gordon, C. Martin, E.C. Blomberg, P. Prommapan, J.Q. Yan, S.L. Bud'ko, P.C. Canfield, Phys. Rev. B **78**, 224506 (2008)
24. P. Das, A.D. Thakur, A.K. Yadav, C.V. Tomy, M.R. Lees, G. Balakrishnan, S. Ramakrishnan, A.K. Grover, Phys. Rev. B **84**, 214526 (2011)
25. M. Bonura, E. Giannini, R. Viennois, C. Senatore, Phys. Rev. B **85**, 134532 (2012)
26. S. Salem-Sugui Jr., J. Mosqueira, A.D. Alvarenga, D. Sóñora, E.P. Herculano, D. Hu, G. Chen, H. Luo, Supercond. Sci. Technol. **28**, 055017 (2015)
27. W. Zhou, X. Xing, W. Wu, H. Zhao, Z. Shi, Sci. Rep. **6**, 22278 (2016)
28. N. Chikumoto, M. Konczykowski, N. Motohira, A.P. Malozemoff, Phys. Rev. Lett. **69**, 1260 (1992)
29. L. Krusin-Elbaum, L. Civale, V.M. Vinokur, F. Holtzberg, Phys. Rev. Lett. **69**, 2280 (1992)
30. L. Klein, E.R. Yacoby, Y. Yeshurun, A. Erb, G. Müller-Vogt, V. Breit, H. Wühl, Phys. Rev. B **49**, 4403 (1994)
31. G.K. Perkins, L.F. Cohen, A.A. Zhukov, A.D. Caplin, Phys. Rev. B **51**, 8513 (1995)
32. M.F. Goffman, J.A. Herbsommer, F. de la Cruz, T.W. Li, P.H. Kes, Phys. Rev. B **57**, 3663 (1998)
33. Y. Kopelevich, P. Esquinazi, Solid State Commun. **122**, 33 (2002)
34. D. Giller, A. Shaulov, R. Prozorov, Y. Abulafia, Y. Wolfus, L. Burlachkov, Y. Yeshurun, E. Zeldov, V.M. Vinokur, J.L. Peng, R.L. Greene, Phys. Rev. Lett. **79**, 2542 (1997)
35. Y. Radzyner, A. Shaulov, Y. Yeshurun, I. Felner, K. Kishio, J. Shimoyama, Phys. Rev. B **65**, 214525 (2002)
36. B. Rosenstein, B. Ya, Shapiro, I. Shapiro, Y. Bruckental, A. Shaulov, Y. Yeshurun, Phys. Rev. B **72**, 144512 (2005)
37. R. Kopeliansky, A. Shaulov, B. Ya, Shapiro, Y. Yeshurun, B. Rosenstein, J.J. Tu, L.J. Li, G. H. Cao, Z.A. Xu, Phys. Rev. B **81**, 092504 (2010)
38. Y. Sun, S. Pyon, T. Tamegai, R. Kobayashi, T. Watashige, S. Kasahara, Y. Matsuda, T. Shibauchi, Phys. Rev. B **92**, 144509 (2015)
39. M. Jirsa, L. Půst, D. Dlouhý, M.R. Koblischka, Phys. Rev. B **55**, 3276 (1997)
40. J. Tao, Q. Deng, Z. Wang, X. Zhu, H.-H. Wen, Phys. Rev. B **91**, 214516 (2015)
41. Z.F. Wu, Z.H. Wang, J. Tao, L. Qiu, S.G. Yang, H.-H. Wen, Supercond. Sci. Technol. **29**, 035006 (2016)
42. T. Nattermann, Phys. Rev. Lett. **64**, 2454 (1990)
43. T. Giamarchi, P. Le Doussal, Phys. Rev. B **55**, 6577 (1997)
44. Y. Abulafia, A. Shaulov, Y. Wolfus, R. Prozorov, L. Burlachkov, Y. Yeshurun, D. Majer, E. Zeldov, H. Wühl, V.B. Geshkenbein, V.M. Vinokur, Phys. Rev. Lett. **77**, 1596 (1996)

45. H. Küpfer, Th. Wolf, C. Lessing, A.A. Zhukov, X. Lançon, R. Maier-Hirmer, W. Schauer, H. Wühl, Phys. Rev. B **58**, 2886 (1998)
46. L. Miu, E. Cimpoiasu, T. Stein, C.C. Almasan, Physica C **334**, 1 (2000)
47. L. Miu, T. Noji, Y. Koike, E. Cimpoiasu, T. Stein, C.C. Almasan, Phys. Rev. B **62**, 15172 (2000)
48. S. Salem-Sugui Jr., L. Ghivelder, A.D. Alvarenga, L.F. Cohen, K.A. Yates, K. Morrison, J. L. Pimentel Jr., H. Luo, Z. Wang, H.-H. Wen, Phys. Rev. B **82**, 054513 (2010)
49. N. Haberkorn, B. Maiorov, I.O. Usov, M. Weigand, W. Hirata, S. Miyasaka, S. Tajima, N. Chikumoto, K. Tanabe, L. Civale, Phys. Rev. B **85**, 014522 (2012)
50. D. Miu, T. Noji, T. Adachi, Y. Koike, L. Miu, Supercond. Sci. Technol. **25**, 115009 (2012)
51. D. Bhoi, P. Mandal, P. Choudhury, J. Appl. Phys. **113**, 183902 (2013)
52. J. Zaanen, O. Gunnarsson, Phys. Rev. B **40**, R7391 (1989)
53. V.J. Emery, S.A. Kivelson, Q.H. Lin, Phys. Rev. Lett. **64**, 475 (1990)
54. P. Abbamonte, A. Rusydi, S. Smadici, G.D. Gu, G.A. Sawatzky, D.L. Feng, Nat. Phys. **1**, 155 (2005)
55. J.M. Tranquada, B.J. Sternlieb, J.D. Axe, Y. Nakamura, S. Uchida, Nature **375**, 561 (1995)
56. J.T. Park et al., Phys. Rev. Lett. **102**, 117006 (2009)
57. B. Lake et al., Nature **415**, 299 (2002)
58. A.R. Moodenbaugh, Y. Xu, M. Suenaga, T.J. Folkerts, R.N. Shelton, Phys. Rev. B **38**, 4596 (1988)
59. K. Kumagai, Y. Nakamura, Y. Watanabe, Y. Nakamichi, H. Nakajima, J. Magn. Magn. Mater. **76–77**, 601 (1988)
60. Y. Koike, A. Kobayashi, T. Kawaguchi, M. Kato, T. Noji, Y. Ono, T. Hikita, Y. Saito, Solid State Commun. **82**, 889 (1992)
61. T. Adachi, K. Omori, Y. Tanabe, Y. Koike, J. Phys. Soc. Japan **78**, 114707 (2009)
62. T. Suzuki, T. Goto, K. Chiba, T. Shinoda, T. Fukase, H. Kimura, K. Yamada, M. Ohashi, Y. Yamaguchi, Phys. Rev. B **57**, R3229 (1988)
63. I. Watanabe, T. Adachi, K. Takahashi, S. Yairi, Y. Koike, K. Nagamine, Phys. Rev. B **65**, 180516 (2002)
64. J. Chang et al., Phys. Rev. B **85**, 134520 (2012), and references therein
65. C.P. Bean, Phys. Rev. Lett. **8**, 250 (1962)
66. E.M. Gyorgy, R.B. van Dover, K.A. Jackson, L.F. Schneemeyer, J.V. Waszczak, Appl. Phys. Lett. **55**, 283 (1989)
67. Y. Yeshurun, A.P. Malozemoff, A. Shaulov, Rev. Mod. Phys. **68**, 911 (1996)
68. C.J. van der Beek, M. Konczykowski, S. Kasahara, T. Terashima, R. Okazaki, T. Shibauchi, Y. Matsuda, Phys. Rev. Lett. **105**, 267002 (2010)
69. Y. Tanabe, T. Adachi, K. Omori, H. Sato, T. Noji, Y. Koike, J. Phys. Soc. Jpn. **76**, 113706 (2007)
70. G.P. Mikitik, E.H. Brandt, Phys. Rev. B **64**, 184514 (2001)
71. R. Gilardi et al., Phys. Rev. Lett. **88**, 217003 (2002)
72. L. Miu, Y. Tanabe, T. Adachi, Y. Koike, D. Miu, G. Jakob, H. Adrian, Phys. Rev. B **78**, 024520 (2008)
73. M.P. Maley, J.O. Willis, H. Lessure, M.E. McHenry, Phys. Rev. B **42**, 2639 (1990)
74. G. Pasquini, L. Civale, H. Lanza, G. Nieva, Phys. Rev. B **59**, 9627 (1999)
75. M. Zehetmayer, Sci. Rep. **5**, 9244 (2015)
76. L. Miu, T. Adachi, K. Omori, Y. Koike, D. Miu, Phys. Rev B **82**, 064520 (2010)
77. R. Khasanov, M. Bendele, A. Amato, K. Conder, H. Keller, H.-H. Klauss, H. Luetkens, E. Pomjakushina, Phys. Rev. Lett. **104**, 087004 (2010)
78. A.P. Malozemoff, Physica C **185–189**, 264 (1991)
79. P.W. Anderson, Y.B. Kim, Rev. Mod. Phys. **36**, 39 (1964)
80. L. Miu, D. Miu, T. Petrisor, A. El-Tahan, G. Jakob, H. Adrian, Phys. Rev. B **78**, 212508 (2008)
81. M.V. Feigel'man, V.B. Geshkenbein, A.I. Larkin, V.M. Vinokur, Phys. Rev. Lett. **63**, 2303 (1989)

82. L. Miu, D. Miu, Supercond. Sci. Technol. **23**, 025033 (2010)
83. L. Miu, A.M. Ionescu, I. Ivan, D. Miu, T. Adachi, K. Omori, Y. Koike, Physica C **519**, 79 (2015)
84. U. Divakar et al., Phys. Rev. Lett. **92**, 237004 (2004)
85. M. Kofu, S.-H. Lee, M. Fujita, H.-J. Kang, H. Eisaki, K. Yamada, Phys. Rev. Lett. **102**, 047001 (2009)
86. S. Kohout, T. Schneider, J. Roos, H. Keller, T. Sasagawa, H. Takagi, Phys. Rev. B **76**, 064513 (2007)
87. Q. Li, M. Suenaga, T. Kimura, K. Kishio, Phys. Rev. B **47**, 2854 (1993)
88. A.A. Schafgans, A.D. LaForge, S.V. Dordevic, M.M. Quazilbash, W.J. Padilla, K.S. Burch, Z.Q. Li, S. Komiya, Y. Ando, D.N. Basov, Phys. Rev. Lett. **104**, 157002 (2010)
89. Y.Q. Zhang, X.Q. Xiang, J.F. Ding, X.G. Li, Supercond. Sci. Technol. **21**, 095005 (2008)
90. G. Campi et al., Nature **525**, 359 (2015)
91. Y.M. Wang, M.S. Fuhrer, A. Zettl, S. Ooi, T. Tamegai, Phys. Rev. Lett. **86**, 3626 (2001)
92. M. Ichioka, M. Takigawa, K. Machida, J. Phys. Soc. Japan **70**, 33 (2001)
93. J.M. Tranquada, J.D. Axe, N. Ichikawa, Y. Nakamura, S. Uchida, B. Nachumi, Phys. Rev. B **54**, 7489 (1996)
94. D. Kuzmanovski, M.G. Vavilov, Supercond. Sci. Technol. **25**, 084001 (2012)
95. L. Miu, Phys. Rev. B **72**, 132502 (2005)
96. T. Klein, H. Grasland, H. Cercellier, P. Toulemonde, C. Marcenat, Phys. Rev. B **89**, 014514 (2014)
97. A.K. Pramanik, L. Harnagea, S. Singh, S. Aswartham, G. Behr, S. Wurmehl, C. Hess, R. Klingeler, B. Büchner, Phys. Rev. B **82**, 014503 (2010)
98. A. Galluzzi, M. Polichetti, K. Buchkov, E. Nazarova, D. Mancusi, S. Pace, Supercond. Sci. Technol. **28**, 115005 (2015)
99. A. Leo et al., Supercond. Sci. Technol. **28**, 125001 (2015)
100. F. Gömöry, Supercond. Sci. Technol. **10**, 523 (1997)
101. L.A. Angurel, F. Amin, M. Polichetti, J. Aarts, P.H. Kes, Phys. Rev. B **56**, 3425 (1997)
102. I. Joumard, T. Klein, J. Marcus, Phys. Rev. Lett. **87**, 167002 (2001)
103. G. Pasquini, D. Pérez, Daroca, C. Chiliotte, G.S. Lozano, V. Bekeris, Phys. Rev. Lett. **100**, 247003 (2008)
104. E. Bartolomé, A. Palau, A. Llordés, T. Puig, X. Obradors, Phys. Rev. B **81**, 184530 (2010)
105. G. Prando, P. Carretta, R. De Renzi, S. Sanna, A. Palenzola, M. Putti, M. Tropeano, Phys. Rev. B **83**, 174514 (2011)
106. R. Villa, V.G. Geshkenbein, R. Prozorov, G. Blatter, Phys. Rev. Lett. **115**, 207001 (2015)
107. L. Miu, P. Mele, I. Ivan, A.M. Ionescu, D. Miu, J. Supercond. Novel Magn. **28**(2), 361 (2015)
108. L. Miu, I. Ivan, A.M. Ionescu, D. Miu, AIP Adv. **6**, 065027 (2016)
109. I. Ivan, A.M. Ionescu, D. Miu, P. Mele, L. Miu, Supercond. Sci. Technol. **29**, 095013 (2016)
110. P. Fabbricatore, S. Farinon, G. Gemme, R. Musenich, R. Parodi, B. Zhang, Phys. Rev. B **50**, 3189 (1994)
111. X.S. Ling, J.I. Budnik, in *Magnetic Susceptibility of Superconductors and other Spin Systems*, ed. by R.A. Hein, T.L. Francavilla, D.H. Liebenberg (Plenum, NY, 1991), p. 377
112. J.R. Clem, A. Sanchez, Phys. Rev. B **50**, 9355 (1994)
113. V.B. Geshkenbein, V.M. Vinokur, R. Fahrenbacher, Phys. Rev. B **43**, 3748 (1991)
114. A.M. Campbell, J. Phys. C **2**, 1492 (1969); **4**, 3186 (1971)
115. S. Monah, J. Sinha, S.S. Banerjee, Y. Myasoedov, Phys. Rev. Lett. **98**, 027003 (2007)
116. S. Salem-Sugui Jr., J. Mosqueira, A.D. Alvarenga, D. Sóñora, E.P. Herculano, D. Hu, G. Chen, H. Luo, Supercond. Sci. Technol. **28**, 055017 (2015)
117. K. Deligiannis, P.A.J. de Groot, M. Oussena, S. Pinfold, R. Langan, R. Gagnon, L. Taillefer, Phys. Rev. Lett. **79**, 2121 (1997)
118. T. Nishizaki, T. Naito, N. Kobayashi, Phys. Rev. B **58**, 11169 (1998)
119. H. Safar, P.L. Gammel, D.A. Huse, D.J. Bishop, W.C. Lee, J. Giapintzakis, D.M. Ginsberg, Phys. Rev. Lett. **70**, 3800 (1993)

120. J. Kierfeld, V. Vinokur, Phys. Rev. B **61**, R14 928 (2000)
121. D. Pal, S. Ramakrishnan, A.K. Grover, D. Dasgupta, B.K. Sarma, Phys. Rev. B **63**, 132505 (2001)
122. A.A. Zhukov, H. Küpfer, H. Claus, H. Wühl, M. Kläser, G. Müller-Vogt, Phys. Rev. B **52**, R9871 (1995)
123. A.D. Thakur, S.S., Benerjee, M.J. Higgins, S. Ramakrishnan, A.K. Grover, Phys. Rev. B **72**, 134524 (2005)
124. L. Miu, Phys. Rev. B **65**, 096501 (2002)
125. D. Pal, S. Ramakrishnan, A.K. Grover, Phys. Rev. B **65**, 096502 (2002)

# Chapter 7
# Emergence of an Interband Phase Difference and Its Consequences in Multiband Superconductors

**Yasumoto Tanaka**

## 7.1 Introduction

The phase of a superconducting wave function (an order parameter) is ordinarily described by a scalar [1, 2]. However, there is not a definite principle preventing the possibility of using a vector or a matrix for the phase [3–6]. The idea of multiple phases was initially considered by Leggett, who discussed their tiny fluctuations in multiband superconductors [3], Then, Leggett applied the idea to superfluid helium-3 [7–10], and the new topology resulting from the multiple phases in multiband superconductors has not been considered. The multiple phases in the superfluid wave function were established as a consequence of spin-triplet $p$-wave pairing, and the condensation of these pairs in superfluid helium-3 was studied in subsequent works by using the Bardeen-Cooper-Schrieffer formalism (BCS theory) [11–15]. This great success initiated the quest for spin-triplet pairing, $p$-wave pairing, other non-spin-singlet non-$s$-wave pairing, and for condensates composed of these unconventional pairs in new superconductors [16–22] where the role of spin fluctuations is spotlighted. Though the multiband approach is not strictly necessary in these non-spin-singlet non-$s$-wave superconductors, the consideration of multiple phases has been reimported from the superfluidity to the superconductivity. In the case of triplet superconductors, it has been proved that degenerate wave functions require multiple phases to describe them, as the Bardeen–Cooper–Schrieffer (BCS) Hamiltonian has degenerate eigenfunctions. When introducing multiple phases, degeneracy had been always discussed, even for the multiband superconductors (Multiple phases had not been considered for nondegenerate

Y. Tanaka (✉)
National Institute of Advanced Industrial Science and Technology,
AIST-Tsukuba-Central-2-30381020, 1-1-1 Umezono, Tsukuba,
Ibaraki 305-8568, Japan
e-mail: y.tanaka@aist.go.jp



A. Crisan (ed.), *Vortices and Nanostructured Superconductors*, Springer Series in Materials Science 261, DOI 10.1007/978-3-319-59355-5_7

multiband superconductors, but only for multiband superconductors with multiple degenerate electronic bands until 2001.) [23].

Multiple phases have been also used to explore the mixing of superconducting order parameters with different symmetries, like, $(d+is)$-waves (the mixing of *d*- and *s*-wave pairing). In this case, the superconducting condensate does not have degenerate wave functions [24–26].

Topology is a key issue of multiple phases both in degenerate systems and when considering the mixing of wave functions with different symmetries. Our understanding of this aspect is based on a topological quantum number (a winding number), which represents the number of rotations around a (singular) point. According to this topological quantum number, the magnetic flux is quantized. It becomes a unit of magnetic flux quantum for a single-component condensate. However, the quantization condition is different for a condensate with multiple quantum phases.

When Leggett brought in the idea of the multiple phases of a superconducting condensate, he concentrated on the small fluctuation of a newly introduced quantum phase of the condensate. His main conclusion was that direct experimental detection was beyond the limits of the then available techniques. Instead, he proposed an indirect method. If we extended the discussion from the small fluctuation to a "topological object," a direct detection method might be automatically deduced. Topological detection is a definite proof of the presence of multiple components. It also changes our basic understanding of superconducting condensates, bringing them together with other condensates that have multiple phases of a macroscopic wave function, such as multicomponent Bose–Einstein condensates [27–37], superfluid helium-3 [38–42], or other systems appearing in particle physics field theories [43–51] and cosmology [52–54].

Now, this challenge of detecting a special topology originating from multiple quantum phases has been elaborated [4, 5, 55] for wave functions that have the same symmetry and are not degenerate. It was necessary to understand the abnormal temperature evolution of superconducting gaps for a multilayer cuprate superconductor, $Cu_xBa_2Ca_3Cu_4O_y$ [56, 57], which has at least two gaps showing different temperature dependence [58–61]. Some typical multilayer cuprate superconductors, of which $Cu_xBa_2Ca_3Cu_4O_y$ is just an example, are illustrated in Fig. 7.1 [62, 63]. The temperature evolution of the superconducting gaps in $Cu_x$-$Ba_2Ca_3Cu_4O_y$ is different from that seen in conventional superconductors. For the latter, the temperature dependence of the superconducting gap is similar for any point in the momentum space and any electronic band [64–66]. A consequence of the challenge of introducing multiple phases is the need to realize a novel internal coordinate space specifying theses multiple phases at higher temperature, as compared with triplet superconductors and other superconducting condensates that have multiple components. It is not necessary to invoke spin-triplet pairing or a rare wave function like a *p*-wave as the orbital function of a pair. It is easier to realize multiple phases by controlling the band structure instead of controlling the pair function, and we have many candidates for such superconductors.

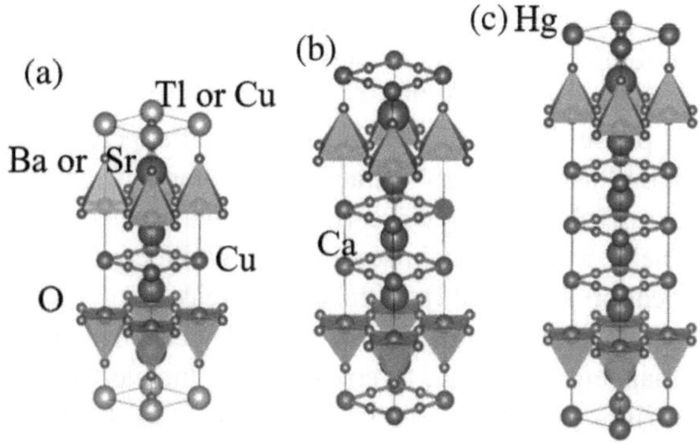


**Fig. 7.1** Crystal structures of some multilayer cuprate superconductors drawn with VESTA 3 [62]. **a** $(Tl,Cu)(Ba,Sr)_2Ca_2Cu_3O_y$, which has three $CuO_2$ planes in a unit cell. **b** $Cu_xBa_2Ca_3Cu_4O_y$, which has four $CuO_2$ planes in a unit cell. **c** $HgBa_2Ca_4Cu_5O_y$, which has five $CuO_2$ planes in a unit cell. Multilayer cuprate superconductors have multiple electronic bands. (Reprinted from [63]. © Y. Tanaka)

Once a multiband superconductor without degenerate components is assigned a superconducting order parameter with multiple phases, the resulting superconductivity in this superconductor is denoted as "multicomponent superconductivity based on a multiband superconductor" [55]. It is considered that each band has each component, and different components are coupled through an "interband Josephson coupling." This superconductivity has salient topological properties that are found in other quantum condensates with multiple components in particle physics, and in cosmology, but that single-component superconductivity never exhibits. An interband phase difference soliton was discovered by the author as a first and prominent topological object [4, 5]. It was accompanied by the fractionalization of the unit magnetic flux quantum [5], which is achieved through the same mechanism as other examples of fractional quantization, like the fractional charge of the quark [45, 67–69]. There have been subsequent works on the topological nature of multicomponent superconductivity based on multiband superconductors [70–79], and the relationship between such superconductivity and other multicomponent quantum condensates has been also investigated in depth [6, 80, 81]. The essential commonalities between a vortex molecule composed of fractional vortices and phase difference solitons and a quark confined in a hadron are mentioned, for example, in [67–69, 82–84]. The quark with a fractional charge corresponds to the fractional vortex, while gluons correspond to solitons.

Introducing multiple nondegenerate components with the same symmetry also changes the nature of the superconducting transition [85–91]. Obviously, we can expect a "second superconducting transition" originating from the internal degrees of freedom related to the interband phase difference [85, 87, 88, 92, 93]. Moreover, we have to account for a collective excitation based on the interband phase difference other than pair breaking [3, 89–91, 94, 95]. The effect of such a collective excitation dramatically changes the nature of the superconducting phase transition.

A superconducting gap does not close at the superconducting transition temperature (with increasing temperature), when there are significant collective excitations (interband phase fluctuations) [86, 96, 97].

The interband phase difference of multicomponent superconductivity based on multiband superconductors is very attractive for practical applications. We can build up a quantum phase of a wave function that does not interact with the electromagnetic field, which is reminiscent of the Weinberg-Salam theory (i.e., the electroweak unification theory) [44]. In the electroweak theory, a field escaped from the Higgs boson becomes the photon, the neutral weak boson $Z^0$ gets separated from the electromagnetic field, and, as consequence, the electromagnetic field does not "see" the neutrinos [44]. Similarly, the electromagnetic field does not see either the interband phase difference solitons [4, 98]. Hence, an analogy can be drawn between an interband phase difference mode and the $Z^0$ boson, and between the interband phase difference soliton and the neutrino. From the point of view of electronics, we may manipulate the soliton without considering the influence of the electromagnetic field [99]. Such manipulation of the electromagnetically inactive interband phase solitons could have the potential to solve problems like the quantum decoherence of the qubit [100].

In this chapter, we introduce the dramatic phenomena expected in multicomponent superconductivity based on multiband superconductors.

## 7.2 Quantum Phase of a Superconducting Wave Function and the Interband Phase Difference Soliton in Multiband Superconductors

When the intraband pairing interaction is considerably larger than the interband interaction (which implies that the interband pair interaction contributes very little —less than a few percent—to the superconducting condensation energy), the pair density is mainly determined by the former. We can take an approximation in which the pair density is constant for each band, and the topology of the superconducting condensate is determined by the phase of the superconducting wave function. (This can be regarded as an extension of the conventional London approximation for single-component superconductivity [101].)

The momentum $\boldsymbol{p}_i$ of a BCS pair on the $i$-th band can be written as

$$\boldsymbol{p}_i = 2m_i^* \boldsymbol{v}_i + 2e^* \boldsymbol{A}, \tag{7.1}$$

where $2m_i^*$, $\boldsymbol{v}_i$, and $2e^*$ are, respectively, the mass, velocity, and charge of the pair, and $\boldsymbol{A}$ is the vector potential of the magnetic field. We may substitute $\left(-i\hbar \frac{\partial}{\partial x_i}\right)$ for $\boldsymbol{p}_i$, and then $\left(-i\hbar \frac{\partial \Psi}{\partial x_i}\right)$ approximates $\left(\hbar \Psi \frac{\partial \theta_i}{\partial x_i}\right)$, where $\Psi$ is the superconducting wave function, $\theta_i$ is the phase of the superconducting component on the $i$-th band,

and $\boldsymbol{x}_i$ gives the position of the $i$-th band. Then, (7.1) adopts the following form. (Note that we can drop $\Psi$ because it appears on both sides of (7.1), so we can divide it by $\Psi$.)

$$\hbar\ grad\ \theta_i = 2m_i^* \boldsymbol{v}_i + 2e^* \boldsymbol{A}. \tag{7.2}$$

The different components are separated within the bands in momentum space, but superimposed at the same location in real space. Equation (7.2) means that the gradient of the quantum phase of each component on a different band can be different, but the vector potential works equally for every band. The phase difference $\theta_i - \theta_j$ is gauge invariant (with regard to the electromagnetic field), though the phase $\theta_i$ itself is not [4].

Let us consider the case where there is no electromagnetic field. In this situation, we can explicitly take a gauge with $\boldsymbol{A} = 0$. Under this gauge, the current density on the $i$-th band approximates

$$\boldsymbol{J}_i = 2e^* n_i \boldsymbol{v}_i, \tag{7.3}$$

where $n_i$ is the density of the pair of the $i$-th band. However, there is no magnetic induction, so the total electromagnetic current must be zero:

$$\textstyle\sum_i \boldsymbol{J}_i = 0, \tag{7.4}$$

which indicates that

$$\textstyle\sum_i \frac{n_i}{m_i^*} grad\ \theta_i = 0. \tag{7.5}$$

As a result, the first component flows in the direction opposite the direction along which the second component flows in the two-band case. The cartoons in Fig. 7.2 show the difference between the real current in single-component superconductivity and a "hidden" current in two-component superconductivity. When there are three or more than three bands, the flow direction is determined by an energy valley [55, 102]. This current flow is similar to an "axial current" [103, 104]. If there is no Josephson coupling, this current may be conserved. However, when there is pair hopping between two bands (as discussed later, this causes the soliton to acquire a mass), this antiparallel current is not conserved like the real current [103, 104], but rather appears somewhere in the sample and disappears at another location.

When there is no interband Josephson interaction, the fluctuation of the interband phase difference enters a Nambu–Goldstone mode [105, 106] that spreads over the whole sample. There is no threshold energy needed to excite the mode. When there is a finite interband interaction, the fluctuation of the massless mode becomes a massive mode [107] known as the Leggett mode [3] (When there is a finite real current, the massless mode is restored, as discussed later.). When there is

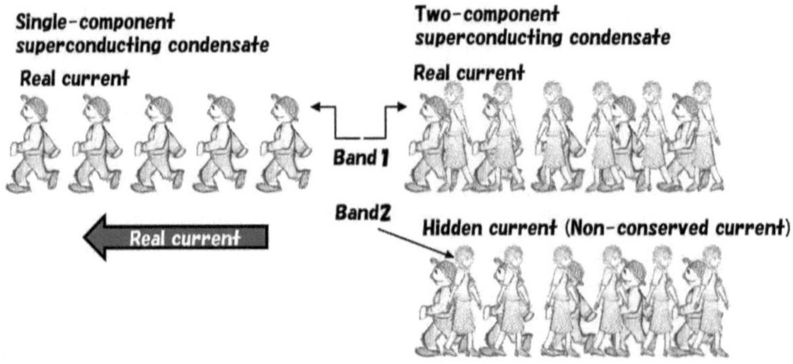


**Fig. 7.2** Schematic cartoons to compare the currents in a single-component and a two-component condensate. In the case of two-component superconductivity, there is another current (as seen in the *bottom panel*) apart from the ordinary real current (seen in the *top panels*). This additional current is electromagnetically inactive because of cancellation. When there is interband pair hopping, this current needs not be conserved. (© Y. Tanaka)

an interband phase difference of $2\pi$ rad between two locations, we can find an interband phase difference soliton [4, 5], a kink in the one-dimensional case, and a domain wall for two or three dimensions. Such a kink or domain wall separates the two superconducting domains having different phase (But there is no interband phase difference within a domain.) [82, 101]. The variation of the interband phase difference as function of the location in the interband phase difference soliton is shown schematically in Fig. 7.3, which also shows, for comparison, a massless mode with a rotation of $2\pi$ rad, for which the rate of change of the interband phase difference with respect to the position constant. This soliton has been extensively studied by many researchers [95, 98, 108–112].

It is important to note that the tail and head of the kink have different phases, as seen in Fig. 7.3. This phase shift comes from the mutual cancellation of the magnetic induction, in other words, from the condition that the total real current should be zero. The current (the gradient of the quantum phase or "gauge field") on one band "bends" the internal phase (coordinate) space of another band. This role played by the current is analogous to that of an ordinary vector potential for the real magnetic field, where the vector potential gives the curvature in the "internal phase space" defined by the quantum phase. For the vector potential, the coupling constant is $2e$; in our case, however, the Josephson interaction itself gives the coupling strength at the kink and the domain wall.

This correspondence originates the fractionalization of the magnetic flux quanta when we connect the head and the tail of the kink. We have to compensate the phase shift generated by the kink structure using the ordinary vector potential for the magnetic field [4, 5]. This compensation can be achieved through an external field or a self-induced current [113], as seen, e.g., in the appearance of half-flux quanta in *d*-wave superconducting films on a tri-crystal substrate [24, 114–116], where the grain boundary of the superconducting film generated by the grain

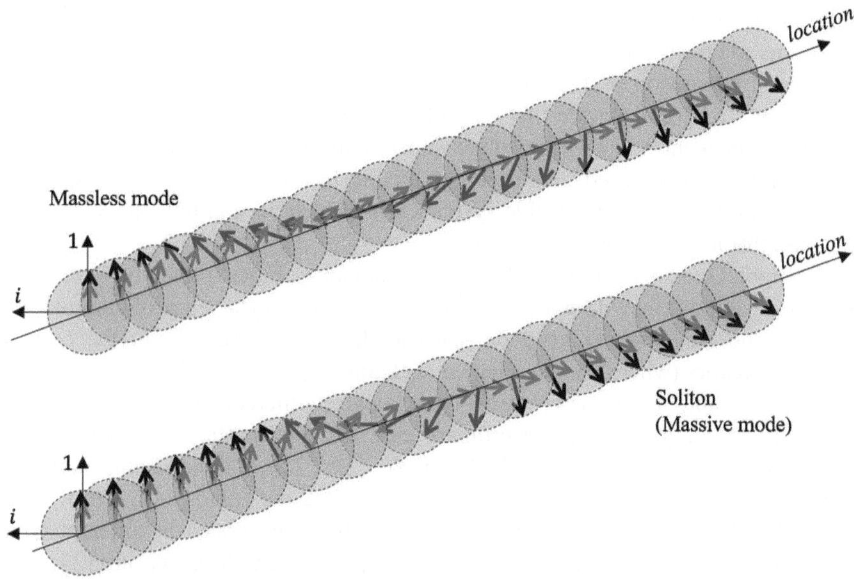


**Fig. 7.3** Interband phase difference soliton (massive mode, *bottom panel*) and the corresponding massless mode (*top panel*). The rate of change of the interband phase difference is confined to the soliton in the center of the figure. This rate of change is constant for the massless mode. (© Y. Tanaka)

boundary of the substrate leads to the phase shift. (The soliton can generate a phase shift on a perfect film, i.e., without grain boundary. It is very useful to consider a basic design of the superconducting electronics using the phase difference soliton.) It should be noted that there are some articles in which these considerations were not properly taken into account [108].

An intercomponent phase difference soliton in a two-component Bose–Einstein condensate formed by ultracold atomic gas was also discussed by Son and Stephanov (Son–Stephanov soliton, [33]). Some remarks concerning the relationship between this phase soliton and the interband phase difference soliton are in order. An ultracold atom does not have charge, and hence, the Son–Stephanov soliton does not have the ability to provoke the fractional quantization of the magnetic flux. This is instructive when we consider the relationship between the quantum number and the quantized quantity (which are not same). A quantum number gives a boundary condition (in other words, a quantization condition), and quantized quantities are a consequence of those boundary conditions. (For example, the reader may consider whether an "elementary charge" gives a boundary condition or is determined by a boundary condition.) Consequently, the unit magnetic flux quantum is determined by a boundary condition [1], and hence, we may find some ways to manipulate fractional flux quanta [117, 118].

For the Son–Stephanov soliton, two components are coupled by the externally applied photon field. Unlike the two-band superconductor, the strength of the interaction can be tuned and turned on/off. This property is very attractive, since it allows us to create various fractional/conventional vortex lattices by modifying the Josephson interaction [32]. To realize the same in a two-band superconductor, we may design a special device that controls the strength of the Josephson coupling through a proximity effect from other superconductor [111]. A mimic multiband superconductor composed of a superconducting bi-layer sandwiching a magnetic layer might be one of the feasible systems with which we can tune the strength of the Josephson interaction [119, 120].

It might be also mentioned that two components rather prefer to have some spatial separation (they do not tend to overlap) in an ordinary ultracold atomic gas. There is no backbone to keep the particle density constant, as in the case of superconductors (where a crystal lattice skeleton helps to keep the total density of carriers constant). The weak repulsive force between different condensates separates the different components in the ordinary case (as condensation tends to collapse under an attractive interaction) [27, 28, 35, 37, 121, 122]. A soliton in this system is normally considered a "dark soliton" [34, 123], i.e., the soliton of a density gradient. The typical interactions between dark solitons generated in different components or in a domain wall between two components are discussed in [124]. The interband phase difference soliton in a two-band superconductor and the solitons in a two-component ultracold atomic gas behave differently when charge and density play crucial roles.

The relationship between the soliton in a spin-triplet condensate [22, 39, 125–128] and the interband phase difference soliton in a two-band superconductor has also been discussed. In a spin-triplet condensate, the components degenerate. Izyumov and Laptex described a soliton for a condensate having multiple degenerate components [129]. Rotation of the intercomponent phase difference is observed between the head and tail of a soliton, but in general it is not of $2\pi$ rad. For superconducting condensates, we could expect to be able to compensate the phase mismatch between two states by sandwiching a single soliton (in other words, between the head and tail of a soliton) with the electromagnetic vector potential, that is, by introducing fractional magnetic quanta. However, we cannot do so. This phase mismatch accompanied by a non-$2\pi$-rad intercomponent rotation is defined by a gauge, that is, not the electromagnetic gauge. To generate a phase mismatch that can be compensated by the electromagnetic vector potential, we need two or more than two solitons. The fractional magnetic flux quantum becomes an integer multiple of 1/n of the unit flux quantum, where n is the number of components [82]. A half-integer flux quantum in the spin-triplet superconductor is the most popular example. In multiband superconductors, the components do not degenerate. Consequently, the phase mismatch accompanied by the interband phase difference soliton can be compensated by the electromagnetic gauge potential, but fractionalization is not given by a simple rational number such as 1/2 and 1/3. It is rather determined by the effective mass and pair density. Babaev applied this fractionalization to a vortex and discussed the possibility of emergence of fractional vortices in $MgB_2$ [71, 130]. A group from the Uppsala University and a Korean researcher tried to find out their signature in the flux noise [131], following Babaev's advice, but they have not reached a conclusive result yet. Selection of an adequate material is one of the most important issues in the quest for fractional vortices because the interband Josephson interaction always tends to confine a fractional vortex in a normal vortex [67, 132]. Thus, the search for fractional vortices has continued in other candidate materials. Multilayer cuprates [132–134] have shown some promising hints and an ultra-underdoped cuprate superconductor [117, 118] provided a direct observation. These successful experimental results will be addressed later in detail.

A special property of nondegenerate systems can be found in the response of the interband phase difference to an applied current. An externally applied current produces a phase difference when the coherence lengths of the two components are not equal [135, 136]. The ability to generate the soliton is an important property of systems with nondegenerate components, as this phenomenon does not occur in multicomponent systems with degenerate components. This will be further discussed later.

## 7.3 The Mass of the Soliton and the Interband Interaction

The origin of the mass of the phase difference soliton is the interband Josephson interaction. The existence of a mass implies that the soliton has a characteristic length. When there is no interband interaction, the length of the soliton increases up to the sample size, as can be seen in Fig. 7.3. The situation is analogous to the twist (torsion) of a string. (Even if we pinch the two ends of the string and twist it, we cannot produce a local kink. On the contrary, a uniform and continuous helix is observed.). Imagine a toy model of the sine-Gordon equation consisting of a rubber hose and paper cards, as shown in Fig. 7.4 [137–139]. The torsion is confined to the hose, and due to the interband Josephson interaction, solitons cause deformations to some parts of the hose. We can draw an analogy between this mechanism and the generation of the quark masses through chiral symmetry breaking [103, 107]. Because of the presence of the interband Josephson interaction, the quantum phases of two bands cannot rotate freely. One component in the two-band superconductor plays the role of a right-handed quark field and another component that of a left-handed quark field. The "mass" of the interband phase difference mode is caused by pair hopping (much like the $\pi$-meson in the Yukawa model [140, 141]), which implies "substitution" of neighborhoods in an internal (coordinate) space. The instantaneous and continuous reassemble of the internal space network (one pair leaves one location specified by an internal gauge and settles at another location) constitutes the mechanism of generation of the mass of the interband phase difference soliton. It is equivalent to the conversion of one particle into another by changing some quantum number.

Interestingly, however, the Josephson interaction does not always inhibit the emergence of the massless mode. In the next section, we describe this mode and frustrated superconductivity as the background for a massless mode.

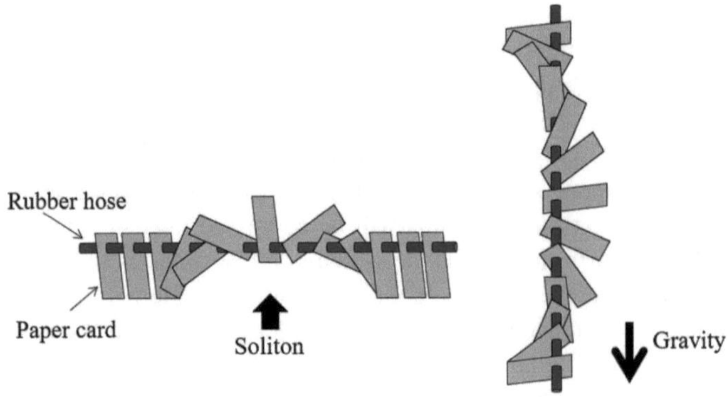


**Fig. 7.4** Toy model of a soliton to show the relationship between the massive (*left panel*, on horizontal holding) and massless modes (*right panel*, on vertical holding). The rotation of a card corresponds to an interband phase difference [137–139]. (© K. Tanaka)

## 7.4 Frustrated Superconductivity and the Massless Leggett Mode

Normally, the sign of the wave function of a pair does not change by pair hopping between bands. This is the case when an electron–phonon interaction mediates pair formation. However, the sign of the wave function can be reversed when an electronic mediator causes the interband pair hopping. This type of interband Josephson interaction is sometimes called a "repulsive interaction" and was independently discovered by Kondo and Peretti in 1962 [142, 143]. They introduced the sign-reversal order parameter, which is now widely known as the $s_{\pm}$-wave. In this kind of superconductivity, the sign of one band is reversed and the interband phase difference is locked to π rad. When there are three bands and the interband Josephson interactions are positive (repulsive), we cannot set all the interband phase differences to π rad. If we allow only two values (0 or π rad), the only possibility is that one interband phase difference becomes π rad and the other two, 0 rad. However, if we allow yet another arbitrary interband phase difference, there is a new possible order parameter. Let us consider that a "superconducting wave function" consisting of three components, $\Psi = \varphi_1 + \varphi_2 + \varphi_3$, $\Psi = |\varphi_1|e^{i\theta_1} + |\varphi_2|e^{i\theta_2} + |\varphi_3|e^{i\theta_3}$ could be this new wave function. Actually, the Ginzburg–Landau formalism suggests that there is a case with a finite intercomponent phase difference that is neither 0 nor π rad, as can be seen in Fig. 7.5a-1 and a-2 [102]. These two states are degenerate chiral states. In a superconductor, one of the two may be selected, and this choice is sometimes called a "time-reversal symmetry breaking." In 1999, Agterberg, Barzykin, and Gor'kov discussed this state for a three-band superconductor with three equivalent symmetrical bands [23]. At that time, it could be considered that these degenerate electronic bands were essential, because it was believed that degenerate electronic bands were necessary to produce the time-reversal symmetry-broken state. Actually,

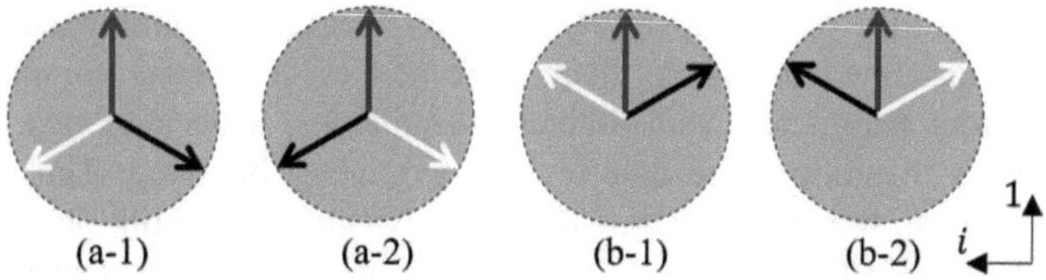


**Fig. 7.5** Order parameter of frustrated superconductivity. **a-1** and **a-2** illustrate two degenerate order parameters which we can call of "Mercedes-Benz-logo" type. When all the interband interactions are positive, there appears an order parameter of Mercedes-Benz-logo type. When one of the interband interactions is negative, there appears an order parameter like those illustrated in (**b-1**) and (**b-2**), that we can call of "dinosaur footprint" type [102]. (© Y. Tanaka)

Tsuei and Kirtley had raised the alarm in their comprehensive review on the order parameter realization in cuprate superconductors [24]. (They wrote "they further proved that time-reversal symmetry can be broken only when the representation is multidimensional," referring to the work of Yip and Garg [26].) They also noted that the time-reversal symmetry breaking can occur through the mixing of order parameters with different symmetries [25, 144]. (Within the Ginzburg–Landau formalism, this chiral instability is caused by biquadratic terms $\eta|\varphi_1|^2|\varphi_2|^2\cos(2(\theta_1-\theta_2))$, where $\eta$ is a constant and $\theta_1-\theta_2=\pi/2$ and $\theta_1-\theta_2=-\pi/2$ sometimes give a minimum value. They correspond to the time-reversal symmetry-broken state, $|\varphi_1|\pm i|\varphi_2|$. Quadratic terms—the Josephson interaction terms, $\gamma|\varphi_1||\varphi_2|\cos(\theta_1-\theta_2)$, where $\gamma$ is a constant and $\theta_1-\theta_2=0$ or $\theta_1-\theta_2=\pi$ gives a minimum value—are treated to inhibit the emergence of the time-reversal symmetry-broken state [145]. It prefers to a time-reversal symmetrical state either $|\varphi_1|+|\varphi_2|$ or $|\varphi_1|-|\varphi_2|$. Hence, the emergence of the time-reversal symmetry breaking had so far been connected with the presence of either degenerate components or mixing of order parameters with different symmetries such as *d*- and *s*-waves [25, 144, 146]. The idea that the mixing of order parameters with different symmetries could induce a time-reversal symmetry breaking at the interface was introduced by Kuboki and Sigrist in a preprint [147]. During the refereeing process (which ended with the rejection of the article) and later, the concept spread and many researchers developed the physics of time-reversal symmetry breaking around the interface [147–154]. According to such formalism, a local symmetry breaking produced a subdominant order parameter (e.g., an *s*-wave order parameter in the case of *d*-wave superconductivity) that triggered the time-symmetry breaking. The roles of frustration at the interface and mixing of order parameters with different symmetries had not been well separated. As a result, we might think that the mixing of order parameters can solely cause the time-reversal symmetry breaking, without the need for frustration. On the other hand, it is not clear whether frustration can solely break the symmetry. In any case, it might have been believed that the mixing of order parameters and not frustration was the origin of the time-reversal symmetry breaking. Discussions of the time-reversal symmetry in the bulk are purely theoretical ones since there is not any established experimental work on the subject.

In 2001, the idea of time-reversal symmetry breaking in multiband superconductors was discussed in [4]. As there were neither degenerate components nor mixing of order parameters with different symmetries, it became clear that frustration suffices to produce the time-reversal symmetry breaking. This discussion referred mainly to the interface, but was later extended to a bulk phase in which layers with a positive and negative interband are alternately stacked. Ng and Nagaosa revisited the discussion about the time-reversal symmetry breaking at the interface of a two-band superconductor in 2009 [155]. Extensive discussion on the time-reversal symmetry breaking in the bulk state has clearly separated the role of frustration from that of the mixing of order parameters with different symmetries [88, 102, 156–159]. Frustration can solely produce the time-reversal symmetry breaking, and degenerate components and various order parameters with different symmetries are not necessary. However, Yip and Garg's principle inhibits the time-reversal symmetry breaking at the superconducting transition in single-component superconductivity, and in the context of the mean field theory described by the Ginzburg–Landau formalism, time-reversal symmetry breaking below the superconducting transition temperature is not discussed at all. The effects of the fluctuation are also beyond their principle. Stanev and Tesanovic discussed the phase transition to the time-reversal symmetry-broken state below the superconducting transition temperature [88], and we will later comment on their results.

Frustration in multiband superconductors can be apparent when we explicitly write the interband Josephson interaction in the Ginsburg–Landau formalism [102, 157].

$$f_{GL} = \alpha_1|\varphi_1|^2 + \frac{1}{2}\beta_1|\varphi_1|^4 + \alpha_2|\varphi_2|^2 + \frac{1}{2}\beta_2|\varphi_2|^4 + \alpha_3|\varphi_3|^2 + \frac{1}{2}\beta_3|\varphi_3|^4 \\ + \gamma_{12}|\varphi_1||\varphi_2|cos(\theta_1 - \theta_2) + \gamma_{23}|\varphi_2||\varphi_3|cos(\theta_2 - \theta_3) + \gamma_{31}|\varphi_3||\varphi_1|cos(\theta_3 - \theta_1). \tag{7.6}$$

$\alpha_i$, $\beta_i$, and $\gamma_{ij}$ are constants. At a superconducting transition temperature, $|\varphi_i| \to 0$ and $|\varphi_i|^4$ can be eliminated. The free energy, $f_{GL}$, becomes

$$f_{GL} = \alpha_1' |\varphi_1' + \varphi_2' + \varphi_3'|^2. \tag{7.7}$$

$\varphi_2'/\varphi_1'$ and $\varphi_3'/\varphi_1'$ are determined by the BCS gap equation, and they are real numbers. $\alpha_1'$ is a constant. The BCS gap equation gives other two solutions (order parameters) resulting in higher energy. These solutions are dropped from $f_{GL}$ at the transition temperature. The information on the interband phase difference is eliminated, and the chiral state (i.e., the time-reversal symmetry-broken state) is never observed. The state observed constitutes a generalization of Yip and Garg's principle. We have to reinstate another order parameter with the same symmetry as the present order parameter (so, after reinstatement, there are two (or three) order parameters with the same symmetry) to discuss the chiral state when there is a finite

$|\varphi_i|$ and $|\varphi_i|^4$ cannot be dropped. By leaving the interband Josephson interaction in the formalism [4, 5], rather than reinstating another order parameter, the effect, role, and consequences of frustration become clear [84, 96, 160–168].

An explicit description of the interband Josephson interaction shows that such interaction does not always lock the interband phase difference. The massless mode can appear even in the presence of an interband interaction. This was initially discussed for the four-band case in [169] and later generalized to other cases in [170] by Hase and Yanagisawa. (In the first report, a massive mode for a five-band superconductor was addressed. According to their study, this case also yields a massless mode.) The presence of a massless mode (which is illustrated in Fig. 7.6) suggests the importance of an interband phase difference fluctuation in frustrated

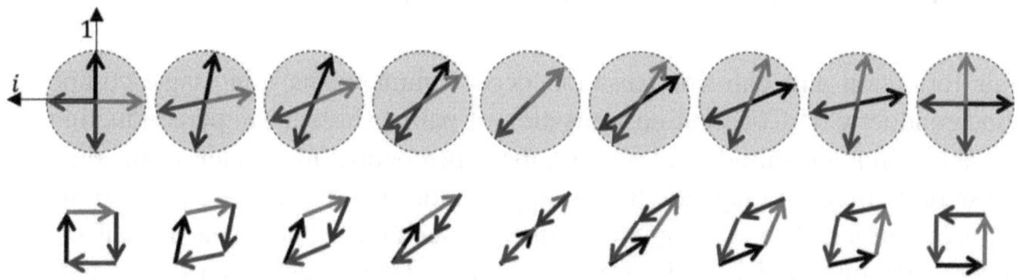


**Fig. 7.6** Massless mode is present even when there is a finite Josephson coupling. When the pair densities of all components are equal and the Josephson coupling constant between any two bands is the same, all the phase configurations illustrated in the upper panel give the same Josephson coupling energy. In the upper panel, a vector designates an order parameter for each component [169]. If we can make a square by connecting these vectors (*left end* of the lower panel), then the Josephson energy is minimum. We can deform that square without increasing the Josephson coupling. This representation is introduced by Yanagisawa and Hase [170]. According to Yanagisawa and Hase, this principle works for any polygons when we increase number of components. (© Y. Tanaka)

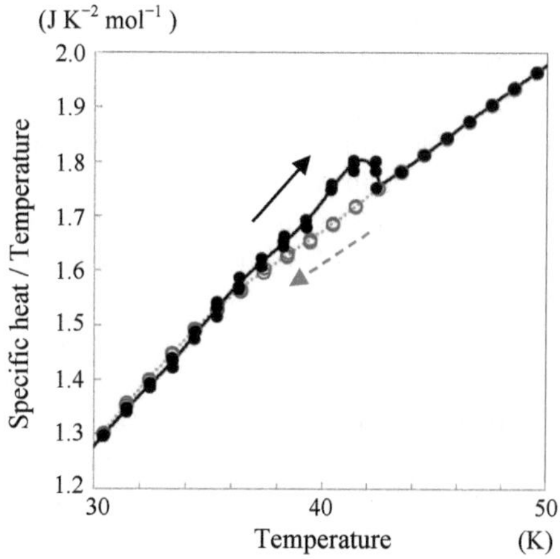


**Fig. 7.7** Additional transition around 40 K (below the “conventional superconducting transition” that occurs at 108 K) in the multilayer cuprate superconductor $HgBa_2Ca_4Cu_5O_y$. We notice that it is a first-order phase transition, as there is a large hysteresis. (Reprinted with permission from [85]. Copyright 2014 by Physical Society of Japan.)

multiband superconductors. The appearance of a massless Leggett mode at the chiral phase transition (which, in the case of multiband superconductors, may be connected with the onset of the time-reversal symmetry-broken state) has been also extensively investigated [89, 90, 94, 95, 171]. As we will discuss in the next section, the presence of the massless mode may change the picture of the superconducting transition [86, 87, 96, 97]. Even though these considerations apply to a superconducting state without an external current, we mention that a massless mode tends to be produced by external currents or thermal fluctuations [91, 136]. Finally, frustration is also attractive due to its potential applications in the fabrication of the memory of a superconducting computer [172].

## 7.5 Entropy and Another Superconducting Transition

Pair formation and pair condensation occur simultaneously in single-component superconductivity. Coherent condensation of pairs causes the superconducting gap to have a finite magnitude and, in turn, the superconducting gap aligns the quantum phase of pairs. If no magnetic field is applied, the superconducting phase transition is second-order. A one-component superconductor is either in a normal or in a superconducting, and it is the free energy that determines which one is realized. Hence, the standard lore among the superconducting community is that the superconducting transition is a second-order phase transition between two states [2]. A first-order phase transition is rare for a quantum condensate described by the BCS formalism. Experimentally, there are only two examples without an applied magnetic field. One is the AB transition between a state having a nodeless gap to one having a nodal gap of the superfluid helium-3 [38, 173–176], and the other is an "extra phase transition" experienced by the multilayer cuprate superconductor $HgBa_2Ca_4Cu_5O_y$ [85].

A theoretical study of multicomponent quantum condensates changes drastically the conventional picture of the superconducting phase transition. There are many different candidates for the wave function of the superconducting condensate. When one wave function (order parameter) is selected among several candidates at the superconducting transition temperature, the others can be metastable states. If we decrease the temperature, it may happen that one of those metastable states has lower free energy than the normal state. Moreover, if we continue decreasing the temperature, its energy may be further suppressed and become lower than the selected condensate just below the superconducting transition temperature. This means that such a candidate is waiting for a chance to appear when the temperature is reduced. The extra superconducting phase transition observed in $HgBa_2Ca_4Cu_5O_y$ that we mentioned above constitutes a typical example. As seen in Fig. 7.7, this superconductor has a first-order phase transition around 40 K, while the superconducting transition temperature is above 100 K. The phase transition that occurs around 40 K is between two different superconducting states, whereas the phase transition that takes place above 100 K is between the normal and the superconducting state.

The extra phase transition can be interpreted as follows. This superconductor has five $CuO_2$ planes in one unit cell, as seen in the crystal structure depicted in Fig. 7.1c [177]. There are several possible superconducting wave functions originating from these multiple $CuO_2$ planes; we may construct them as linear combinations of the superconducting components on each plane, that is, $\Psi_{+++++} = \varphi_1 + \varphi_2 + \varphi_3 + \varphi_4 + \varphi_5$, $\Psi_{+-+-+} = \varphi_1 - \varphi_2 + \varphi_3 - \varphi_4 + \varphi_5$, and so on, where $\varphi_j = |\varphi_j| exp(i\theta_j)$ is the superconducting component on the $j$-th plane. It can possible to change between these different order parameters, which are separated by a potential barrier. Once one state is selected, it can survive after the free energy of another state becomes lower than its free energy. When some agitation or quantum effect helps the condensates to go over the potential barrier, there is a change of the order parameters. The metastable state protected by the potential barrier might be either a supercooled or superheated state, and the phase transition from it to the ground state is first order. Such a phase transition is also realized in a Josephson junction composed of two superconducting layers that sandwich a magnetic layer [120]. The wave function of this system can be denoted by $\Psi = \varphi_1 + \varphi_2$, where $\varphi_1$ and $\varphi_2$ are the superconducting wave functions on two different superconducting layers. The phase transition between $\Psi_{++} = |\varphi_1| + |\varphi_2|$ and $\Psi_{+-} = |\varphi_1| - |\varphi_2|$ was clearly observed through an experimental measurement of the critical current through this Josephson junction, where the critical current is suppressed at the transition temperature.

So far, we have considered that all candidates other than the normal state are superconducting states. We implicitly assume no other "exotic" state which is neither a superconducting nor the normal state exists. However, it is not sure that there are only the normal and superconducting states. We encounter this question in the context of frustrated multicomponent superconductivity based on multiband superconductors with nondegenerate components, as discussed in the previous section. We mentioned that the BCS gap equation at the superconducting transition temperature requires the intercomponent phase difference to be either 0 or $\pi$ rad, which can be symbolically denoted by $\Psi = |\varphi_1| \pm |\varphi_2| \pm |\varphi_3|$ when the superconducting wave function does not have any degenerate component and mean field theory can be applied. When the chiral state denoted by $\Psi = |\varphi_1|e^{i\theta_1} + |\varphi_2|e^{i\theta_2} + |\varphi_3|e^{i\theta_3}$ appears at low temperatue, it might suggest there is another phase transition between $\Psi = |\varphi_1| \pm |\varphi_2| \pm |\varphi_3|$ and $\Psi = |\varphi_1|e^{i\theta_1} + |\varphi_2|e^{i\theta_2} + |\varphi_3|e^{i\theta_3}$ at a lower temperature than the superconducting transition between the normal state and $\Psi = |\varphi_1| \pm |\varphi_2| \pm |\varphi_3|$ state. Even though this is a scenario considered by some researchers [88–90, 92, 161, 164, 165, 178, 179], such a phase transition suffers from a serious inconsistency in the entropy counting.

When $\Psi = |\varphi_1| + |\varphi_2|e^{i\theta_2} + |\varphi_3|e^{i\theta_3}$ (as illustrated in Fig. 7.5a-1 or b-1, for example) is selected at the lower transition (chiral transition) temperature, $\Psi = |\varphi_1| + |\varphi_2|e^{-i\theta_2} + |\varphi_3|e^{-i\theta_3}$ (as illustrated in Fig. 7.5a-2 or b-2) might be also a candidate. Hence, there are two chiral wave functions and these are degenerate although originally there was no degenerate superconducting component. This means that, just above the chiral transition temperature, the system has an entropy if there are two states. All the entropy of an unbounded particle is lost through pair formation within the BCS formalism. (Strictly speaking, the state occupied by the pair does not have an entropy.) The wave function of the pair is determined at the higher transition temperature, so it is not possible that the pair retains the corresponding entropy just above the chiral transition temperature. Introducing an elementary excitation other than pair breaking is inevitable to consider the chiral transition. The interband phase difference fluctuation (Leggett mode) is this elementary excitation that provides the extra entropy necessary for the chiral phase transition [89, 90]. It is natural to consider that this fluctuation is already present at the higher phase transition temperature (corresponding to the transition between the normal state and one superconducting state) and continues to be present between this temperature and the lower transition temperature (corresponding to the transition between a nonchiral and a chiral superconducting state). This scenario poses the question if a state between these two phase transitions can be called a "superconducting state." [86, 87, 96, 97].

Allowing for the interband phase fluctuation implies that the pair retains the entropy after pair formation. This scenario is not contemplated by the BCS formalism. In the BCS scheme, the gain in condensation energy generated by the opening gap compensates the cost of giving up the entropy necessary to form the pair. The gap is opened by the pair formation, in turn [180]. However, if the pair retains the entropy after its formation, the superconducting gap can open without paying the (partial or full) cost of the missing entropy [86, 87, 96, 97].

The gap still opens above the transition temperature specified by a mean field BCS theory [86]. The phase of the pair is aligned in a band and helps to open the gap. However, the interband phase difference is unsettled and this is what causes the entropy. We can call it a "partial coherent state" or "partially correlated system." Understanding this state is beyond the scope of conventional quantum condensation and, of course, also beyond the scope of Landau's Fermi liquid theory. Multicomponent superconductivity based on multiband superconductors sheds a light on the limitations of our conventional understanding of condensed matter.

The BCS formalism implicitly assumes that all the pair phases get aligned at the end. This is automatically attained when there is only negative (ordinary) pairing interaction. Then, the interband phase difference fluctuation does not have any room to grow up. Actually, if we admit pair formation without alignment of the phase of the pair, the phase is never settled [181]. This situation corresponds to disordered bosons. (One concrete example is an atomic gas above the Bose–Einstein condensate critical temperature.) This completely departs from the concept of a quantum condensate, which should be characterized by a limited number of characteristic values, such as a superconducting phase, instead of using the degrees of freedom attributed to each particle (and whose number is of the order of Avogadro's number). Our discussion on frustrated multicomponent superconductors tells us how fragile phase alignment becomes when a positive pair hopping interaction is introduced. Because *d*-wave superconductivity has already been presented [24], we may find some relationship between the abnormal properties in the normal state of this superconductor and the interband phase difference fluctuation predicted in frustrated superconductivity based on multiband superconductors [182]. Of course, there have been discussion and understanding on the schema under the crossover between the Bose-Einstein phase transition and BCS phase transition [183–186]. The pair formation does not require a coherent phase against other particle and pre-formed pairs, which are created at much higher temperature than the superconducting phase transition. This is possible when one particle hardly changes its partner. This situation is fulfilled when the coherence length (a size of the pair) is much smaller than the distance between the pairs. It is BEC limit in the conventional BCS-BEC crossover scenario. In case of the multiband superconductor, the particle frequently changes its partner after pair formation. The discussion of the unsettled interband phase difference and its curious consequence is beyond the conventional BCS-BEC crossover schema. It can be called as "momentum space" BCS-BEC crossover, where a Leggett mode condensed into the chiral state [87, 91, 187]. It might be stressed again that the contrast in the strength between interband interaction and intraband interaction is essential. Without this contrast, the frustration occurred between pairs leads only highly correlated pairs having random phases. Macroscopic quantum phase never has a chance to appear.

Reducing the degrees of freedom by coherence, but not shirking into one degree of freedom and keeping few degrees of freedoms, we can apply these few degrees of freedom to a new superconducting electronics [100, 172, 188]. From this merits for an industrial application, properties of the superconducting state having large entropy originating from the interband phase difference mode should be explored.

## 7.6 Current-Induced Interband Phase Difference

I devote the final section of this chapter to explaining a new direction in the experimental research of multicomponent superconductivity based on multiband superconductors.

In 2006, a group from Stanford succeeded in detecting the interband phase difference soliton experimentally using a mimic multiband superconductor that consisted of an ultrathin superconducting double layer [108, 113]. They observed the effect of the interband phase difference soliton in the periodic response of the current induced in the ring geometry. The same year, an AIST (National Institute of Advanced Industrial Science and Technology) group also reported an indication of the formation of an interband phase difference soliton in the multilayer cuprate superconductor $CuBa_2Ca_2Cu_3O_y$ [189], which is an intrinsic multiband superconductor [82, 133]. They detected an AC magnetization signal that could be due to the rotation of the vortex molecule formed by fractional vortices and the interband phase difference soliton [134].

Later, the Stanford group observed plenty of fractional vortices (which are the direct consequence of the intercomponent phase difference) in an ultra-underdoped cuprate superconductor, which can be considered as a mimic multiband superconductor [117, 118]. Figure 7.8 shows an instructive diagram that explains their

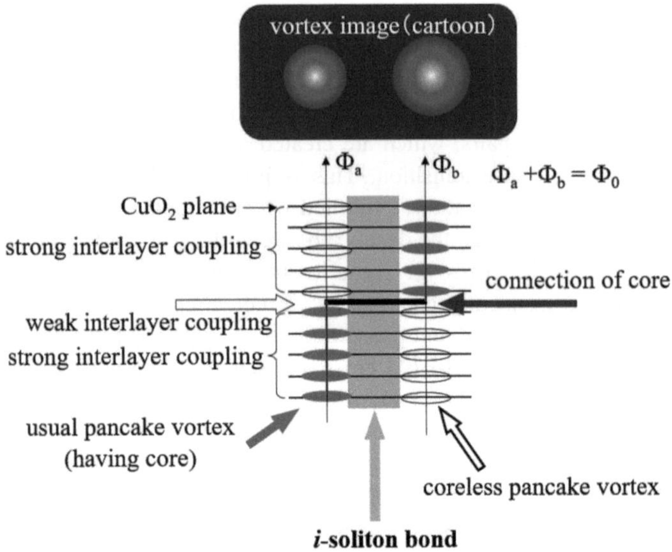


**Fig. 7.8** Schematic of a plausible scenario for fractional vortex formation in a cuprate superconductor. When there is a strong and a weak interlayer coupling, multicomponent superconductivity can emerge. Each single winding vortex carries a fractional flux quantum denoted by $\Phi_a$ or $\Phi_b$. The sum of these becomes $\Phi_0$ (Reprinted with permission from [55]. Copyright 2015 by IOP publishing)

situation. According to their interpretation, the column consisting of pancake vortices induces fractional vortices when it is cut inside the sample. One end of this cut is connected to the end of another column by a kink parallel to the $CuO_2$ planes. The kink corresponds to the core of a Josephson vortex. They interpreted that this Josephson vortex is originating from the pinning of the pancake vortex. From a multicomponent point of view, this Josephson vortex corresponds to the intercomponent phase difference soliton, and the column corresponds to the fractional vortices. A Josephson vortex cannot carry a full unit magnetic flux quantum because the thickness of the sample is comparable to or smaller than the magnetic penetration depth, and the size of the sample in the direction parallel to the $CuO_2$ planes is comparable to the effective screening length. This fact provides a mechanism for the appearance of a fractional vortex since a unit magnetic flux cannot be supported due to incomplete shielding. In the limit where the thickness of the sample becomes small and the carried magnetic flux can be neglected, the Josephson vortex becomes the pure intercomponent phase difference soliton. The Stanford experiment suggests that this property of the soliton (the fact that it can produce fractional vortices) can be observed even when it carries some amount of magnetic flux (the complete abandonment of the magnetic flux is not always necessary). The same type of fractional vortices was also discussed by De Col, Geshkenbein, and Blatter from viewpoint of the dislocation of the pancake vortex lattice (or the loss of a pancake vortex) [77].

In 2015, a collaboration between NIMS (National Institute for Materials Science), TUS (Tokyo University of Science), and AIST found an indication of the interband phase difference soliton in the quantum oscillations of a Little-Parks circuit using a $(Tl,Cu)(Ba,Sr)_2Ca_2Cu_3O_y$ film. Their observations are similar to those by the Stanford group mentioned above [190] and will be described below.

The experimental realization of multicomponent superconductivity based on multiband superconductors is still problematic. In particular, the emergence of the interband phase difference soliton has not been well controlled yet. There are many theoretical proposals to detect the soliton experimentally [95, 98, 109–112, 135, 191] that are being implemented gradually. We add some comments about these experimental procedures.

The interband phase difference soliton can present solely [82, 83, 192] but, unfortunately, it is electromagnetically inactive. As a result, there has not been any proposal yet to detect the interband phase difference soliton directly. However, the soliton produces fractional flux quanta. Moreover, a fractional vortex is always accompanied by a soliton. Thus, observing fractional magnetic flux quanta constitutes a definite proof of the presence of an interband phase difference soliton. This approach was proposed by the author along with the prediction of the interband phase difference soliton in [4]. After that work, many modifications have been suggested [67, 71, 74, 77–79, 130, 193, 194]. Figure 7.9 shows the basic ideas behind these proposals by means of a cartoon.

Gurevich and Vinokur [98] suggested at a very early stage how to generate interband phase difference solitons. According to their proposal, an injected nonequilibrium current should be converted into an interband phase difference

**Fig. 7.9** Cartoons illustrating a basic mechanism to create the fractional vortices. We should divide the conventional vortex having the unit magnetic flux quantum, $\Phi_0$ into fractional vortices having fractional magnetic flux quanta $\Phi_a$ and $\Phi_b$, where $\Phi_0 = \Phi_a + \Phi_b$. (© Y. Tanaka)

soliton. Later the same authors proposed another method [135] with a very similar experimental implementation but based on a completely different physical concept. In their second proposal, they argue that the state having the interband phase difference appears when the applied current density reaches the "intrinsic critical value" for a "weaker band," that is, a band with longer coherence length than the others. The experimental realization of these two different ideas may actually induce some confusion because the role of the applied current is substantially different in each of them. When comparing both methods, Gurevich and Vinokur did not discuss their merits, demerits, or feasibility; they only said that the new method was based on the equilibrium current, as opposed to their first proposal, which made use of the nonequilibrium current. Hence, their motivation to propose a new method was not clear. From an experimental point of view, there is no difference, and experimentalists would probably have a hard time distinguishing if the current is converted into an equilibrium current or works as the equilibrium current. Moreover, other researchers, Fenchenko and Yerin, presented a newer theoretical work adopting a similar scheme. Their conclusions seem to be completely different from those of Gurevich and Vinokur: According to Gurevich and Vinokur, the interband phase difference should occur under the external applied current. However, Fenchenko and Yerin used newer and more realistic calculations to conclude that no interband phase different occurs in that situation [195].

This theoretical confusion is a paradigmatic example and the reason why experimental work on the interband phase difference dynamics remains stuck.

I revisited the question if the interband phase difference is induced by an external current in 2015 [136]. When there are two condensates with different coherence length, the interband phase difference is inevitably induced along a superconducting line modeled by a one-dimensional line terminated by the natural boundary condition (we admit a finite interband phase difference at an end of the line). The mechanism is schematically drawn in Fig. 7.10. A larger current can flow if both components survive as compared with the case where only one component remains. The pair density decreases with increasing current. (Particles never disappear in the classical dynamic system; however, the pair is broken and disappears in the superconductivity.) If two components align the phase (the pairs have equal

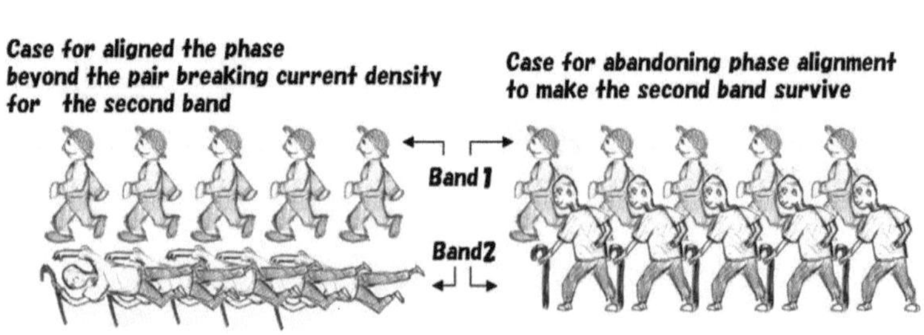


**Fig. 7.10** Cartoons illustrating the origin of the interband phase difference under an externally applied current. The current is larger when the second band survives and contributes to the current, even though there is an interband phase difference. (© Y. Tanaka)

momentum for all bands), which means there is no interband phase difference, the pair density in the weaker band decreases much faster than that in the stronger band. (Here, "stronger band" means that the condensation energy contributed by the intraband pair interaction from this band is much larger than that from another band. In other words, the coherence length of the stronger band is shorter than that of the weaker band.) Then the stronger band solely supports the current. If we admit the interband phase difference, the weaker band can also contribute to support the current. Then, we expect larger current density when we admit the interband phase difference. This is a basic mechanism for the generation of the interband phase difference. A rigorous analysis of the one-dimensional circuit tells us that the interband phase difference inevitably appears in two-component superconductivity based on two-band superconductors. There is no threshold on the external applied current. The finite current is surely accompanied by a finite interband phase difference. This result is slightly different from the Gurevich and Vinokur scenario, in which the interband phase difference appears at some threshold current.

The interband Josephson interaction cannot inhibit the emergence of the interband phase difference in a thin line, though it suppresses the extent of the interband phase difference. This finding sounds curious since there does not seem to be any interband phase difference in the conventional superconducting circuits, as shown by Fenchenko and Yerin's calculation. However, the superconducting line in the circuit is usually terminated by a large bank at the current lead; however, the role of the large banks in the calculation is not well understood in the calculation. This large bank becomes a boundary condition (Dirichlet boundary condition (fixed boundary condition)). The interband phase difference is locked by this large bank and becomes 0 rad. Fenchenko and Yerin might not see any interband phase difference generation because they attached the large superconducting banks to both end of the superconducting line. They also assume the superconducting coherence can be kept above the pair breaking current as seen in the standard textbook describing the properties of the standard Ginzburg–Laudau theory, which is valid near superconducting transition temperature [196]. However, Gurevich and Vinokur considered the superconducting coherence should be broken above the pair breaking current as mentioned in the standard textbook, of which conclusion derived by the BCS formalism and might be valid at lower temperature [197]. Though there is some minor difference on assumptions between these two groups, I consider the essential origin bearing the different conclusion, which one group concluded the emergence of the interband phase difference is possible and other group concluded the interband phase difference does not appear is the difference in the boundary condition.

The large superconducting bank keeps the state in the connected superconducting line so as not to invoke a finite interband phase difference under the externally applied current, though this state is metastable state. This means this bank can also trap an interband phase difference in the line under no current. The interband phase difference of the multiples of $2\pi$ rad between one end and another end matches the Dirichlet boundary condition. It can be trapped. The interband phase difference of the rotation of $2\pi$ rad becomes the interband phase difference

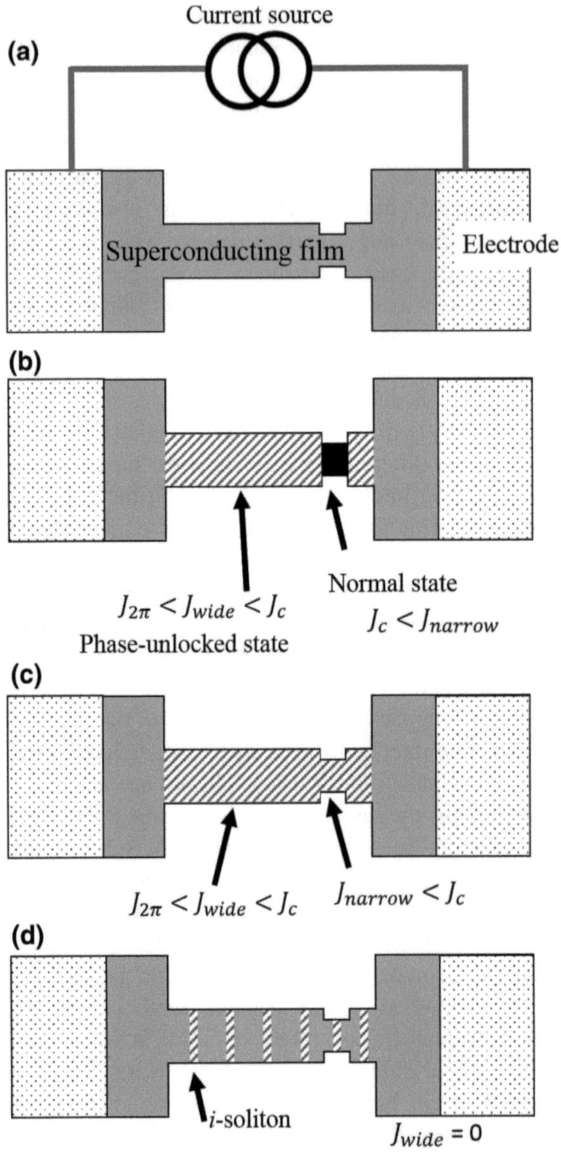


soliton under zero current. Then, we can generate the interband phase difference soliton controlling the boundary condition. On the contrary, the interband phase difference texture discussed by Gurevich and Vinokur always requires the continuously applied current in both of method they proposed [135]. This continuous applied current tends to introduce the difficulty and ambiguity on the experimental measurement. The applied current induced the extra magnetic induction, and we

**Fig. 7.11** Method for trapping an interband phase difference soliton (*i*-soliton) in a circuit by controlling the boundary conditions. **a** Circuit design. The *light-gray shaded* area indicates the superconducting film. The *dotted hatch* indicates the electrode on the superconducting film. A superconducting circuit is composed of the *wide-line* region and *narrow-line* region. We can break the superconducting state at the narrow-line region while maintaining superconductivity at the wide-line region. The end of the wide-line region attached to the normal state becomes the natural boundary condition where the interband phase difference is not locked. When the current is high enough to generate a rotation of $2\pi$ rotations, this rotation enters into the circuit as shown in (**b**). The narrow-line region can return to the superconducting state by decreasing the current. When the current is still large enough to maintain of $2\pi$ rotations at this current, the rotations are trapped by the two Dirichlet boundary conditions of the two large superconducting banks that terminate the superconducting line (**c**). After switching off the external current, the rotation of the $2\pi$ radians becomes the *i*-soliton (**d**). The *i*-soliton is trapped inside the superconducting line. In the figures, $J_c$ denotes the critical current density and $J_{2\pi}$ denotes the current density by which a rotation of $2\pi$ rotations of the interband phase difference is generated. $J_{wide}$ and $J_{narrow}$ are current densities at the wide-line region and the narrow-line region, respectively. In (**b**)–(**d**), we omit the external current source. (Reprinted with permission from [136]. Copyright 2015 by Elsevier.)

should distinguish this magnetic induction and the magnetic induction due to the fractional magnetic flux quanta, for example. If the fractional magnetic flux quanta can be kept under the silent condition without the externally applied current, measurement, analysis, and interpretation become more definite and certain.

Controlling of the boundary condition becomes a key issue on this measurement. This technique is schematically described in Fig. 7.11. When we make a weak location in the line, we can simply switch on the emergence of the superconductivity by the current (of course, we can design other circuit in which we break the superconductivity by other method, for example, a local heating, local irradiation, and so on). The current higher than a critical current at the weak location converts this location to the normal state. Then, the natural boundary condition is achieved (the current is conserved through this location, but the interband phase difference is not locked). When the state having more than 2 $\pi$ interband phase difference rotation is more stable than the interband phase difference locked state at this current density, interband phase difference rotation goes into the line. After feeding the interband phase difference texture, we decrease the current density and recover the superconductivity at the weak location. The Dirichlet boundary condition is achieved. If the state in which an interband phase difference rotation is more than $2\pi$ rad is still a ground state with this decreased current, the interband phase texture is trapped in the line. Finally, it becomes interband phase difference soliton after switching off the externally applied current.

If we can put the interband phase soliton between sample edge and the rim of the hole inside the line as shown in Fig. 7.12, we can generate the fractional flux quanta inside this hole. These are one of the ideal scenarios to trap the fractional flux quanta.

On the way to realize this scenario, we measured the Little-Parks oscillation on the multilayer cuprate $(Tl,Cu)(Sr,Ba)_2Ca_2Cu_3O_y$ film [190]. The narrow line having hole was prepared, and the current-voltage response of this line is measured. Because we measure the voltage, there are some locations converted into the normal

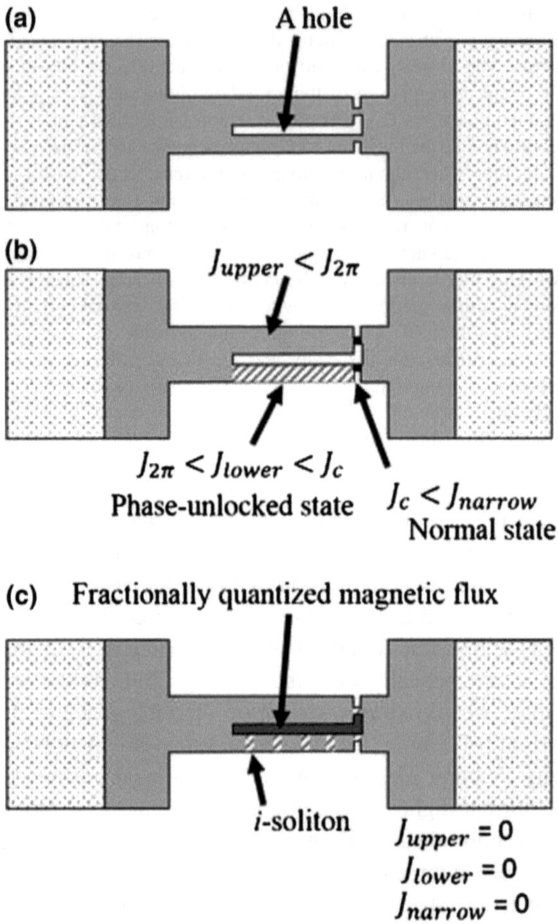


**Fig. 7.12** Circuit to detect the *i*-soliton. We add a hole to the circuit shown in Fig. 7.11. The *upper part* above the hole is wider than that of the part below the hole in the wide-line region (**a**). There are two bridges at narrow region. The current density in the wide-line region above the hole is higher than that in the wide-line region below the hole when two bridges at narrow region give the same electrical resistance. When the current density in the wide-line region below the hole is large enough to generate a rotation of $2\pi$ radians, but that of the wide-line region above the hole is not sufficient to do so, the rotation of $2\pi$ radians only enters into the wide-line region below the hole (**b**). After switching off the external current, the *i*-soliton is trapped only in the wide-line region below the hole. To compensate for the phase shift that accompanies the *i*-solitons, the fractionally quantized magnetic flux appears in the hole (**c**). $J_{above}$ and $J_{below}$ are current densities at the wide-line region above the hole and the wide-line region below the hole, respectively. $J_c$ and $J_{2\pi}$ are the same as the quantities described in Fig. 7.11. We omit the external current source described in Fig. 7.10a. (Reprinted with permission from [136]. Copyright 2015 by Elsevier.)

state. This location corresponds to the natural condition discussed above. We expect the DC-SQUID (superconducting quantum interference device) like situation appears around the hole. If it works, we can measure the oscillation in the voltage

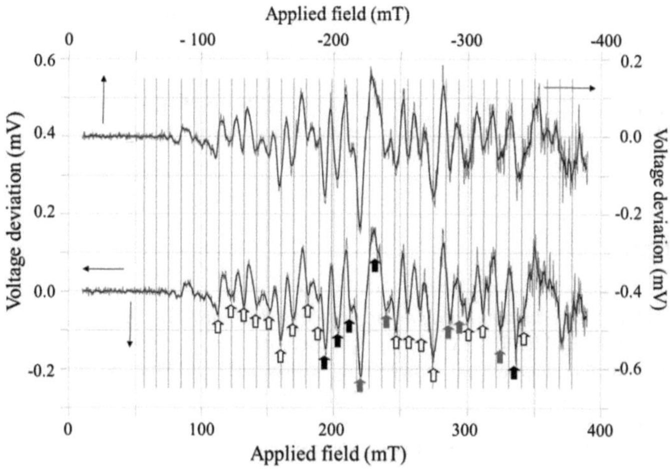


**Fig. 7.13** Oscillations in the detected signal, extracted by subtracting the base. For the top signal, the applied field increased from 0 to −400 mT and for the bottom signal, from 0 to 400 mT. The vertical lines are drawn at intervals of 9.45 mT, a value determined by trial and error to reproduce as many dip locations as possible. The dips indicated by the *unfilled arrows* are formed by the usual Little-Parks oscillations, while the dips indicated by the black (*gray*) filled arrows shift the location from the expected positions by the usual Little-Parks oscillations by 0.55 (0.33) of the period. (Reprinted with permission from [190]. Copyright 2015 by Elsevier.)

varying the externally applied field. This is a basic/conventional principle of the Little-Parks experimental technique. We succeeded to measure the quantum oscillation that may correspond to increment of the flux quanta inside the hole. After some regular oscillation keeping the equal interval with increasing the magnetic field, we found a shorten interval. After this shorten interval, the equal interval is recovered as shown in Fig. 7.13. A similar tendency had been reported as the generation of the interband phase difference soliton in the mimic multiband superconductor by Stanford group [108, 113]. We consider the same situation is also reproduced in Little-Parks circuit consisting of $(Tl,Cu)(Sr,Ba)_2Ca_2Cu_3O_y$ film where the current induces the interband phase difference. One interpretation is that the soliton generated at some locations traveled to the ring location and might be trapped in the SQUID.

## 7.7 Summary

The method and technique to give a definite proof of multicomponent superconductivity based on multiband superconductors needs still to be further explored. I expect we will be able to find out many multiband superconductors in which the multicomponent superconductivity emerges, after the complete establishment of methods for the generation and detection of interband phase difference solitons.

These methods should be based on the current-induced interband phase difference and the detection of the fractional flux quantization [55].

**Acknowledgements** This work was partially supported by JSPS KAKENHI Grant Number JP 16K06275. We would like to thank Editage (www.editage.jp) for English language editing.

## References

1. M. Tinkham, *Introduction to Superconductivirty* (McGraw-Hill, Inc. New York, 1996)
2. J.F. Annett, N. Goldenfeld, A.J. Leggett, Experimental constrains on the paring state of the cuprate superconductors: an emergening consensus, in *Physical Properties of High Temperature Superconductors*, ed. by D.M. Ginsberg (World Scientific, Singapore, 1996), pp 375–461
3. A.J. Leggett, Number-phase fluctuations in two-band superconductors. Prog. Theor. Phys. 36, 901–930 (1966)
4. Y. Tanaka, Phase instability in multi-band superconductors. J. Phys. Soc. Jpn. **70**, 2844–2847 (2001)
5. Y. Tanaka, Soliton in two-band superconductor. Phys. Rev. Lett. **88**, 017002 (2001)
6. T. Yanagisawa, Quarks and fractionally quantized vortices in superconductors: an analogy between two worlds, in *Recent Advances in Quarks Research*, ed. by H Fujikage, K Hyobanshi (Nova Science, New York, 2013), pp 113–46
7. A.J. Leggett, Interpretation of recent results on $He^3$ below 3 mK: a new liquid phase? Phys. Rev. Lett. **29**, 1227–1230 (1972)
8. D.D. Osheroff, R.C. Richardson, D.M. Lee, Evidence for a new phase of solid $He^3$. Phys. Rev. Lett. **28**, 885–888 (1972)
9. A.J. Leggett, A theoretical description of the new phases of liquid $^3He$. Rev. Mod. Phys. **47**, 331–414 (1975)
10. A.J. Leggett, Nobel lecture: superfluid $He^3$: the early days as seen by a theorist. Rev. Mod. Phys. **76**, 999–1011 (2004)
11. J. Bardeen, L.N. Cooper, J.R. Schrieffer, Microscopic theory of superconductivity. Phys. Rev. **106**, 162–164 (1957)
12. J. Bardeen, L.N. Cooper, J.R. Schrieffer, Theory of superconductivity. Phys. Rev. **108**, 1175–1204 (1957)
13. P.W. Anderson, P. Morel, Generalized Bardeen-Cooper-Schrieffer states and aligned orbital angular momentum in the proposed low-temperature phase of liquid $He^3$. Phys. Rev. Lett. **5**, 136–138 (1960)
14. P.W. Anderson, P. Morel, Generalized Bardeen-Cooper-Schrieffer states and the proposed low-temperature phase of liquid $He^3$. Phys. Rev **123**, 1911–1934 (1961)
15. R. Balian, N.R. Werthamer, Superconductivity with pairs in a relative p wave. Phys. Rev. **131**, 1553–1564 (1963)
16. F. Steglich, J. Aarts, C.D. Bredl, W. Lieke, D. Meschede, W. Franz, H. Schäfer, Superconductivity in the presence of strong Pauli paramagnetism: $CeCu_2Si_2$. Phys. Rev. Lett. **43**, 1892–1896 (1979)
17. H.R. Ott, H. Rudigier, Z. Fisk, J.L. Smith, $UBe_{13}$: an unconventional actinide superconductor. Phys. Rev. Lett. **50**, 1595–1598 (1983)
18. H.R. Ott, H. Rudigier, T.M. Rice, K. Ueda, Z. Fisk, J.L. Smith, *p*-wave superconductivity in $UBe_{13}$. Phys. Rev. Lett. **52**, 1915–1918 (1984)
19. R. Joynt, L. Taillefer, The superconducting phases of $UPt_3$. Rev. Mod. Phys **74**, 235–294 (2002)

20. M. Sigrist, K. Ueda, Phenomenological theory of unconventional superconductivity. Rev. Mod. Phys. **63**, 239–311 (1991)
21. A.P. Mackenzie, Y. Maeno, The superconductivity of $Sr_2RuO_4$ and the physics of spin-triplet pairing. Rev. Mod. Phys. **75**, 657–712 (2003)
22. L.-F. Zhang, V.F. Becerra, L. Covaci, M.V. Milošević, Electronic properties of emergent topological defects in chiral $p$-wave superconductivity. Phys. Rev. B **94**, 24520 (2016)
23. D.F. Agterberg, V. Barzykin, L.P. Gor'kov, Conventional mechanisms for exotic superconductivity. Phys. Rev. B **60**, 14868–14871 (1999)
24. C.C. Tsuei, J.R. Kirtley, Pairing symmetry in cuprate superconductors. Rev. Mod. Phys. **72**, 969–1016 (2000)
25. Q.P. Li, B.E.C. Koltenbah, R. Joynt, Mixed $s$-wave and $i$ $d$-wave superconductivity in high-$T_c$ systems. Phys. Rev. B **48**, 437–455 (1993)
26. S. Yip, A. Garg, Superconducting states of reduced symmetry: general order parameters and physical implications. Phys. Rev. B **48**, 3304–3308 (1993)
27. C.J. Myatt, E.A. Burt, R.W. Ghrist, E.A. Cornell, C.E. Wieman, Production of two overlapping Bose-Einstein condensates by sympathetic cooling. Phys. Rev. Lett. **78**, 586–589 (1997)
28. D.S. Hall, M.R. Matthews, J.R. Ensher, C.E. Wieman, E.A. Cornell, Dynamics of component separation in a binary mixture of Bose-Einstein condensates. Phys, Rev. Lett **81**, 1539–1542 (1998)
29. M.R. Matthews, B.P. Anderson, P.C. Haljan, D.S. Hall, C.E. Wieman, E.A. Cornell, Vortices in a Bose-Einstein condensate. Phys. Rev. Lett. **83**, 2498–2501 (1999)
30. K. Doi, Y. Natsume, Calculation of Bose-Einstein condensations and characteristic features of fluctuations for systems with and without a vortex in two-component alkali atom gases. J. Phys. Soc. Jpn. **70**, 167–172 (2001)
31. K. Kasamatsu, M. Tsubota, M. Ueda, Vortex phase diagram in rotating two-component Bose-Einstein condensates. Phys. Rev. Lett. **91**, 150406 (2003)
32. M. Cipriani, M. Nitta, Crossover between integer and fractional vortex lattices in coherently coupled two-component Bose-Einstein condensates. Phys. Rev. Lett. **111**, 170401 (2013)
33. D.T. Son, M.A. Stephanov, Domain walls of relative phase in two-component Bose-Einstein condensates. Phys. Rev. A **65**, 63621 (2002)
34. P. Öhberg, L. Santos, Dark solitons in a two-component Bose-Einstein condensate. Phys. Rev. Lett. **86,** 2918–2921 (2001)
35. B.D. Esry, C.H. Greene, J.J.P. Burke, J.L. Bohn, Hartree-Fock theory for double condensates. Phys. Rev. Lett. **78**, 3594–3597 (1997)
36. C.K. Law, H. Pu, N.P. Bigelow, J.H. Eberly, "Stability signature" in two-species dilute Bose-Einstein condensates. Phys. Rev. Lett. **79**, 3105–3108 (1997)
37. H. Pu, N.P. Bigelow, Properties of two-species Bose condensates. Phys. Rev. Lett. **80**, 1130–1133 (1998)
38. D.R. Tilley, J. Tilley, Superfluidity and superconductivity 3rd edn. (Adams Hilger, Bristol and New York, 1990)
39. O.V. Lounasmaa, E. Thuneberg, Vortices in rotating superfluid $^3$He. Proc. Natl. Acad. Sci. U.S.A **96**, 7760–7767 (1990)
40. M.M. Salomaa, G.E. Volovik, Quantized vortices in superfluid $^3$He. Rev. Mod. Phys. **59**, 533–613 (1987)
41. T.A. Tokuyasu, D. Hess, J.A. Sauls, Vortex states in an unconventional superconductor and the mixed phases of $UPt_3$. Phys. Rev. B **41**, 8891–8903 (1990)
42. J.A. Sauls, M. Eschrig, Vortices in chiral, spin-triplet superconductors and superfluids. New J. Phys. **11**, 75008 (2009)
43. C.N. Yang, R.L. Mills, Conservation of isotopic spin and isotopic gauge invariance. Phys. Rev. **96**, 191–195 (1954)
44. S. Weinberg, A model of leptons. Phys. Rev. Lett. **19**, 1264–1266 (1967)
45. M. Gell-Mann, A schematic model of baryons and mesons. Phys. Lett. **8**, 214–215 (1964)

46. M. Eto, Y. Hirono, M. Nitta, S. Yasui, Vortices and other topological solitons in dense quark matter. Prog. Theor. Exp. Phys. **2014**, 012D01 (2014)
47. S. Sasaki, H. Suganuma, H. Toki, Dual Ginzburg-Landau theory with QCD-monopoles for dynamical chiral-symmetry breaking. Prog. Theor. Phys. **94**, 373–384 (1995)
48. Y. Koma, H. Toki, Weyl invariant formulation of the flux-tube solution in the dual Ginzburg-Landau theory. Phys. Rev. D **62**, 54027 (2000)
49. Y. Koma, M. Koma, D. Ebert, H. Toki, Effective string action for the U(1) × U(1) dual Ginzburg–Landau theory beyond the London limit. Nuclear Physics B **648**, 189–202 (2003)
50. M. Nitta, Fractional instantons and bions in the O(N) model with twisted boundary conditions. JHEP **2015**, 1–38 (2015)
51. J. Ashcroft, M. Eto, M. Haberichter, M. Nitta, M.B. Paranjape, Head butting sheep: kink collisions in the presence of false vacua. J. Phys. A: Math. Theory **49**, 365203 (2016)
52. T.W.B. Kibble, Topology of cosmic domains and strings. J. Phys. A: Math. Gen. **9**, 1387–1398 (1976)
53. A. Dey, S. Mahapatra, T. Sarkar, Very general holographic superconductors and entanglement thermodynamics. JHEP **12**, 1–32 (2014)
54. G.L. Giordano, A.R. Lugo, Holographic phase transitions from higgsed, non abelian charged black holes. JHEP **7**, 1–29 (2015)
55. Y. Tanaka, Multicomponent superconductivity based on multiband superconductors. Supercond. Sci. Technol. **28**, 034002 (2015)
56. H. Ihara, K. Tokiwa, H. Ozawa, M. Hirabayashi, H. Matuhata, A. Negishi, Y.S. Song, New high-$T_c$ superconductor $Ag_{1-x}Cu_xBa_2Ca_{n-1}Cu_nO_{2n+3-\delta}$ family with $T_c > 117$ K. Jpn. J. Appl. Phys **33**, L300–L303 (1994)
57. H. Ihara, K. Tokiwa, H. Ozawa, M. Hirabayashi, A. Negishi, H. Matuhata, Y.S. Song, New high-$T_c$ superconductor family of Cu-Based $Cu_{1-x}Ba_2Ca_{n-1}Cu_nO_{2n+4-\delta}$ with $T_c > 116$ K. Jpn. J. Appl. Phys **33**, L503–L506 (1994)
58. N. Hamada, H. Ihara, Electronic band structure of $CuBa_2Ca_3Cu_4O_{10+x}(x = 0, 1)$. Physica B **284**, 1073–1074 (2000)
59. N. Hamada, H. Ihara, Electronic band structure of $CuBa_2Ca_{n-1}Cu_nO_{2n+2}$ and $CuBa_2Ca_{n-1}Cu_nO_{2n+1}F$ (n = 3 − 5). Physica C **357**, 108–111 (2001)
60. Y. Tokunaga, K. Ishida, Y. Kitaoka, K. Asayama, K. Tokiwa, A. Iyo, H. Ihara, Effect of carrier distribution on superconducting characteristics of the multilayered high-$T_c$ cuprate $(Cu_{0.6}C_{0.4})Ba_2Ca_3Cu_4O_{12+y}$: $^{63}$Cu-NMR study. Phys. Rev. B **61**, 9707–9710 (2000)
61. Y. Tanaka, A. Iyo, N. Shirakawa, M. Ariyama, M. Tokumoto, S.I. Ikeda, H. Ihara, Specific heat study on $Cu_xBa_2Ca_{n-1}Cu_nO_y$. Physica C **357**, 222–225 (2001)
62. K. Momma, F. Izumi, VESTA 3 for three-dimensional visualization of crystal, volumetric and morphology data. J. Appl. Crystallogr. **44**, 1272–1276 (2011)
63. Y. Tanaka, Tl-based, Hg-based and multilayer (Cu-base and F-based) cuprate superconductors, in *Physics of Vortex State in Superconductors*, ed. by K. Kadowaki (Shokabo, Tokyo, in Japanese, 1996) in press
64. H. Suhl, B.T. Matthias, L.R. Walker, Bardeen-Cooper-Schrieffer theory of superconductivity in the case of overlapping bands. Phys. Rev. Lett. **3**, 552–554 (1959)
65. A. Vargunin, T. Örd, K. Rägo, Thermal fluctuations of order parameters in two-gap superconductors. J. Supercond. Nov. Magn. **24**, 1127–1131 (2011)
66. A. Vargunin, K. Rägo, T. Örd, Two-gap superconductivity: interband interaction in the role of an external field. Supercond. Sci. Technol. **26**, 65008 (2013)
67. J. Goryo, S. Soma, H. Matsukawa, Deconfinement of vortices with continuously variable fractions of the unit flux quanta in two-gap superconductors. EPL **80**, 17002 (2007)
68. M. Nitta, M. Eto, T. Fujimori, K. Ohashi, Baryonic bound state of vortices in multicomponent superconductors. J. Phys. Soc. Jpn. **81**, 84711 (2012)
69. S.B. Gudnason, M. Nitta, Fractional Skyrmions and their molecules. Phys. Rev. D **91**, 85040 (2015)
70. E. Babaev, L.D. Faddeev, A.J. Niemi, Hidden symmetry and knot solitons in a charged two-condensate Bose system. Phys. Rev. B **65**, 067001 (2002)

71. E. Babaev, Vortices with fractional flux in two-gap superconductors and in extended Faddeev model. Phys. Rev. Lett. **89**, 100512 (2002)
72. E. Babaev, J. Carlström, J. Garaud, M. Silaev, J.M. Speight, Type-1.5 superconductivity in multiband systems: magnetic response, broken symmetries and microscopic theory—a brief overview. Physica C **479**, 2–14 (2012)
73. E. Babaev, M. Speight, Semi-meissner state and neither type-I nor type-II superconductivity in multicomponent superconductors. Phys. Rev. B **72**, 180502 (2005)
74. P.J. Pereira, L.F. Chibotaru, V.V. Moshchalkov, Vortex matter in mesoscopic two-gap superconductor square. Phys. Rev. B **84**, 144504 (2011)
75. L.F.C.L.F. Chibotaru, V.H.D.V.H. Dao, A. Ceulemans, Thermodynamically stable noncomposite vortices in mesoscopic two-gap superconductors. EPL **78** (2007) 47001
76. L.F. Chibotaru, V.H. Dao, Stable fractional flux vortices in mesoscopic superconductors. Phys. Rev. B **81**, 20502 (2010)
77. A. De Col, V.B. Geshkenbein, G. Blatter, Dissociation of vortex stacks into fractional-flux vortices. Phys. Rev. Lett. **94**, 97001 (2005)
78. R. Geurts, M.V. Milošević, F.M. Peeters, Vortex matter in mesoscopic two-gap superconducting disks: influence of Josephson and magnetic coupling. Phys. Rev. B **81**, 214514 (2010)
79. R.M. da Silva, M.V. Milošević, D. Dominguez, F.M. Peeters, J.A. Aguiar, Distinct magnetic signatures of fractional vortex configurations in multiband superconductors. Appl. Phys. Lett. **105**, 232601
80. T. Yanagisawa, Theory of multi-band superconductivity, in *Recent Advances in Superconductivity Research* ed. by C.B. Taylor (New York: Nova Science, 2013), pp. 219–48
81. T. Yanagisawa, Chiral sine-Gordon model. EPL **113**, 41001 (2016)
82. Y. Tanaka, A. Crisan, D.D. Shivagan, A. Iyo, K. Tokiwa, T. Watanabe, Interpretation of abnormal AC loss peak based on vortex-molecule model for a multicomponent cuprate superconductor. Jpn. J. Appl. Phys **46**, 134–135 (2007)
83. Y. Tanaka, A. Crisan, Ambiguity in the statistics of single-component winding vortex in a two-band superconductor. Physica B **404**, 1033–1039 (2009)
84. T. Yanagisawa, Y. Tanaka, I. Hase, K. Yamaji, Vortices and chirality in multi-band superconductors. J. Phys. Soc. Jpn. **81**, 24712 (2012)
85. Y. Tanaka, A. Iyo, S. Itoh, K. Tokiwa, T. Nishio, T. Yanagisawa, Experimental observation of a possible first-order phase transition below the superconducting transition temperature in the multilayer cuprate superconductor $HgBa_2Ca_4Cu_5O_y$. J. Phys. Soc. Jpn. **83**, 74705 (2014)
86. Y. Tanaka, T. Yanagisawa, T. Nishio, Fluctuation-assisted gap evolution in frustrated multiband superconductors. Physica C **483**, 86–90 (2012)
87. Y. Tanaka, T. Yanagisawa, T. Nishio, Unlocking interband phase difference in multiband superconductors. Physica C **485**, 64–70 (2013)
88. V. Stanev, Z. Tešanović, Three-band superconductivity and the order parameter that breaks time-reversal symmetry. Phys. Rev. B **81**, 134522 (2010)
89. V. Stanev, Model of collective modes in three-band superconductors with repulsive interband interactions. Phys. Rev. B **85**, 174520 (2012)
90. S.-Z. Lin, X. Hu, Massless leggett mode in three-band superconductors with time-reversal-symmetry breaking. Phys. Rev. Lett. **108**, 177005 (2012)
91. T. Yanagisawa, Y. Tanaka, Fluctuation-induced Nambu-Goldstone bosons in a Higgs–Josephson model. New J. Phys. **16**, 123014 (2014)
92. V. Stanev, A.E. Koshelev, Complex state induced by impurities in multiband superconductors. Phys. Rev. B **89**, 100505 (2014)
93. M. Silaev, E. Babaev, Unusual mechanism of vortex viscosity generated by mixed normal modes in superconductors with broken time reversal symmetry. Phys. Rev. B **88**, 220504 (2013)
94. K. Kobayashi, M. Machida, Y. Ota, F. Nori, Massless collective excitations in frustrated multiband superconductors. Phys. Rev. B **88**, 224516 (2013)

95. S.-Z. Lin, Ground state, collective mode, phase soliton and vortex in multiband superconductors. J. Phys.: Cond. Matt. **26**, 493202 (2014)
96. T.A. Bojesen, A. Sudbø, Fluctuation effects in phase-frustrated multiband superconductors. J. Supercond. Nov. Magn. **28**, 3193–3204 (2015)
97. T.A. Bojesen, E. Babaev, A. Sudbø, Phase transitions and anomalous normal state in superconductors with broken time-reversal symmetry. Phys. Rev. B **89**, 104509 (2014)
98. A. Gurevich, V.M. Vinokur, Interband phase modes and nonequilibrium soliton structures in two-gap superconductors. Phys. Rev. Lett. **90**, 047004 (2003)
99. Y. Tanaka, A. Iyo, D. Shivagan. P. Shirage, K. Tokiwa, T. Watanabe, N. Terada, Method for controlling inter-component phase difference soliton and inter-component phase difference soliton circuit device. US Patent 8902018 (2014)
100. Y. Tanaka, A. Iyo, N. Terada, S. Kawabata, A. Sundaresan, T. Watanabe, K. Tokiwa, Quantum turing machine. US Patent 7,400,282 (2008)
101. Y. Tanaka, D.D. Shivagan, A. Crisan, A. Iyo, P.M. Shirage, K. Tokiwa, T. Watanabe, N. Terada, Vortex molecule, fractional flux quanta, and interband phase difference soliton in multi-band superconductivity and multi-component superconductivity. J. Phys.: Conf. Ser. **150**, 052267 (2009)
102. Y. Tanaka, T. Yanagisawa, Chiral ground state in three-band superconductors. J. Phys. Soc. Jpn. **79**, 114706 (2010)
103. J.S. Bell, R. Jackiw, A PCAC puzzle: $\pi 0 \rightarrow \gamma\gamma$ in the $\sigma$-model in the sigma model. Il Nuovo Cimento A **60**, 47–61 (1969)
104. S.L. Adler, Axial-vector vertex in spinor electrodynamics. Phys. Rev. **177**, 2426–2438 (1969)
105. Y. Nambu, Quasi-particles and gauge invariance in the theory of superconductivity. Phys. Rev. **117**, 648–663 (1960)
106. Field theories with ≪Sperconductor≫ solutions. J. Goldstone. Il Nuovo Cimento **19**, 154–164 (1961)
107. I.Y. Nambu, G. Jona-Lasinio, Dynamical model of elementary particles based on an analogy with superconductivity. Phys. Rev. **122**, 345–358 (1961)
108. H. Bluhm, Magnetic response measurements of mesoscopic superconducting and normal metal rings. Doctoral Degree Thesis, Stanford University, USA, 2013
109. S.V. Kuplevakhsky, A.N. Omelyanchouk, Y.S. Yerin, Soliton states in mesoscopic two-band-superconducting cylinders. Low Temp. Phys. **37**, 667–677 (2011)
110. K.V. Samokhin, Phase solitons and subgap excitations in two-band superconductors. Phys. Rev. B **86**, 064513 (2012)
111. V. Vakaryuk, V. Stanev, W.-C. Lee, A. Levchenko, Topological defect-phase soliton and the pairing symmetry of a two-band superconductor: role of the proximity effect. Phys. Rev. Lett. **109**, 227003 (2012)
112. S.-Z. Lin, X. Hu, Phase solitons in multi-band superconductors with and without time-reversal symmetry. New J. Phys. **14**, 063021 (2012)
113. H. Bluhm, N.C. Koshnick, M.E. Huber, K.A. Moler, Magnetic response of mesoscopic superconducting rings with two order parameters. Phys. Rev. Lett. **97**, 237002 (2006)
114. J.R. Kirtley, C.C. Tsuei, J.Z. Sun, C.C. Chi, L.S. Yu-Jahnes, A. Gupta, M. Rupp, M.B. Ketchen, Symmetry of the order parameter in the high-Tc superconductor $YBa_2Cu_3O_{7\text{-}d}$. Nature **373**, 225–228 (1995)
115. C.C. Tsuei, J.R. Kirtley, C.C. Chi, L.S. Yu-Jahnes, A. Gupta, T. Shaw, J.Z. Sun, M.B. Ketchen, Pairing symmetry and flux quantization in a tricrystal superconducting ring of $YBa_2Cu_3O_{7-\delta}$. Phys. Rev. Lett. **73**, 593–596 (1994)
116. C.C. Tsuei, J.R. Kirtley, Z.F. Ren, J.H. Wang, H. Raffy, Z.Z. Li, Pure $dx^2 - y^2$ order-parameter symmetry in the tetragonal superconductor $TI_2Ba_2CuO_{6+\delta}$. Nature **387**, 481–483 (1997)
117. J.W. Guikema, H. Bluhm, D.A. Bonn, R. Liang, W.N. Hardy, K.A. Moler, Two-dimensional vortex behavior in highly underdoped $YBa_2Cu_3O6_{+x}$ observed by scanning Hall probe microscopy. Phys. Rev. B **77**, 104515 (2008)

118. L. Luan, O.M. Auslaender, D.A. Bonn, R. Liang, W.N. Hardy, K.A. Moler, Magnetic force microscopy study of interlayer kinks in individual vortices in the underdoped cuprate superconductor $YBa_2Cu_3O_{6+x}$. Phys. Rev. B **79**, 214530 (2009)
119. H. Sickinger, A. Lipman, M. Weides, R.G. Mints, H. Kohlstedt, D. Koelle, R. Kleiner, E. Goldobin, Experimental evidence of a $\varphi$ Josephson junction. Phys. Rev. Lett. **109**, 227003 (2012)
120. V.V. Ryazanov, V.A. Oboznov, A.Y. Rusanov, A.V. Veretennikov, A.A. Golubov, J. Aarts, Coupling of two superconductors through a ferromagnet: Evidence for a $\pi$ junction. Phys. Rev. Lett. **86**, 2427–2430 (2001)
121. T.-L. Ho, V.B. Shenoy, Binary mixtures of Bose condensates of akali atoms. Phys. Rev. Lett. **77**, 3276–3279 (1996)
122. H. Pu, N.P. Bigelow, Collective excitations, metastability, and nonlinear response of a trapped two-species Bose-Einstein condensate. Phys. Rev. Lett. **80**, 1134–1137 (1998)
123. D.L. Feder, M.S. Pindzola, L.A. Collins, B.I. Schneider, C.W. Clark, Dark-soliton states of Bose-Einstein condensates in anisotropic traps. Phys. Rev. A **62**, 53606 (2000)
124. Domain wall solitons in binary mixtures of Bose-Einstein condensates. Phys. Rev. Lett. **87** (2001) 140401
125. K. Maki, T. Tsuneto, Magnetic resonance and spin waves in the A phase of superfluid $He^3$. Phys. Rev. B **11**, 2539 (1975)
126. K. Maki, H. Ebisawa, Magnetic excitations in superfluid $^3He$. J. Low Temp. Phys. **23**, 351–365 (1976)
127. K. Maki, P. Kumar, Magnetic solitons in superfluid $He^3$. Phys. Rev. B **14**, 118 (1976)
128. M. Przedborski, Planar topological defects in unconventional superconductors. Master's Degree Thesis, Brock University, Canada, 2013
129. Y.A. Izyumov, V.M. Laptev, Vortex structure in superconductors with a many-component order parameter. Phase Transitions A Multinatl. J. **20**, 95–112 (1990)
130. E. Babaev, Phase diagram of planar U (1) × U (1) superconductor: Condensation of vortices with fractional flux and a superfluid state. Nucl Phys. B **686**, 397–412 (2004)
131. O. Festin, P. Svedlindh, S.-I. Lee, Flux noise in $MgB_2$ film, in *SPIE's First International Symposium on Fluctuations and Noise*, vol. 5112 (2003), pp. 338–345
132. A. Crisan, Y. Tanaka, A. Iyo, L. Cosereanu, K. Tokiwa, T. Watanabe, Anomalous vortex melting line in the two-component superconductor $(Cu,C)Ba_2Ca_3Cu_4O_{10+\delta}$. Phys. Rev. B **74,** 184517 (2006)
133. A. Crisan, Y. Tanaka, D.D. Shivagan, A. Iyo, L. Cosereanu, K. Tokiwa, T. Watanabe, Anomalous AC susceptibility response of $(Cu,C)Ba_2Ca_2Cu_3O_y$: experimental indication of two-component vortex matter in multi-layered cuprate superconductors. Jpn. J. Appl. Phys **46**, L451–L453 (2007)
134. D.D. Shivagan, A. Crisan, P.M. Shirage, A. Sundaresan, Y. Tanaka, A. Iyo, K. Tokiwa, T. Watanabe, N. Terada, Vortex molecule and *i*-soliton studies in multilayer cuprate superconductors. J. Phys. Conf. Ser. **97**, 012212 (2008)
135. A. Gurevich, V.M. Vinokur, Phase textures induced by dc-current pair breaking in weakly coupled multilayer structures and two-gap superconductors. Phys. Rev. Lett. **97**, 137003 (2006)
136. Y. Tanaka, I. Hase, T. Yanagisawa, G. Kato, T. Nishio, S. Arisawa, Current-induced massless mode of the interband phase difference in two-band superconductors. Physica C **516**, 10–16 (2015)
137. Y. Tanaka, K. Tanaka, K. Tanaka, K. Tonooka, N. Kikuchi, K. Mashiko, A. Iyo, Y. Shimoi, Topological soliton model. Japan patent 5098946 (2012)
138. Y. Tanaka, K. Mashiko, A. Iyo, D.D. Shivagan, P.M. Shirage, N. Kikuchi, K. Tonooka, N. Terada, K. Tokiwa, T. Watanabe, Topological soliton model. Japan patent 5099483, 2012
139. K. Tanaka, Chaos, cosmos, and diversity (Research of soliton 5) (2012). https://www.shizecon.net/award/detail.html?id=244
140. I.H. Yukawa, On the interaction of elementary particles. Proc. Phys.-Math. Soc. Jpn. **17**, 48–57 (1935)

141. H. Yukawa, Models and methods in the meson theory. Rev. Mod. Phys. **21**, 474–479 (1949)
142. J. Kondo, Superconductivity in transition metals. Prog. Theor. Phys. **29**, 1–9 (1963)
143. J. Peretti, Superconductivity of transition elements. Phys. Lett. **2**, 275–276 (1962)
144. F. Wenger, S. Östlund, d-wave pairing in tetragonal superconductors. Phys. Rev. B **47**, 5977–5983 (1993)
145. Y. Imry, On the statistical mechanics of coupled order parameters. J. Phys. C: Solid State Phys. 8, 567–577 (1975)
146. W.-C. Lee, S.-C. Zhang, C. Wu, Paring state with a time-reversal symmetry breaking in FeAs-Based superconductors. Phys. Rev. Lett. **102**, 217002 (2009)
147. K. Kuboki, M. Sigrist, Proximity-induced time-reversal symmetry breaking at Josephson junctions between unconventional superconductors. J. Phys. Soc. Jpn. **65**, 361–364 (1996)
148. M. Matsumoto, H. Shiba, Coexistence of different symmetry order parameters near a surface in *d*-wave superconductors I. J. Phys. Soc. Jpn. **64**, 3384–3396 (1995)
149. M. Matsumoto, H. Shiba, Coexistence of different symmetry order parameters near a surface in *d*-wave superconductors II. J. Phys. Soc. Jpn. **64**, 4867–4881 (1995)
150. M. Matsumoto, H. Shiba, Coexistence of different symmetry order parameters near a surface in *d*-wave superconductors III. J. Phys. Soc. Jpn. **65**, 2194–2203 (1996)
151. M. Fogelström, S.-K. Yip, Time-reversal symmetry-breaking states near grain boundaries between d-wave superconductors. Phys. Rev. B **57**, R14060–R14063 (1998)
152. A. Huck, A. van Otterlo, M. Sigrist, Time-reversal symmetry breaking and spontaneous currents in s-wave/normal-metal/d-wave superconductor sandwiches. Phys. Rev. B **56**, 14163–14167 (1997)
153. M. Sigrist, D.B. Bailey, R.B. Laughlin, Fractional vortices as evidence of time-reversal symmetry breaking in high-temperature superconductors. Phys. Rev. Lett. **74**, 3249–3252 (1995)
154. D.B. Bailey, M. Sigrist, R.B. Laughlin, Fractional vortices on grain boundaries: the case for broken time-reversal symmetry in high-temperature superconductors. Phys. Rev. B **55**, 15239–15247 (1997)
155. T.K. Ng, N. Nagaosa, Broken time-reversal symmetry in Josephson junction involving two-band superconductors. EPL **87**, 17003 (2009)
156. Y. Tanaka, T. Yanagisawa, Chiral state in three-gap superconductors. Solid State Commun. **150**, 1980–1982 (2010)
157. Y. Tanaka, P.M. Shirage, A. Iyo, Time-reversal symmetry-breaking in two-band superconductors. Physica C **470**, 2023–2026 (2010)
158. J. Garaud, E. Babaev, Domain walls and their experimental signatures in $s + is$ superconductors. Phys. Rev. Lett. **112**, 017003 (2014)
159. J. Garaud, J. Carlström, E. Babaev, M. Speight, Chiral CP2 skyrmions in three-band superconductors. Phys. Rev. B **87**, 014507 (2013)
160. S. Gillis, J. Jäykkä, M.V. Milošević, Vortex states in mesoscopic three-band superconductors. Phys. Rev. B **89**, 024512 (2014)
161. N.V. Orlova, A.A. Shanenko, M.V. Milošević, F.M. Peeters, A.V. Vagov, V.M. Axt, Ginzburg-Landau theory for multiband superconductors: Microscopic derivation. Phys. Rev. B **87**, 134510 (2013)
162. Y. Yerin, S.-L. Drechsler, G. Fuchs, Ginzburg-landau analysis of the critical temperature and the upper critical field for three-band superconductors. J. Low. Temp. Phys. **173**, 247–263 (2013)
163. A. Vagov, A.A. Shanenko, M.V. Milošević, V.M. Axt, V.M. Vinokur, J.A. Aguiar, F.M. Peeters, Superconductivity between standard types: multiband versus single-band materials. Phys. Rev. B **93**, 174503 (2016)
164. X. Hu, Z. Wang, Stability and Josephson effect of time-reversal-symmetry-broken multicomponent superconductivity induced by frustrated intercomponent coupling. Phys. Rev. B **85**, 064516 (2012)
165. Y. Takahashi, Z. Huang, X. Hu, H–T phase diagram of multi-component superconductors with frustrated inter-component couplings. J. Phys. Soc. Jpn. **83**, 034701 (2014)

166. Y. Takagashi, Theoretical study on vortex states in unconventional superconductors. Doctoral Degree of Thesis, Tsukuba Univ. Japan, 2014
167. D. Weston, E. Babaev, Classification of ground states and normal modes for phase-frustrated multicomponent superconductors. Phys. Rev. B **88**, 214507 (2013)
168. T.A. Bojesen, E. Babaev, A. Sudbø, Time reversal symmetry breakdown in normal and superconducting states in frustrated three-band systems. Phys. Rev. B **88**, 220511 (2013)
169. Y. Tanaka, T. Yanagisawa, A. Crisan, P.M. Shirage, A. Iyo, K. Tokiwa, T. Nishio, A. Sundaresan, N. Terada, Domains in multiband superconductors. Physica C **471**, 747–750 (2011)
170. T. Yanagisawa, I. Hase, Massless modes and abelian gauge fields in multi-band superconductors. J. Phys. Soc. Jpn. **82**, 124704 (2013)
171. K. Kobayashi, Y. Ota, M. Machinda, Analysis of collective excitation for multi band superconductor: frustrated spin model approach. Physica C **494**, 13–16 (2013)
172. Y. Tanaka. Y. Tanaka, Superconducting frustration bit. Physica C **505**, 55–64 (2014)
173. J.C. Wheatley, Experimental properties of the extraordinary phases of liquid $^3He$ at millikelvin temperatures. Physica **69**, 218–244 (1973)
174. T.J. Greytak, R.T. Johnson, D.N. Paulson, J.C. Wheatley, Heat flow in the extraordinary phases of liquid $He^3$. Phys. Rev. Lett. **31**, 452 (1973)
175. R.T. Johnson, D.N. Paulson, C.B. Pierce, J.C. Wheatley, Measurements along the melting curve of $He^3$ at millikelvin temperatures. Phys. Rev. Lett. **30**, 207–210 (1973)
176. R.L. Kleinberg, D.N. Paulson, R.A. Webb, J.C. Wheatley, Supercooling and superheating of the AB transition in superfluid $^3He$ near the polycritical point. J. Low. Temp. Phys. **17**, 521–528 (1974)
177. K. Tokiwa, A. Iyo, T. Tsukamoto, H. Ihara. Czech, Synthesis of $HgBa_2Ca_3Cu_4O_{10+\delta}$ (Hg-1234) and $HgBa_2Ca_4Cu_5O_{12+\delta}$(Hg-1245) from oxygen controlled precursors under high pressure. J. Phys. **46 [Suppl. 3]**, 1491–1492 (1996)
178. R.G. Dias, A.M. Marques, Frustrated multiband superconductivity. Supercond. Sci. Technol. **24**, 085009 (2011)
179. B.J. Wilson, M.P. Das, Time-reversal-symmetry-broken state in the BCS formalism for a multi-band superconductor. J. Phys. Condens. Matter **25**, 425702 (2013)
180. J. Bardeen, J.R. Schrieffer, Recent developments in superconductivity. Prog. Low Temp. Phys. **3**, 170–287 (1961)
181. A.J. Leggett, Diatomic molecules and cooper pairs. Modern trends in the theory of condensed matter **115**, 13–27 (1980)
182. V.J. Emery, S.A. Kivelson, Importance of phase fluctuations in superconductors with small superfluid density. Nature **374**, 434–437 (1995)
183. P. Nozieres, S. Schmitt-Rink, Bose condensation in an attractive fermion gas: from weak to strong coupling superconductivity. J. Low Temp. Phys. **59**, 195–211 (1985)
184. D.M. Eagles, Possible pairing without superconductivity at low carrier concentrations in bulk and thin-film superconducting semiconductors. Phys. Rev. **186**, 456–463 (1969)
185. M. Holland, S.J.J.M.F. Kokkelmans, M.L. Chiofalo, R. Walser, Resonance superfluidity in a quantum degenerate Fermi gas. Phys. Rev. Lett. **87**, 120406 (2001)
186. C. Regal, Experimental realization of BCS-BEC crossover physics. PhD Thesis, University of Colorado, USA, 2006
187. T. Yanagisawa, Nambu-Goldstone-Leggett modes in multi-condensate superconductors. Nov. Supercond. Mater. **1**, 95–106 (2015)
188. Y. Tanaka, A. Iyo, A. Crisan, K. Tokiwa, T. Watanabe, N. Terada, Method of generation and method of detection of interband phase difference soliton and interband phase difference circuir. US Patent 7,522,078, 2009
189. M.N. Regueiro, M. Jaime, M.A.A. Franco, J.J. Capponi, C. Chaillout, J.L. Tholence, A. Sulpice, P. Lejay, Pressure effects in high temperature superconductors. Physica C **235**, 2093–2094 (1994)

190. Y. Tanaka, G. Kato, T. Nishio, S. Arisawa, Observation of quantum oscillations in a narrow channel with a hole fabricated on a film of multiband superconductors. Solid State Commun. **201**, 95–97 (2015)
191. S.-Z. Lin, C. Reichhardt, Stabilizing fractional vortices in multiband superconductors with periodic pinning arrays. Phys. Rev. B **87**, 100508 (2013)
192. A. Crisan, Y. Tanaka, A. Iyo, Exotic vortex matter: pancake vortex molecules and fractional-flux. J. Super. Nov. Magn. **24**, 1–6 (2011)
193. E. Babaev, J. Jäykkä, M. Speight, Magnetic field delocalization and flux inversion in fractional vortices in two-component superconductors. Phys. Rev. Lett. **103**, 237002 (2009)
194. M.A. Silaev, Stable fractional flux vortices and unconventional magnetic state in two-component superconductors. Phys. Rev. B **83**, 144519 (2011)
195. V.N. Fenchenko, Y.S. Yerin, Phase slip centers in a two-band superconducting filament: application to $MgB_2$. Physica C **480**, 129–136 (2012)
196. Ref. [1], pp. 215–224
197. J. Bardeen, Critical fields and currents in superconductors. Rev. Mod. Phys. **34**, 667–681 (1962)

# Chapter 8
# Fluctuation Modes in Multi-gap Superconductors

**Takashi Yanagisawa**

## 8.1 Introduction

The study of multiband superconductors has a long history and is started from works by Moskalenko [1], Suhl et al. [2], Peretti [3], and Kondo [4], as a generalization of the Bardeen-Cooper-Schrieffer (BCS) theory [5] to a multi-gap superconductor. The first observed two-band superconductor is Nb-doped $SrTiO_3$ [6, 7]. The critical field $H_{c2}(0)$ and the sizable positive curvature of $H_{c2}(T)$ in $YNi_2B_2C$ and $LuNi_2B_2C$ were analyzed within an effective two-band model on the basis of multiband Eliashberg theory [8]. $MgB_2$ [9] and iron-based superconductors [10] were discovered later. It was pointed out from a theoretical point of view that the sign of gap function depends on the sign of the pair-transfer interaction between two bands, and the signs of two gaps are opposite to each other when the pair-transfer interaction is repulsive.

There are many interesting properties in multi-gap superconductors. We show some of them in the following.
(1) Multiband superconductors have a possibility to exhibit high critical temperature $T_c$. $T_c$ is always enhanced in the presence of interband interactions for $s$-wave superconductors. $MgB_2$ [9] and iron-based superconductors [10] are multiband superconductors with relatively high $T_c$. We also mention that layered cuprates [11–13] can be regarded as a multi-gap superconductor. The mechanism of superconductivity in high-temperature cuprates has been studied intensively since its discovery [14–21]. The ladder model is also related with a two-gap superconductor [22–24].
(2) Unusual isotope effect has been observed in multiband superconductors. This depends on the nature of the attractive interaction in the pairing mechanism [25–28] The isotope exponent $\alpha$ of (Ba,K)$Fe_2As_2$ takes values even in the range of $\alpha < 0$ and $\alpha > 0.5$, depending on the property of glue, especially strength and the range of attractive interactions [27, 28]. It is sometimes difficult to determine the pair symmetry of a multiband superconductor. It is still controversial whether the symmetry of

T. Yanagisawa (✉)
Electronics and Photonics Research Institute, National Institute of Advanced Industrial Science and Technology, 1-1-1 Umezono, Tsukuba, Ibaraki 305-8568, Japan
e-mail: t-yanagisawa@aist.go.jp



A. Crisan (ed.), *Vortices and Nanostructured Superconductors*, Springer Series in Materials Science 261, DOI 10.1007/978-3-319-59355-5_8

the electron pair is $s_\pm$ or $s_{++}$, or there is a line node in the gap function of iron-based superconductors.

The pairing symmetry of noncentrosymmetric $LaNiC_2$ [29, 30] and $LaNiGa_2$ [31–33] is also not confirmed yet. It has been suggested that the time-reversal symmetry is broken in these materials by a muon spin relaxation measurement [29]. The non-unitary triplet pairing was proposed theoretically [29], whereas there are experimental data suggesting that the superconductivity in $LaNiC_2$ is BCS-like [34–36]. It is indicated that $LaNiC_2$ is highly correlated material with strong electronic interactions [37].

(3) In $N$-gap superconductors, the gap functions are written as $\Delta_j = |\Delta_j|e^{i\theta_j}$ for $j = 1, \ldots, N$. The $U(1)^N$ phase invariance at most can be spontaneously broken. The Coulomb repulsive interaction turns the one-phase mode $\Phi = c_1\theta_1 + \cdots + c_N\theta_N$ into a gapped plasma mode. Thus, there are at most $N - 1$ modes, and they can be low-energy excitation modes in superconductors. These modes are in general massive due to Josephson interactions. There is, however, a possibility that some of these modes become massless Nambu-Goldstone modes when the Josephson couplings are frustrated.

The Josephson couplings between different bands will bring about attractive phenomena; they are (a) time-reversal symmetry breaking (TRSB) [38–50], (b) the existence of massless (gapless) modes [51–57] and low-lying excited states, and (c) the existence of kinks and fractionally quantized flux vortices [58–62]. The phase-difference mode between two gaps is sometimes called the Leggett mode [63]. This mode will yield new excitation modes in multi-gap superconductors. The Leggett mode is realized as a Josephson plasma oscillation in layered superconductors.

(4) The amplitude mode of the gap function is represented by a Higgs boson in a superconductor. The effective action is given by the time-dependent Ginzburg-Landau (TDGL) model when the temperature $T$ is near the critical temperature $T_c$. The TDGL model includes the dissipation effect on the amplitude, and thus the Higgs mode may not be defined clearly. In contrast, at low temperature $T \ll T_c$, the effective action is given by the quadratic form of the Higgs boson and the mass of the Higgs boson are defined definitely.

(5) The existence of fractionally quantized flux vortices is very significant and interesting. The kink (soliton) solution of phase difference leads to a new mode and the existence of half-quantum flux vortices in two-gap superconductors. A generalization to a three-gap superconductor is not trivial and results in very attractive features, that is, chiral states with time-reversal symmetry breaking and the existence of fractionally quantized vortices [38–40, 42]. Further, in the case with more than four gaps, a new state is predicted with a gapless excitation mode [64].

(6) A new type of superconductors, called the 1.5 type as an intermediate of types I and II, was proposed for two-gap superconductors [65, 66]. The 1.5-type state suggests that an attractive interaction works between vortices. This state may be realized as a result of a multiband effect and does not occur in a single-band superconductor.

(7) There is an interesting and profound analogy between particles physics and superconductivity. For example, there is a similarity between the Dirac equation and the

gap equation of superconductivity [67, 68]. Nambu first noticed this property and brought the idea of spontaneous symmetry breaking into the particle physics.

The mass of the Higgs particle corresponds to the inverse of the coherence length, and the masses of gauge bosons W and Z correspond to the inverse of the penetration depth. When we use $m_W \sim 80.41\,\mathrm{GeV}/c^2$, $m_Z \sim 91.19\,\mathrm{Gev}/c^2$, and $m_H \sim 126\,\mathrm{GeV}/c^2$, the Ginzburg-Landau parameter $\kappa$ is roughly

$$\kappa = \frac{\lambda}{\xi} \sim \frac{m_{W,Z}}{m_H} \sim 1.5. \tag{8.1}$$

This suggests that the universe corresponds to a type-II superconductor.

In this paper, we discuss several interesting properties of the Nambu-Goldstone mode, the Leggett mode, and the Higgs mode in multi-gap superconductors. We focus on superconductors in the clean limit, and impurity effects are left for future studies. This paper is organized as follows: We show the basic formulation of the BCS theory and discuss an analogy between the theory of superconductivity and the mass generation in the particle physics in Sect. 8.2. In Sect. 8.3, we give a survey on a history of multi-gap superconductivity. We give a formula for the effective action on the basis of the functional integral method in Sect. 8.4. We discuss the Nambu-Goldstone and Leggett modes in Sect. 8.5. The dispersion relation of the Higgs mode is examined in Sect. 8.6. Section 8.7 is devoted to a discussion on time-reversal symmetry breaking. In Sect. 8.8, we show that the half-quantized flux vortex can be regarded as a monopole in a multi-gap superconductor. In Sect. 8.9, we discuss the emergency of massless Nambu-Goldstone mode when there is a frustration between Josephson couplings. We give a discussion on the sine-Gordon model in Sect. 8.10. We investigate a chiral symmetry breaking where fluctuations restore time-reversal symmetry from the ground state with time-reversal symmetry breaking in the subsequent Sect. 8.11. In Sect. 8.12, we show an *SU*(*N*) sine-Gordon model which is a generalization. This model is a generalization of the conventional sine-Gordon model to that with multiple variables and is regarded as a model of *G*-valued fields for a Lie group *G*. This model is reduced to a unitary matrix model in some limit. We give a summary in the last section.

## 8.2 Gap Equation and an Analogy to the Particle Physics

### 8.2.1 BCS Theory

Let us consider the BCS Hamiltonian:

$$H = \int d\mathbf{r} \sum_\sigma \psi_\sigma^\dagger(\mathbf{r}) \left( \frac{\mathbf{p}^2}{2m} - \mu \right) \psi_\sigma(\mathbf{r}) - g \int d\mathbf{r} \psi_\uparrow^\dagger(\mathbf{r}) \psi_\downarrow^\dagger(\mathbf{r}) \psi_\downarrow(\mathbf{r}) \psi_\uparrow(\mathbf{r}), \tag{8.2}$$

where $\sigma$ is the spin index $\uparrow$ and $\downarrow$, $\mu$ is the chemical potential, and $g > 0$ is the coupling constant of the attractive interaction. In the momentum space, this is written as

$$H = \sum_{k\sigma} \xi_k c_{k\sigma}^\dagger c_{k\sigma} - g\frac{1}{V}\sum_{kk'q} c_{k'\uparrow}^\dagger c_{-k'+q\downarrow}^\dagger c_{-k+q\downarrow} c_{k\uparrow}, \tag{8.3}$$

where $\xi_k = \epsilon_k - \mu$ for the electron dispersion $\epsilon_k$. The corresponding Lagrangian density is

$$\mathcal{L} = \sum_\sigma \psi_\sigma^\dagger(x)\left(i\hbar\frac{\partial}{\partial t} + \frac{\hbar^2}{2m}\nabla^2 + \mu\right)\psi_\sigma(x) + g\psi_\uparrow^\dagger(x)\psi_\downarrow^\dagger(x)\psi_\downarrow(x)\psi_\uparrow(x). \tag{8.4}$$

Using the Nambu notation [67],

$$\psi(x) = \begin{pmatrix} \psi_\uparrow(x) \\ \psi_\downarrow^\dagger(x) \end{pmatrix}, \tag{8.5}$$

the Lagrangian density becomes

$$\mathcal{L} = \psi^\dagger\left(\sigma_0 i\hbar\frac{\partial}{\partial t} - \sigma_3\xi(\nabla)\right)\psi - \frac{g}{4}\left[(\psi^\dagger\psi)^2 - (\psi^\dagger\sigma_3\psi)^2\right], \tag{8.6}$$

where $\sigma_0$ is the unit matrix and $\xi(\nabla) = -\hbar^2\nabla^2/(2m) - \mu = \mathbf{p}^2/(2m) - \mu$. The vacuum partition function is represented by a functional integral,

$$Z = \int d\psi^\dagger d\psi \exp\left(\frac{i}{\hbar}\int d^dx\mathcal{L}\right). \tag{8.7}$$

$d$ is the space-time dimension. This can be written in a bilinear form by applying a Hubbard-Stratonovich transformation,

$$\exp\left(\frac{i}{\hbar}g\int d^dx\psi_\uparrow^\dagger\psi_\downarrow^\dagger\psi_\downarrow\psi_\uparrow\right) =$$
$$\int d\Delta^* d\Delta \exp\left[-\frac{i}{\hbar}\int d^dx\left(\Delta^*\psi_\downarrow\psi_\uparrow + \Delta\psi_\uparrow^\dagger\psi_\downarrow^\dagger + \frac{1}{g}|\Delta|^2\right)\right], \tag{8.8}$$

where $\Delta^*$ and $\Delta$ are auxiliary fields, and an overall normalization factor is excluded. The partition function has the form

$$Z = \int d\psi^\dagger d\psi \int d\Delta^* d\Delta \exp\left(\frac{i}{\hbar}\int d^dx\mathcal{L}_{eff}\right), \tag{8.9}$$

where

$$\mathcal{L}_{eff} = \psi^\dagger \left[ \sigma_0 i\hbar \frac{\partial}{\partial t} - \sigma_3 \xi(\nabla) - \begin{pmatrix} 0 & \Delta \\ \Delta^* & 0 \end{pmatrix} \right] \psi - \frac{1}{g} |\Delta|^2. \tag{8.10}$$

The field equations obtained by variation of the Lagrangian are

$$\left[ i\hbar \frac{\partial}{\partial t} - \sigma_3 \xi(\nabla) - \begin{pmatrix} 0 & \Delta \\ \Delta^* & 0 \end{pmatrix} \right] \psi = 0, \tag{8.11}$$

$$\Delta = g\psi_\uparrow \psi_\downarrow. \tag{8.12}$$

The equation for $\Delta$ shows that $\Delta$ describes a pair of electrons that forms a spin singlet. It is clear that there is a similarity between this equation and the Dirac equation. If we approximate $\Delta$ by its average $\bar{\Delta} = g\langle \psi_\uparrow \psi_\downarrow \rangle$, we obtain a self-consistency equation for $\bar{\Delta}$. By performing the Grassmann integration over the fields $\psi^\dagger$ and $\psi$, we obtain the effective action

$$S(\Delta^*, \Delta) = -\frac{1}{g} \int d^d x |\Delta(x)|^2 - i\hbar \mathrm{Tr} \ln \begin{pmatrix} p_0 - \xi(\mathbf{p}) & -\Delta(x) \\ -\Delta^*(x) & p_0 + \xi(\mathbf{p}) \end{pmatrix}, \tag{8.13}$$

for which the partition function is

$$Z = \int d\Delta^* d\Delta \exp\left( \frac{i}{\hbar} S(\Delta^*, \Delta) \right). \tag{8.14}$$

Now, the averaged value $\bar{\Delta}$ of the gap function $\Delta$ is determined by adopting the saddle point approximation. The field equation reads

$$\frac{\delta S(\bar{\Delta}^*, \bar{\Delta})}{\delta \bar{\Delta}^*} = 0. \tag{8.15}$$

We obtain a solution assuming that $\bar{\Delta} > 0$ is a constant. This yields

$$\frac{1}{g} \bar{\Delta} = i\hbar \mathrm{Tr} G_0(p) \begin{pmatrix} 0 & 0 \\ 1 & 0 \end{pmatrix}, \tag{8.16}$$

where $G_0(p)$ is the Green function including $\bar{\Delta}$,

$$\begin{aligned} G_0(p) &= \begin{pmatrix} p_0 - \xi(\mathbf{p}) & -\bar{\Delta} \\ -\bar{\Delta}^* & p_0 + \xi(\mathbf{p}) \end{pmatrix}^{-1} \\ &= \frac{1}{p_0^2 - E(\mathbf{p})^2 + i\delta} \begin{pmatrix} p_0 + \xi(\mathbf{p}) & \bar{\Delta} \\ \bar{\Delta}^* & p_0 - \xi(\mathbf{p}) \end{pmatrix}. \end{aligned} \tag{8.17}$$

Here,

$$E(\mathbf{p}) = \sqrt{\xi(\mathbf{p})^2 + \bar{\Delta}^2} \tag{8.18}$$

is the single-particle excitation energy. Then, we obtain the gap equation

$$\frac{1}{g} = \frac{1}{2} \int \frac{d^3k}{(2\pi)^3} \frac{1}{E(\mathbf{k})}. \tag{8.19}$$

The superconducting gap is

$$\bar{\Delta} = 2\hbar\omega_D \exp\left(-\frac{1}{\rho g}\right), \tag{8.20}$$

with the energy cutoff $\hbar\omega_D$ and the density of states $\rho$ at the Fermi energy.

### *8.2.2 Nambu-Jona-Lasinio Model*

The Nambu-Jona-Lasinio model is

$$\mathcal{L} = \bar{\psi} i\gamma^\mu \partial_\mu \psi + g[(\bar{\psi}\psi)^2 - (\bar{\psi}\gamma_5\psi)^2], \tag{8.21}$$

which has the form similar to the BCS model. We set $\hbar = 1$ in this section. $\gamma_\mu$ and $\gamma_5$ are Dirac gamma matrices. This Lagrangian is invariant under the particle number and chiral transformations,

$$\psi \to \exp(i\alpha)\psi, \quad \bar{\psi} \to \bar{\psi}\exp(-i\alpha) \tag{8.22}$$

$$\psi \to \exp(i\gamma_5\alpha)\psi, \quad \bar{\psi} \to \bar{\psi}\exp(i\gamma_5\alpha). \tag{8.23}$$

In a similar way after the spontaneous symmetry breaking, the fermion (nucleon) acquires a mass $m \propto 2g\langle\bar{\psi}\psi\rangle$.

Using an identity

$$1 = \text{const.} \int d\sigma' d\pi' \exp i \int d^4x \Big[ -\frac{1}{4g}(\sigma'^2 + \pi'^2) \Big], \tag{8.24}$$

the partition function is written as

$$Z = \int d\bar{\psi} d\psi d\sigma' d\pi' \exp i \int d^4x \Big[ \bar{\psi} i\gamma^\mu \partial_\mu \psi + g((\bar{\psi}\psi)^2 - (\bar{\psi}\gamma_5\psi)^2) - \frac{1}{4g}(\sigma'^2 + \pi'^2) \Big]. \tag{8.25}$$

We define new $\sigma$ and $\pi$ fields by

$$\sigma' = \sigma + 2g\bar{\psi}\psi, \tag{8.26}$$

$$\pi' = \pi + 2gi\bar{\psi}\gamma_5\psi, \tag{8.27}$$

then, we have

$$Z = \int \bar{\psi} d\psi d\sigma d\pi \exp\left( i \int d^4x \mathcal{L}_{eff} \right), \tag{8.28}$$

where

$$\begin{aligned} \mathcal{L}_{eff} &= \mathcal{L} - \frac{1}{4g}[(\sigma + 2g\bar{\psi}\psi)^2 + (\pi + 2gi\bar{\psi}\gamma_5\psi)^2] \\ &= \bar{\psi}[i\gamma^\mu \partial_\mu - (\sigma + i\gamma_5\pi)]\psi - \frac{1}{4g}(\sigma^2 + \pi^2). \end{aligned} \tag{8.29}$$

Then, we obtain the effective action

$$\begin{aligned} S_{eff}(\sigma, \pi) &= -i \ln \det[i\gamma^\mu \partial_\mu - (\sigma + i\gamma_5\pi)] - \frac{1}{4g} \int d^4x (\sigma^2 + \pi^2) \\ &= -i \mathrm{Tr} \ln[i\gamma^\mu \partial_\mu - (\sigma + i\gamma_5\pi)] - \frac{1}{4g} \int d^4x (\sigma^2 + \pi^2). \end{aligned} \tag{8.30}$$

The saddle point approximation leads to a solution such that $\sigma = \sigma_0$ is a constant and $\pi = 0$. The equation for $\sigma_0$ is

$$\sigma_0 = 2ig \mathrm{Tr} \frac{1}{i\gamma^\mu \partial_\mu - \sigma_0}. \tag{8.31}$$

Because $\sigma_0$ is the mass $m$ of the fermion $\psi$, the mass $m$ is determined by

$$1 = 8gi \int \frac{d^4k}{(2\pi)^4} \frac{1}{k^2 - m^2} = 8g \int \frac{d^4k_E}{(2\pi)^4} \frac{1}{k_E^2 + m^2}, \tag{8.32}$$

which has a nontrivial solution $m \neq 0$ when

$$0 < \frac{2\pi^2}{g\Lambda^2} < 1. \tag{8.33}$$

In this case, the mass $m$ is determined by

$$\frac{2\pi^2}{g\Lambda^2} = 1 - \frac{m^2}{\Lambda^2} \ln\left(1 + \frac{\Lambda^2}{m^2}\right). \tag{8.34}$$

We define the field $h_\sigma$ by

$$\sigma = \sigma_0 + h_\sigma, \tag{8.35}$$

then $h_\sigma$ is a massive boson whose mass is $2m$. The field $\pi$ represents a massless boson which is the Nambu-Goldstone boson. The result that the field $h_\sigma$ acquires the mass $2m$ is well understood by an analogy to superconductivity, where the excitation energy is $2\bar{\Delta}$ when a pair of electrons is excited above the Fermi energy. The model has been generalized to a more realistic two-flavor model:

$$\mathcal{L} = \bar{\psi} i\gamma^\mu \partial_\mu \psi + g[(\bar{\psi}\psi)^2 - \sum_i (\bar{\psi}\gamma_5 \tau_i \psi)(\bar{\psi}\gamma_5 \tau_i \psi)]. \tag{8.36}$$

As is obvious from the discussion here, the nucleon mass generation is very analogous to the gap generation in superconductors. We note that the mass is finite only when $0 < 2\pi^2/g\Lambda^2 < 1$ holds in the Nambu-Jona-Lasinio model, while the superconducting gap always exists as far as $g > 0$.

## 8.3 Multi-gap Superconductivity

We give a brief survey on the research of multi-gap superconductivity [69]. Two years after the BCS theory was proposed [5], an extension to two overlapping bands was considered by Moskalenko [1] and Suhl, Matthias and Walker [2]. After these works, Peretti [3], Kondo [4], and Geilikman [70] reconsidered superconductors with multiple bands. The motivation of Kondo's work is to understand the small isotope effect observed for some transition metal superconductors. Kondo investigated the exchange-like integral between different bands, which is a non-phonon effective attractive interaction, and proposed a possibility of small, being less than 0.5, or vanishing of the isotope effect of the critical temperature $T_c$ using the two-band model. It was found by early works that the critical temperature is enhanced higher than both of critical temperatures of uncoupled superconductors due to the interband coupling. The critical field $H_{c2}(0)$ and the sizable positive curvature of $H_{c2}(T)$ in $YNi_2B_2C$ and $LuNi_2B_2C$ were analyzed on the basis of an effective two-band model. The Ginzburg-Landau model was extended to include two conduction bands [42, 71–73]. Kondo, at the same time, introduced different phases assigned to two different gaps with phase difference $\pi$. This indicates that we can take the phase difference $\varphi$ to be 0 or $\pi$ for the two-band model. A simple generalization to a three-band model was investigated much later than Kondo's work. It was shown independently [38–40] that the phase difference other than 0 or $\pi$ is possible. It was indicated that the intermediate value of the phase difference $\varphi$ leads to time reversal symmetry breaking, which is a new state in three-band superconductors. There have been many works for a pairing state with time-reversal symmetry breaking [41–48, 51, 52, 57, 74–76] with relation to iron-based superconductors [77], and also from the viewpoint of holographic superconductors [78–80].

Leggett [63] considered small fluctuation of phase difference, which yields fluctuation in the density of Cooper pairs. This indicates a possibility of a collective excitation of phase difference mode. Leggett examined the Josephson term $-J\cos(\varphi)$

using the expansion $\cos(\varphi) = 1 - (1/2)\varphi^2 + \cdots$. In the presence of large fluctuation of $\varphi$, we are not allowed to use this approximation. In this situation, we must employ a sine-Gordon model. This model has a kink solution [81] with fluctuation from $\varphi = 0$ to $2\pi$, which results in a new collective mode [60, 82–86].

An intensive study of multi-gap superconductivity started since the discovery of $MgB_2$, and especially iron-based superconductors. A new kind of superconductivity, called the type 1.5, was proposed for $MgB_2$ [65] where it seems that there is an attractive inter-vortex interaction preventing the formation of Abrikosov vortex lattice. A theoretical prediction was given based on the model with vanishing Josephson coupling [87]. There are some controversial on this subject [88–90]. We expect that the Higgs mode plays a role in this issue, because Higgs mode will produce an attractive force between vortices. A three-band model is now considered as a model for iron-based superconductors, and the time-reversal symmetry breaking is investigated intensively.

## 8.4 Effective Action of Multi-gap Superconductors

Let us consider the Hamiltonian for multi-gap superconductors:

$$H = \sum_{i\sigma} \int d\mathbf{r}\psi^{\dagger}_{i\sigma}(\mathbf{r})K_i(\mathbf{r})\psi_{i\sigma}(\mathbf{r}) - \sum_{ij} g_{ij} \int d\mathbf{r}\psi^{\dagger}_{i\uparrow}(\mathbf{r})\psi^{\dagger}_{i\downarrow}(\mathbf{r})\psi_{j\downarrow}(\mathbf{r})\psi_{j\uparrow}(\mathbf{r}), \quad (8.37)$$

where $i$ and $j$ (=1, 2,...) are band indices. $K_i(\mathbf{r})$ stands for the kinetic operator: $K_i(\mathbf{r}) = p^2/(2m_i) - \mu \equiv \xi_i(\mathbf{p})$ where $\mu$ is the chemical potential. We assume that $g_{ij} = g^*_{ji}$. The second term indicates the pairing interaction with the coupling constants $g_{ij}$. This model is a simplified version of multiband model where the coupling constants $g_{ij}$ are assumed to be constants.

In the functional-integral formulation, using the Hubbard-Stratonovich transformation, the partition function is expressed as follows:

$$Z = \int d\psi_\uparrow d\psi_\downarrow \int d\Delta^* d\Delta \exp\left( -\int_0^\beta d\tau d^d x \sum_{ij} \Delta^*_i (G^{-1})_{ij}\Delta_j \right) \times \exp\left( -\sum_j \int d\tau d^d x (\psi^*_{j\uparrow}\psi_{j\downarrow}) \begin{pmatrix} \partial_\tau + \xi_j(\mathbf{p}) & \Delta_j \\ \Delta^*_j & \partial_\tau - \xi_j(\mathbf{p}) \end{pmatrix} \begin{pmatrix} \psi_{j\uparrow} \\ \psi^*_{j\downarrow} \end{pmatrix} \right), \quad (8.38)$$

where $G = (g_{ij})$ is the matrix of coupling constants. $(G^{-1})_{ij}$ $(i \neq j)$ indicates the Josephson coupling. The condition for the matrix $G$ has been discussed in [91].

In order to obtain the effective action for phase variables $\theta_j$, we perform the gauge transformation

$$\begin{pmatrix} \psi_{j\uparrow} \\ \psi^*_{j\downarrow} \end{pmatrix} \rightarrow \begin{pmatrix} e^{i\theta_j}\psi_{j\uparrow} \\ e^{-i\theta_j}\psi^*_{j\downarrow} \end{pmatrix}, \tag{8.39}$$

so that $\Delta_j$ are real and positive. The effective action is written in the form

$$S = \sum_{ij} \int d\tau d^d x \Delta_i (G^{-1})_{ij} \Delta_j \cos(2(\theta_i - \theta_j))$$
$$-\mathrm{Tr}\ln \begin{pmatrix} \partial_\tau + i\partial_\tau\theta_j + \xi_j(\mathbf{p} + \nabla\theta_j) & \Delta_j \\ \Delta_j & \partial_\tau + i\partial_\tau\theta_j - \xi_j(\mathbf{p} + \nabla\theta_j) \end{pmatrix}. \tag{8.40}$$

We define the fluctuation mode (Higgs mode) $h_j$ of the amplitude of $\Delta_j$ as

$$\Delta_j = \bar{\Delta}_j + h_j, \tag{8.41}$$

where $\bar{\Delta}_j$ is the gap function given by the saddle point approximation. We define

$$\mathbf{a}_j = \nabla\theta_h, \;\; a_{0j} = i\partial_\tau\theta_j. \tag{8.42}$$

Then, the effective action is written in the form:

$$S = \sum_{ij} \int d\tau d^d x \Delta_i (G^{-1})_{ij} \Delta_j \cos(2(\theta_i - \theta_j))$$
$$- \sum_j \mathrm{Tr}\ln \left( S_{Fj}^{-1} + \begin{pmatrix} 0 & h_j \\ h_j & 0 \end{pmatrix} + V_j \right), \tag{8.43}$$

where

$$V_j = \frac{1}{2m_j}(\mathbf{p}\cdot\mathbf{a}_j + \mathbf{a}_j\cdot\mathbf{p}) + \left( a_{0j} + \frac{1}{2m_j}\mathbf{a}_j^2 \right)\sigma_3, \tag{8.44}$$

and $S_{Fj}$ is defined by

$$S_{Fj}^{-1}(i\omega_n, \mathbf{p}) = \begin{pmatrix} -i\omega_n + \xi_j(\mathbf{p}) & \bar{\Delta}_j \\ \bar{\Delta}_j & -i\omega_n - \xi_j(\mathbf{p}) \end{pmatrix}, \tag{8.45}$$

n the momentum space where $\omega_n$ is the Matsubara frequency. $G$ is the matrix of coupling constants $g_{ij}$: $G = (g_{ij})$. This action is expanded in the form:

$$S = \sum_{ij} \int d\tau d^d x \Delta_i (G^{-1})_{ij} \Delta_j \cos(2(\theta_i - \theta_j))$$
$$- \sum_j \mathrm{Tr}\ln S_{Fj}^{-1} + \sum_j \mathrm{Tr} \sum_{\ell=1}^{\infty} \frac{1}{\ell}(-1)^\ell \left[ S_{Fj} \begin{pmatrix} o & h_j \\ h_j & 0 \end{pmatrix} + S_{Fj}V_j \right]^\ell. \tag{8.46}$$

## 8.5 Nambu-Goldstone and Leggett Modes

### 8.5.1 *Effective Action*

The effective action for phase modes $\theta_j$ is given by the usual quadratic form with the Josephson coupling. The lowest-order contribution is

$$S^{(\ell=1)}[\theta] = \sum_j \mathrm{Tr} n_j \left( i\partial_\tau \theta_j + \frac{1}{2m_j}(\nabla\theta_j)^2 \right), \tag{8.47}$$

where

$$n_j = \int \frac{d^d k}{(2\pi)^d} \left[ 1 - \frac{\xi_j(\mathbf{k})}{E_j(\mathbf{k})}(1 - 2f(E_j)) \right], \tag{8.48}$$

with $E_j = \sqrt{\xi_j^2 + \bar{\Delta}_j^2}$. The second term with $\ell = 2$ is

$$S^{(\ell=2)}[\theta] = \frac{1}{2} \sum_j \mathrm{Tr} S_{Fj}\sigma_3 S_{Fj}\sigma_3 \left( i\partial_\tau \theta_j + \frac{1}{2m_j}(\nabla\theta_j)^2 \right)^2. \tag{8.49}$$

Then, the quadratic terms of the Nambu-Goldstone-Leggett modes are given by

$$\begin{aligned} S^{(2)}[\theta] &= \sum_j \int d\tau d^d x \left[ \rho_j(\partial_\tau\theta_j)^2 + n_j \frac{1}{2m_j}(\nabla\theta_j)^2 \right] \\ &+ \sum_{ij} \int d\tau d^d x \bar{\Delta}_i (G^{-1})_{ij} \bar{\Delta}_j \cos(2(\theta_i - \theta_j)), \end{aligned} \tag{8.50}$$

where $\rho_j$ is the density of states in the $j$-th band. The Nambu-Goldstone-Leggett modes become massive due to the Josephson term. The gap of the Leggett mode (phase-difference mode) is determined by Josephson couplings. In general, the dynamics of the Leggett mode are described by the sine-Gordon model [42, 82]. A generalization of the sine-Gordon model has also been discussed recently [92].

### 8.5.2 *Nambu-Goldstone-Leggett Mode for Neutral Superconductors*

Let us consider a two-band neutral superconductor, where the action density reads

$$\mathcal{L}_E[\theta] = \rho_1(\partial_\tau\theta_1)^2 + \rho_2(\partial_\tau\theta_2)^2 + \frac{n_1}{2m_1}(\nabla\theta_1)^2 + \frac{n_2}{2m_2}(\nabla\theta_2)^2 + 2\gamma_{12}\bar{\Delta}_1\bar{\Delta}_2\cos(2(\theta_1 - \theta_2)). \tag{8.51}$$

Here, $\gamma_{ij} = (G^{-1})_{ij}$. We assume that $\gamma_{12}$ is negative and $\theta_i$ are small, so that we expand the potential $\cos(2(\theta_1 - \theta_2))$ in terms of $\theta_1 - \theta_2$. The dispersion relations of the Nambu-Goldstone mode and the Leggett mode are determined by [98, 101]:

$$\det \begin{pmatrix} \rho_1\omega^2 - \frac{n_1}{2m_1}k^2 - 4|\gamma_{12}|\bar{\Delta}_1\bar{\Delta}_2 & 4|\gamma_{12}|\bar{\Delta}_1\bar{\Delta}_2 \\ 4|\gamma_{12}|\bar{\Delta}_1\bar{\Delta}_2 & \rho_2\omega^2 - \frac{n_2}{2m_2}k^2 - 4|\gamma_{12}|\bar{\Delta}_1\bar{\Delta}_2 \end{pmatrix} = 0, \quad (8.52)$$

where we performed an analytic continuation $i\omega_n \to \omega$. The dispersion relations of the Nambu-Goldstone and Leggett modes are, respectively, given by

$$\omega^2 = \frac{1}{\rho_1 + \rho_2}\left(\frac{n_1}{2m_1} + \frac{n_2}{2m_2}\right)k^2 = v_N^2 k^2, \quad (8.53)$$

$$\begin{aligned} \omega^2 &= 4\frac{\rho_1 + \rho_2}{\rho_1\rho_2}|\gamma_{12}|\bar{\Delta}_1\bar{\Delta}_2 + \frac{1}{\rho_1 + \rho_2}\left(\frac{n_1\rho_2}{2m_1\rho_1} + \frac{n_2\rho_1}{2m_2\rho_2}\right)k^2 \\ &= \omega_J^2 + v_L^2 k^2, \end{aligned} \quad (8.54)$$

where

$$v_N^2 = \frac{1}{3}\frac{\rho_1 v_{F1}^2 + \rho_2 v_{F2}^2}{\rho_1 + \rho_2}, \quad (8.55)$$

$$v_L^2 = \frac{1}{3}\frac{\rho_2 v_{F1}^2 + \rho_1 v_{F2}^2}{\rho_1 + \rho_2}, \quad (8.56)$$

$$\omega_J^2 = 4\frac{\rho_1 + \rho_2}{\rho_1\rho_2}|\gamma_{12}|\bar{\Delta}_1\bar{\Delta}_2. \quad (8.57)$$

$v_{Fj}$ is the Fermi velocity of the $j$-th band.

### *8.5.3 Plasma and Leggett Modes*

For charged superconductors, we introduce the scalar potential $\Phi$. One mode of the Nambu-Goldstone modes becomes a massive plasma mode in the presence of the Coulomb potential $\Phi$. Let us consider the action density given as

$$\begin{aligned} \mathcal{L}_E[\theta] &= \sum_j \left[\rho_j(\partial_\tau\theta_j - e\Phi)^2 + n_j\frac{1}{2m_j}(\nabla\theta_j)^2\right] + \frac{1}{8\pi}(\nabla\Phi)^2 \\ &\quad + \sum_{ij}\bar{\Delta}_i(G^{-1})_{ij}\bar{\Delta}_j\cos(2(\theta_i - \theta_j)), \end{aligned} \quad (8.58)$$

where $e$ is the charge of the electron. We integrate out the field $\Phi$ to obtain the effective action for the fields $\theta_j$:

$$\mathcal{L}_E[\theta] = \frac{1}{8\pi e^2 \rho(0)^2} \sum_{jj'a} \rho_j \rho_{j'} \partial_\tau \zeta_{ja} \partial_\tau \zeta_{j'a} + \sum_{ja} \frac{n_j}{2m_j} \zeta_{ja}^2 + \sum_j \rho_j \left(\partial_\tau \theta_j\right)^2$$
$$-\frac{1}{\rho(0)} \left( \sum_j \rho_j \partial_\tau \theta_j \right)^2 + \sum_{ij} \bar{\Delta}_i (G^{-1})_{ij} \bar{\Delta}_j \cos(2(\theta_i - \theta_j)) + \cdots, \quad (8.59)$$

where we put $\zeta_{ja} = \nabla_a \theta_j$ and $\rho(0) = \sum_j \rho_j$, and the index $a$ takes $x$, $y$, and $z$. $\zeta_{ja}$ represents the massive mode called the plasma mode. The Nambu-Goldstone mode was absorbed by the Coulomb potential to be the massive plasma mode. There are three plasma modes $\zeta_{ja}$ for $a = x, y$, and $z$ for each band.

In the single-band case, the plasma frequency is

$$\omega_{pl,a}^2 = 4\pi e^2 n / m_a, \quad (8.60)$$

where $n$ is the electron density $n = n_{j=1}$. In the case with large anisotropy such as $m_z \gg m_x, m_y$, the one mode $\zeta_z$ has a small plasma frequency.

In the two-band model with equivalent bands, i.e., $\xi_1 = \xi_2 = p^2/(2m) - \mu$ for simplicity, the Lagrangian reads

$$\mathcal{L}_E = \rho_F(\partial_\tau \theta_1 - e\Phi)^2 + \rho_F(\partial_\tau \theta_2 - e\Phi)^2 + \frac{n}{2m}\left((\nabla\theta_1)^2 + (\nabla\theta_2)^2\right) + \frac{1}{8\pi}(\nabla\Phi)^2$$
$$+2\gamma\bar{\Delta}_1\bar{\Delta}_2 \cos(2(\theta_1 - \theta_2)) + \cdots$$
$$= 2\rho_F\left(\frac{1}{2}\partial_\tau \phi - e\Phi\right)^2 + \frac{n}{4m}(\nabla\phi)^2 + \frac{1}{8\pi}(\nabla\Phi)^2$$
$$+\frac{1}{2}\rho_F(\partial_\tau \varphi)^2 + \frac{n}{4m}(\nabla\varphi)^2 + 2\gamma\bar{\Delta}_1\bar{\Delta}_2\cos(2\varphi) + \cdots, \quad (8.61)$$

where $\cdots$ indicates higher order terms including the coupling terms between amplitude modes and phase modes. We introduced the scalar potential $\Phi$ which represents the Coulomb interaction and defined

$$\phi = \theta_1 + \theta_2, \quad \varphi = \theta_1 - \theta_2. \quad (8.62)$$

$\gamma$ denotes the Josephson coupling strength given by $\gamma = \gamma_{12} \equiv (G^{-1})_{12}$. $\rho_F$ is the density of states at the Fermi level. The derivative of the total phase $\nabla\phi$ represents the plasma mode with the plasma frequency $\omega_p^2 = 4\pi n e^2/m$. This is seen by writing the terms of $\phi$ in the following form by integrating out the scalar potential $\Phi$:

$$\frac{1}{2}\rho_F\left(\frac{\omega_n^2}{\mathbf{k}^2 + 16\pi\rho_F e^2} + \frac{n}{2m\rho_F}\right)\mathbf{k}^2|\phi(i\omega_n, \mathbf{k})|^2, \quad (8.63)$$

after the Fourier transformation. This indicates that the plasma mode is described by the derivative of the total phase $\nabla\phi$. By performing the analytic continuation $i\omega_n \to \omega + i\delta$, we obtain the dispersion relation as

$$\omega^2 = \omega_{pl}^2 + c_s^2 \mathbf{k}^2, \tag{8.64}$$

where $\omega_{pl}^2 = 8\pi ne^2/m$ and $c_s^2 = n/(2m\rho_F)$. $\omega_{pl}^2$ for the two-band model is twice that of the single-band model.

The Lagrangian of the phase-difference mode (Leggett mode) $\varphi$ is given by the sine-Gordon model. This mode is a massless mode if the Josephson coupling $\gamma$ vanishes. When $\varphi$ is small, the sine-Gordon model describes an oscillation mode, by using $\cos\varphi = 1 - \varphi^2/2 + \cdots$. We assume that $\gamma$ is positive so that $\varphi$ describes a stable oscillation mode. The frequency of this mode is proportional to the gap amplitude:

$$\omega_J = 2\sqrt{\frac{2|\gamma|}{\rho_F}}\bar{\Delta}. \tag{8.65}$$

The dispersion relation is given as

$$\omega^2 = \omega_J^2 + \frac{1}{3} v_F^2 \mathbf{k}^2. \tag{8.66}$$

In the general case where the two bands are not equivalent, the dispersion of the Leggett mode is given by

$$\begin{aligned} \omega^2 &= \omega_J^2 + \frac{\rho_1 + \rho_2}{\rho_1 \rho_2} \frac{1}{n_{s1}/2m_1 + n_{s2}/2m_2} \frac{n_{s1}}{2m_1} \frac{n_{s2}}{2m_2} k^2 \\ &= \omega_J^2 + \frac{1}{9} \frac{1}{v_N^2} v_{F1}^2 v_{F2}^2 k^2. \end{aligned} \tag{8.67}$$

This kind of oscillation mode is known as the Josephson plasma mode [93–97]. In $MgB_2$, the frequency of the oscillation mode (Leggett mode) was estimated to be 1.6 or 2 THz [98]. There are two superconducting gaps in $MgB_2$; their magnitudes are given by $\Delta_1 \simeq 1.2$–$3.7$ meV ($\pi$ band, smaller gap) and $\Delta_2 \simeq 6.4$–$6.8$ meV ($\sigma$ band, larger gap) [99]. Thus, the frequency of the Leggett mode is larger than $2\Delta_1$. The observation of the Leggett mode in $MgB_2$ was recently reported by Raman scattering measurements [100].

In an $N$-gap superconductor, the plasma frequency is given by the formula

$$\omega_{pl,a}^2 = 4\pi e^2 \frac{n_1 \ldots n_N}{m_{1a} \ldots m_{Na}} \frac{\rho(0)^2}{\rho_1^2 \frac{n_2 \ldots n_N}{m_{2a} \ldots m_{Na}} + \cdots + \rho_N^2 \frac{n_1 \ldots n_{N-1}}{m_{1a} \ldots m_{N-1,a}}}. \tag{8.68}$$

When $N$ gaps are equivalent, this formula reduces to

$$\omega_{pl,a}^2 = 4\pi e^2 \frac{n}{m_a} N. \tag{8.69}$$

When one conduction band has an effective heavy mass compared to other bands, the plasma frequency is determined by its heavy mass.

In general, in an N-gap superconductor, there are $N - 1$ Leggett modes because one mode becomes a massive mode with the plasma frequency by coupling to the Coulomb potential. When N bands are equivalent, the Josephson term is invariant under an $S_N$ group action. When there is an anisotropy that breaks the equivalence among several bands, we have lower symmetry than $S_N$.

## 8.6 Higgs Mode

Let us discuss the fluctuation of the amplitude of gap functions, which is called the Higgs mode. Recently, there has been an increasing interest in a role of the Higgs mode in superconductors [102–106]. In the relativistic model considered by Nambu and Jona-Lasinio, the Higgs mass is just twice the magnitude of superconducting gap $\bar{\Delta}$ [107, 108]. This results in the mass ratio given as [109]

$$m_{NG} : \bar{\Delta} : m_H = 0 : 1 : 2 \tag{8.70}$$

where $m_{NG}$ is the mass of the Nambu-Goldstone boson, and $m_H$ is that of the Higgs boson.

The action up to the second order of $h$ is

$$S^{(2)}[h] = \frac{1}{2}\sum_j \mathrm{Tr}\left[S_{Fj}\sigma_1 h_j S_{Fj}\sigma_1 h_j\right] + \sum_{ij}\int d\tau d^d x h_i (G^{-1})_{ij} h_j \cos(2(\theta_i - \theta_j)). \tag{8.71}$$

### 8.6.1 *Effective Action Near $T_c$*

When the temperature $T$ is near $T_c$, the effective action is given by the time-dependent Ginzburg-Landau (TDGL) action. For the single-band case, we have

$$S^{(2)}[h] = \int d\tau d^d x \rho\Big[\frac{\pi}{8k_B T_c} h i \frac{\partial h}{\partial \tau} + \frac{1}{4\epsilon_F}\ln\left(\frac{2e^{\gamma}\omega_c}{\pi k_B T_c}\right) h\frac{\partial h}{\partial \tau} + \frac{7\zeta(3)}{48\pi^2}\frac{v_F^2}{(k_B T_c)^2}(\nabla h)^2 + \frac{7\zeta(3)}{4\pi^2 (k_B T_c)^2}\bar{\Delta}^2 h^2\Big]. \tag{8.72}$$

The time dependence gives the dissipation effect so that the Higgs mode is not defined in this region.

### 8.6.2 Effective Action at Low Temperature

At low temperature, in contrast, the Higgs mode can be defined clearly. We employ the approximation that the density of states is constant. Then, the action in the single-band case is

$$S^{(2)}[h] = \int d\tau d^d x \rho \Big[ \frac{1}{12\bar{\Delta}^2} \left( \frac{\partial h}{\partial \tau} \right)^2 + \frac{v_F^2}{36\bar{\Delta}^2} (\nabla h)^2 + h^2 \Big]. \tag{8.73}$$

This is obtained by evaluating the Higgs boson one-loop contribution given by

$$\Pi(\mathbf{q}, i\epsilon) = \frac{1}{\beta} \sum_n \frac{1}{V} \sum_{\mathbf{p}} \mathrm{tr} \Big[ S_F(\mathbf{p}+\mathbf{q}, i\omega_n + i\epsilon) \sigma_1 S_F(\mathbf{p}, i\omega_n) \sigma_1 \Big]. \tag{8.74}$$

At absolute zero, $\Pi(\mathbf{q}=0, q_0)$ (where $q_0 = i\epsilon$) is calculated as [110]

$$\Pi(\mathbf{q}=0, q_0) = -\rho(0) \int d\xi \frac{1}{E(\xi)} + 2\rho(0) \Big[ 1 - \left( \frac{q_0}{2\bar{\Delta}} \right)^2 \Big] F\left( \frac{q_0}{2\bar{\Delta}} \right), \tag{8.75}$$

where $E(\xi) = \sqrt{\xi^2 + \bar{\Delta}^2}$. The second term in the action $S^{(2)}[h]$ in (8.71) cancels the first term of $\Pi(\mathbf{q}=0, q_0)$ by the gap equation. When $q_0/2\bar{\Delta} < 1$, $F(q_0/2\bar{\Delta})$ is given by

$$F(x) = \frac{1}{x\sqrt{1-x^2}} \tan^{-1} \left( \frac{x}{\sqrt{1-x^2}} \right), \tag{8.76}$$

for $x < 1$. When $q_0/2\bar{\Delta} > 1$, we obtain by the analytic continuation

$$F(x) = \frac{1}{2x\sqrt{x^2-1}} \log \left| \frac{x - \sqrt{x^2-1}}{x + \sqrt{x^2-1}} \right| + i \frac{\pi}{2x\sqrt{x^2-1}}, \tag{8.77}$$

for $x > 1$. The real part of $F(x)$ is shown in Fig. 8.1, and the behavior of $\Pi(\mathbf{q}=0, q_0)$ is shown for $q_0 < 2\bar{\Delta}$ in Fig. 8.2. Then, for small $q_0/2\bar{\Delta}$, we have

$$\frac{1}{g} + \frac{1}{2}\Pi(\mathbf{q}=0, q_0) = \rho \Big[ 1 - \frac{1}{3} \left( \frac{q_0}{2\bar{\Delta}} \right)^2 + \cdots \Big], \tag{8.78}$$

where $g = g_{11}$. This results in (8.73). $\frac{1}{g} + \frac{1}{2}\Pi(\mathbf{q}=0, q_0)$ has a zero at

$$q_0 = 2\bar{\Delta} \tag{8.79}$$

In the relativistic model, the same calculation leads to $1 - (q_0/2\bar{\Delta})^2$, giving the mass $m_H = 2\bar{\Delta}$.

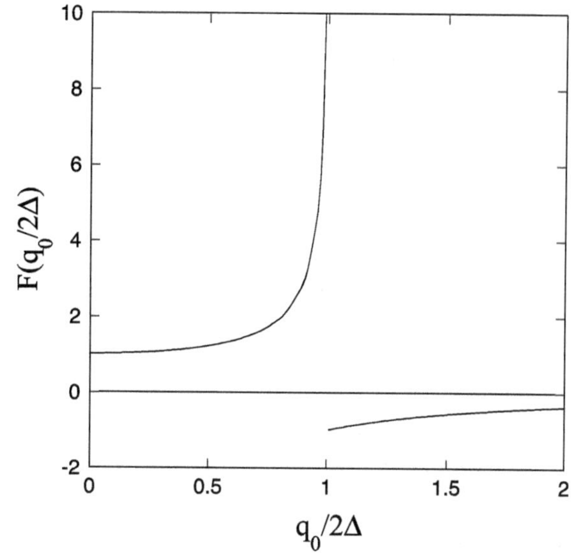


**Fig. 8.1** Re$F(x)$ as a function of $x = q_0/2\Delta$. $F(x)$ has a singularity at $x = 1$

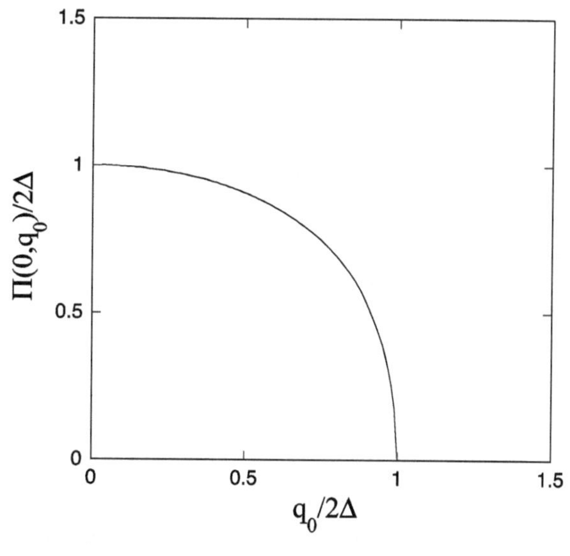


**Fig. 8.2** $pi(\mathbf{q} = 0, q_0)$ as a function of $x = q_0/2\Delta$. The constant which is canceled by $1/g$ is neglected

In the multiband case, the action is given by the quadratic form,

$$S^{(2)}[h] = \int d\tau d^d x \sum_{j\ell} \eta_j H_{j\ell} \eta_\ell. \tag{8.80}$$

The excitation spectra is determined from the condition $\det H = 0$. At low temperatures, the spectrum has a gap being proportional to the mean-field gap amplitude.

For $N = 2$ (two-band superconductor), the matrix $H_{j\ell}$ is written as

$$\begin{pmatrix} \gamma_{11} + \frac{1}{2}\Pi_1 & \gamma_{12} \\ \gamma_{21} & \gamma_{22} + \frac{1}{2}\Pi_2 \end{pmatrix}, \tag{8.81}$$

where $\Pi_\ell$ is

$$\Pi_\ell(\mathbf{q}, i\epsilon) = \frac{1}{\beta}\sum_n \frac{1}{V}\sum_{\mathbf{p}} \mathrm{tr}\left[S_{F\ell}(\mathbf{p}+\mathbf{q}, i\omega_n + i\epsilon)\sigma_1 S_{F\ell}(\mathbf{p}, i\omega_n)\sigma_1\right]. \tag{8.82}$$

Then, the dispersion relation of the Higgs mode $\omega = \omega(\mathbf{q})$ is given by a solution of the equation

$$1 + \frac{1}{2}g_{11}\Pi_1(\mathbf{q},\omega) + \frac{1}{2}g_{22}\Pi_2(\mathbf{q},\omega) + \frac{1}{4}\mathrm{det}G \cdot \Pi_1(\mathbf{q},\omega)\Pi_2(\mathbf{q},\omega) = 0, \tag{8.83}$$

where $\mathrm{det}G = g_{11}g_{22} - g_{12}g_{21}$.

In the multi-gap case, the action becomes

$$S^{(2)}[h] = \int d\tau d^d x \sum_j \left[h_i(-\rho_j f_j + \rho_j \nu_j)h_j + \frac{1}{12\bar{\Delta}_j^2}\left(\frac{\partial h_j}{\partial \tau}\right)^2 + \frac{v_{Fj}^2}{36\bar{\Delta}_j^2}\left(\nabla h_j\right)^2\right] + \int d\tau d^d x \sum_{ij} h_i (G^{-1})_{ij} h_j, \tag{8.84}$$

where we defined

$$\rho_j f_j = \int \frac{d^d k}{(2\pi)^d}\frac{1}{2E_j}\left(1 - 2f(E_j)\right), \tag{8.85}$$

$$\rho_j \nu_j = \int \frac{d^d k}{(2\pi)^d}\frac{1}{\beta}\sum_n \frac{2\bar{\Delta}_j^2}{(\omega_n^2 + \xi_j^2 + \bar{\Delta}_j^2)^2}. \tag{8.86}$$

We set the phase variables $\{\theta_j\}$ to take their equilibrium values that are assumed to be zero here. The excitation gaps of Higgs modes in the multi-gap model are obtained by diagonalizing the Higgs kinetic terms. The Higgs masses in multi-gap superconductors are obtained as a function of Josephson couplings by diagonalizing the quadratic terms in the action. In general, multiple Higgs masses are different to each other. Since the constant term in Eq. (8.84) is closely related to the critical field $H_{c2}$, because this term $\propto 1/\xi^2$ where $\xi$ is the coherence length, we expect that the large upper critical field $H_{c2}(0)$ observed in iron-based superconductors [111] is understood by means of a multiband model of Higgs fields.

## 8.7 Time-Reversal Symmetry Breaking

The gap function, defined as $\Delta_i(\mathbf{r}) = -\sum_j g_{ij}\langle\psi_{j\downarrow}(\mathbf{r})\psi_{j\uparrow}(\mathbf{r})\rangle$, satisfies the gap equation

$$\Delta_i = \sum_j g_{ij}N_j\Delta_j \int d\xi_j \frac{1}{E_j}\tanh\left(\frac{E_j}{2k_BT}\right), \tag{8.87}$$

where $N_j$ is the density of states at the Fermi surface in the $j$-th band and $E_j = \sqrt{\xi_j^2 + |\Delta_j|^2}$. $\Delta_i$ in this section is the mean-field solution in Sect. 8.3 which is obtained by a saddle-point approximation. We set

$$\zeta_j = \int_0^{\omega_{Dj}} d\xi_j \frac{1}{E_j}\tanh\left(\frac{E_j}{2k_BT}\right), \tag{8.88}$$

and $\gamma_{ij} = (G^{-1})_{ij}$ where $G = (g_{ij})$. We write the gap equation in the following form,

$$\begin{pmatrix} \gamma_{11} - N_1\zeta_1 & \gamma_{12} & \gamma_{13} & \cdots \\ \gamma_{21} & \gamma_{22} - N_2\zeta_2 & \gamma_{23} & \cdots \\ \gamma_{31} & \gamma_{32} & \gamma_{33} - N_3\zeta_3 & \cdots \\ \cdots & \cdots & \cdots & \cdots \end{pmatrix}\begin{pmatrix} \Delta_1 \\ \Delta_2 \\ \Delta_3 \\ \cdots \end{pmatrix} = 0. \tag{8.89}$$

$\gamma_{ij}$ $(i \neq j)$ gives the interband Josephson coupling between bands $i$ and $j$ [42].

When the gap functions $\Delta_j$ are complex-valued functions, the time-reversal symmetry is broken. The condition for TRSB is that the following equation for the imaginary part $\mathrm{Im}\Delta_j$ has a nontrivial solution:

$$\begin{pmatrix} \gamma_{12} & \gamma_{13} & \cdots \\ \gamma_{22} - N_2\zeta_2 & \gamma_{23} & \cdots \\ \gamma_{32} & \gamma_{33} - N_3\zeta_3 & \cdots \\ \cdots & \cdots & \cdots \end{pmatrix}\begin{pmatrix} \mathrm{Im}\Delta_2 \\ \mathrm{Im}\Delta_3 \\ \cdots \end{pmatrix} = 0, \tag{8.90}$$

where we adopt that $\Delta_1$ is real for simplicity and $\gamma_{ij}$ are real. We assume that $\gamma_{ij} = \gamma_{ji}$. In the case of $N = 3$, the condition for TRSB has been obtained [43, 47]. We have a necessary condition $\gamma_{12}\gamma_{23}\gamma_{13} > 0$ [39, 40]. The determinant of each $2 \times 2$ matrix in (8.90) should vanish so that nontrivial solution $\mathrm{Im}\Delta_j$ $(j = 2, 3)$ exist. Then, we have

$$\gamma_{12}\gamma_{23} - (\gamma_{22} - N_2\zeta_2)\gamma_{13} = 0, \tag{8.91}$$

$$(\gamma_{22} - N_2\zeta_2)(\gamma_{33} - N_3\zeta_3) - \gamma_{23}^2 = 0, \tag{8.92}$$

$$\gamma_{12}(\gamma_{33} - N_3\zeta_3) - \gamma_{12}\gamma_{23} = 0. \tag{8.93}$$

When we assume $\gamma_{13} \neq 0$, we obtain

$$\gamma_{22} - N_2\zeta_2 = \gamma_{12}\gamma_{23}/\gamma_{13}. \tag{8.94}$$

Similarly, we have by assuming $\gamma_{12} \neq 0$

$$\gamma_{33} - N_3\zeta_3 = \gamma_{23}\gamma_{13}/\gamma_{12}. \tag{8.95}$$

From the gap equation $\gamma_{21}\Delta_1 + (\gamma_{22} - N_2\zeta_2)\Delta_2 + \gamma_{23}\Delta_3 = 0$, we obtain the relation

$$\frac{\Delta_1}{\gamma_{23}} + \frac{\Delta_2}{\gamma_{31}} + \frac{\Delta_3}{\gamma_{12}} = 0. \tag{8.96}$$

The complex numbers $\Delta_1/\gamma_{23}, \ldots$ form a triangle in the TRSB state. The transition from TRSB to the state with time-reversal symmetry takes place when the triangle relation is broken. From (8.94) and (8.95), the critical temperature $T_c$ should satisfy

$$N_j \ln\left(\frac{2e^{\gamma_E}\omega_{Dj}}{\pi k_B T_c}\right) = \gamma_{jj} - \frac{\gamma_{jn}\gamma_{jm}}{\gamma_{nm}}, \tag{8.97}$$

where $j$, $n$, and $m$ are different to one another and $\gamma_E$ is the Euler constant. The stability of TRSB state has been examined by evaluating the free energy [38, 47, 91]

In the simplest case where all the bands are equivalent and $\gamma_{ij}$ $(i \neq j)$ are the same, the chiral state in Fig. 8.3 is realized. We have $(\theta_1, \theta_2, \theta_3) = (0, 2\pi/3, 4\pi/3)$ for Fig. 8.1a and $(\theta_1, \theta_2, \theta_3) = (0, 4\pi/3, 2\pi/3)$ for Fig. 8.1b. The two states are degenerate and have chirality $\kappa = 1$ and $\kappa = -1$, respectively, where the chirality is defined by $\kappa = (2/3\sqrt{3})[\sin(\theta_1 - \theta_2) + \sin(\theta_2 - \theta_3) + \sin(\theta_3 - \theta_1)]$. In the chiral state $\Delta_1/\gamma_{23}, \ldots$ form an equilateral triangle. In this case, the eigenvalues of the gap equation are degenerate and the chiral TRSB state is realized.

For $N > 3$, it is not straightforward to derive the condition for TRSB. We consider here a separable form for the Josephson couplings:

$$\gamma_{ij} = \gamma_i\gamma_j \text{ for } i \neq j, \tag{8.98}$$

where $\gamma_j (\neq 0)$ $(j = 1, \ldots, N)$ are real constants. The condition $\gamma_{12}\gamma_{23}\gamma_{31} = \gamma_1^2\gamma_2^3\gamma_3^2 > 0$ is satisfied. For $N = 4$, we obtain from (8.90)

$$\frac{\Delta_1}{\gamma_2\gamma_3\gamma_4} + \frac{\Delta_2}{\gamma_3\gamma_4\gamma_1} + \frac{\Delta_3}{\gamma_4\gamma_1\gamma_2} + \frac{\Delta_4}{\gamma_1\gamma_2\gamma_3} = 0. \tag{8.99}$$

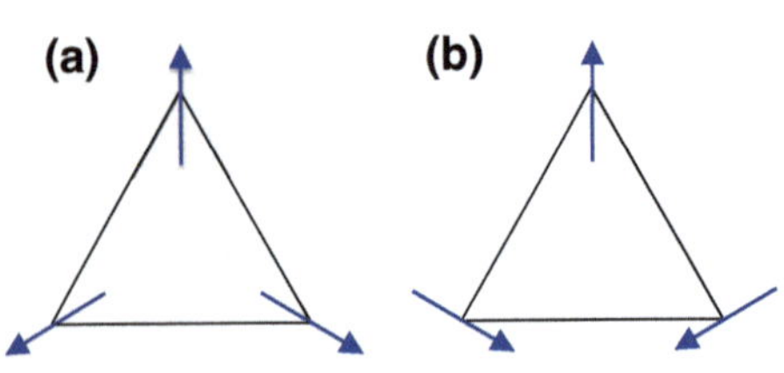


**Fig. 8.3** Chiral state with time-reversal symmetry breaking. Two states have the chirality $\kappa = +1$ for (**a**) and $\kappa = -1$ for (**b**)

Then, the triangle condition in (8.96) is generalized to the polygon condition for general $N \geq 3$:

$$\frac{\Delta_1}{\gamma_2\gamma_3\cdots\gamma_N} + \frac{\Delta_2}{\gamma_3\gamma_4\cdots\gamma_N\gamma_1} + \cdots + \frac{\Delta_N}{\gamma_1\gamma_2\cdots\gamma_{N-1}} = 0. \quad (8.100)$$

We assume that the polygon is not crushed to a line, which means, in the case $N = 3$, the triangle inequality holds. Under these conditions, the solution with time-reversal symmetry breaking exists and massless excitation modes also exist at the same. The existence of massless modes will be examined in next section.

## 8.8 Half-Quantum Flux Vortex and a Monopole

The sine-Gordon model has been studied to investigate a new dynamics of multi-gap superconductors [60, 61, 112]. When the oscillation of phase difference $\varphi$ is small, we can expand the potential around a minimum. This results in the Leggett mode as described in Sect. 8.3. In the presence of large oscillation, we cannot use a perturbative method and we must consider a non-perturbative kink solution. This leads to a half-quantum flux vortex.

The sine-Gordon model has a kink solution [81]. If we impose the boundary condition such that $\varphi \to 0$ as $x \to -\infty$ and $\varphi \to 2\pi$ as $x \to \infty$, we have a kink solution like $\varphi = \pi + 2\sin^{-1}(\tanh(\sqrt{\kappa}x))$ for a constant $\kappa$. The phase difference $\varphi$ should be changed from 0 to $2\pi$ to across the kink. This means that $\theta_1$ changes from 0 to $\pi$ and at the same time $\theta_2$ changes from 0 to $-\pi$. In this case, a half-quantum flux vortex exists at the edge of the kink. This is shown in Fig. 8.4 where the half-quantum vortex is at the edge of the cut (kink). A net change of $\theta_1$ is $2\pi$ by a counterclockwise encirclement of the vortex, and that of $\theta_2$ vanishes. Then, we have a half-quantum flux vortex.

The phase-difference gauge field $\mathbf{B}$ is defined as [51]

$$\mathbf{B} = -\frac{\hbar c}{2e^*}\nabla\varphi. \quad (8.101)$$

The half-quantum vortex can be interpreted as a monopole [42]. Let us assume that there is a cut, namely, kink on the real axis for $x > 0$. The phase $\theta_1$ is represented by

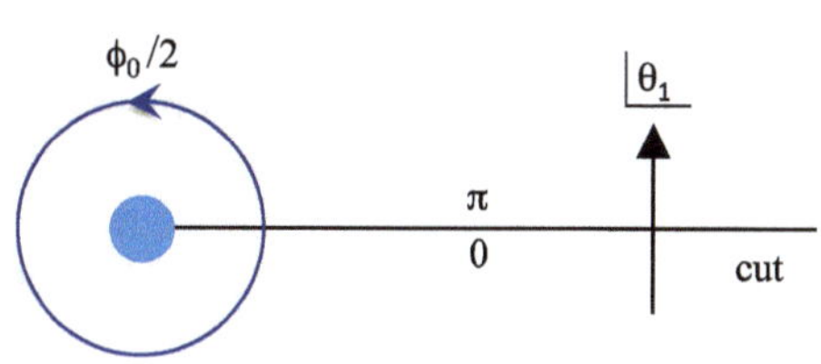


**Fig. 8.4** Half-quantum flux vortex with a line singularity (kink). The phase variables $\theta_1$ changes from 0 to $\pi$ when crossing a singularity

$$\theta_1 = -\frac{1}{2}\mathrm{Im}\log\zeta + \pi, \tag{8.102}$$

where $\zeta = x + iy$. The singularity of $\theta_j$ can be transferred to a singularity of the gauge field by a gauge transformation. We consider the case $\theta_2 = -\theta_1$: $\varphi = 2\theta_1$. Then, we have

$$\mathbf{B} = -\frac{\hbar c}{2e^*}\nabla\varphi = -\frac{\hbar c}{e^*}\frac{1}{2}\left(\frac{y}{x^2+y^2}, -\frac{x}{x^2+y^2}, 0\right). \tag{8.103}$$

Thus, when the gauge field **B** has a monopole-type singularity, the vortex with half-quantum flux exists in two-gap superconductors.

Let us consider the fictitious z-axis perpendicular to the x-y plane. The gauge potential (1-form) is given by

$$\Omega_\pm = -\frac{1}{2}\frac{1}{r(z \pm r)}(ydx - xdy) = \frac{1}{2}(\pm 1 - \cos\theta)d\phi, \tag{8.104}$$

where $r = \sqrt{x^2+y^2+z^2}$, and $\theta$ and $\phi$ are Euler angles. $\Omega_\pm$ correspond to the gauge potential in the upper and lower hemisphere $H_\pm$, respectively. $\Omega_\pm$ are connected by $\Omega_+ = \Omega_- + d\phi$. The components of $\Omega_+$ are

$$\Omega_\mu = \frac{1}{2}(1 - \cos\theta)\partial_\mu\phi. \tag{8.105}$$

At $z = 0$, $\Omega_\mu$ coincides with the gauge field for half-quantum vortex. If we identify $\varphi$ with $\phi$, we obtain

$$\mathbf{B} = \frac{\hbar c}{e^*}\mathbf{\Omega}, \tag{8.106}$$

at $\theta = \pi/2$. $\{\Omega_\pm\}$ is the U(1) bundle $P$ over the sphere $S^2$. The Chern class is defined as

$$c_1(P) = -\frac{1}{2\pi}F = -\frac{1}{2\pi}d\Omega_+. \tag{8.107}$$

The Chern number is given as

$$C_1 = \int_{S^2} c_1 = -\frac{1}{2\pi}\int_{S^2} F = -\frac{1}{2\pi}\left(\int_{H_+} d\Omega_+ + \int_{H_-} d\Omega_-\right) = 1. \tag{8.108}$$

In general, the gauge field **B** has the integer Chern number: $C_1 = n$. For $n$ odd, we have a half-quantum flux vortex.

The half-flux vortex has been investigated in the study of $p$-wave superconductivity [59, 113, 114]. In the case of chiral $p$-wave superconductivity, the singularity of $U(1)$ phase is, however, canceled by the kink structure of the $d$-vector. This is the difference between two-band superconductivity and $p$-wave superconductivity.

As we can expect easily, a fractional quantum flux vortex state is not stable because the singularity (kink, domain wall) costs energy being proportional to the square root of the Josephson coupling. Thermodynamic stability was discussed in [62]. Two vortices form a molecule by two kinks. This state may have lower energy than the vortex state with a single quantum flux $\phi_0$ because the magnetic energy of two fractional vortices is smaller than $\phi_0^2$ of the unit quantum flux. The energy of kinks is proportional to the distance $R$ between two fractional vortices when $R$ is large. Thus, the attractive interaction works between them when $R$ is sufficiently large. There is an interesting analogy between quarks and fractional flux vortices [115].

## 8.9 Massless Nambu-Goldstone Modes

We examined the phase modes that are Nambu-Goldstone modes by nature emerging due to a spontaneous symmetry breaking in Sect. 8.3. There, one mode becomes massive by coupling to the scalar potential, called the plasma mode, and the other modes become massive due to Josephson couplings, called the Leggett modes. In this section, we show that massive modes change into massless modes when some conditions are satisfied.

The Josephson potential is given as

$$V \equiv \sum_{i \neq j} \gamma_{ij} \bar{\Delta}_i \bar{\Delta}_j \cos(\theta_i - \theta_j), \tag{8.109}$$

where $\gamma_{ij} = \gamma_{ji}$ are chosen real. Obviously, the phase difference modes $\theta_i - \theta_j$ acquire masses. This would change qualitatively when $N$ is greater than 3 or equal to 3. We discuss this in this section.

We show that massless modes exist for an $N$-equivalent frustrated band superconductor. Let us consider the potential for $N \geq 4$ given by

$$\begin{aligned} V = \Gamma[&\cos(\theta_1 - \theta_2) + \cos(\theta_1 - \theta_3) + \cdots + \cos(\theta_1 - \theta_N) \\ &+ \cdots + \cos(\theta_{N-1} - \theta_N)]. \end{aligned} \tag{8.110}$$

For $\Gamma > 0$, there are two massive modes and $N - 3$ massless modes, near the minimum $(\theta_1, \theta_2, \theta_3, \theta_4, \ldots) = (0, 2\pi/N, 4\pi/N, 6\pi/N, \ldots)$. This can be seen by writing the potential in the form

$$V = \frac{\Gamma}{2}\left[\left(\sum_{i=1}^{N} \mathbf{S}_i\right)^2 - N\right], \tag{8.111}$$

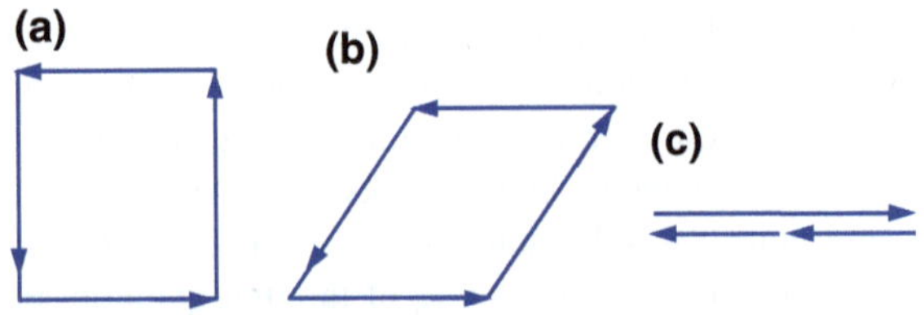


**Fig. 8.5** *Polygon* state satisfying $\sum_j \mathbf{S}_j = 0$ for $N = 4$ in (**a**) and (**b**), where time-reversal symmetry is broken and a massless mode exists. A *linear* state with $\sum_j \gamma_j \Delta_j = 0$ is shown in (**c**), where a massless mode exists but the time-reversal symmetry is not broken

where $\mathbf{S}_i$ $(i = 1, \ldots, N)$ are two-component vectors with unit length $|\mathbf{S}_i| = 1$. $V$ has a minimum $V_{min} = -\Gamma N/2$ for $\sum_i \mathbf{S}_i = 0$. Configurations under this condition have the same energy and can be continuously mapped to each other with no excess energy. At $(\theta_1, \theta_2, \ldots) = (0, 2\pi/N, 4\pi/N, \ldots)$ with $V = -\Gamma N/2$, satisfying $\sum_i \mathbf{S}_i = 0$, the vectors $\mathbf{S}_i$ form a polygon. The polygon can be deformed with the same energy (see Fig. 8.5a, b). The existence of massless modes was examined numerically for the multi-gap BCS model [53]. It has been shown that there is a large region in the parameter space where massless modes exist.

Let us discuss the Josephson potential in a separable form. This is given by

$$V = \sum_{i \neq j} \gamma_{ij} \Delta_i^* \Delta_j = \sum_{i \neq j} \gamma_i \gamma_j \Delta_i^* \Delta_j. \tag{8.112}$$

This is written as

$$V = |P|^2 - \sum_j \gamma_j^2 |\Delta_j|^2, \tag{8.113}$$

where $P = \sum_j \gamma_j \Delta_j$. $V$ has a minimum when $P = 0$ is satisfied. $P = 0$ is equivalent to the polygon condition in (8.100). Because the polygon for $N > 3$ can be deformed continuously without finite excitation energy, a massless mode exists [51] (Fig. 8.5a, b). We have one massless mode for $N = 4$ and two massless modes for $N = 5$. A spin model, corresponding to the Josephson model considered here, also has gapless excitation modes.

When the polygon is crushed to a line, the time-reversal symmetry is not broken. A massless mode, however, exists when $P = 0$. An example is shown in Fig. 8.5c called a linear model. In this model, there are two independent modes and the quadratic term of one mode vanishes as can be shown by explicit calculations. A mode called the scissor mode becomes massless.

Although we did not consider an effect of the amplitude mode (Higgs mode) $\eta_j$, this mode may be important when discussing the stability of massless modes. This is a future problem.

## 8.10 Sine-Gordon Model

### 8.10.1 (d+1)D Sine-Gordon Model

The phase-difference mode is described by the sine-Gordon model. For the two-band model with equivalent bands, the Lagrangian density is

$$\mathcal{L}_E = \frac{1}{2}\rho_F(\partial_\tau\varphi)^2 + \frac{n}{4m}(\nabla\varphi)^2 + 2\gamma_{12}\bar{\Delta}_1\bar{\Delta}_2\cos(2\varphi), \tag{8.114}$$

where $\varphi$ is the half of the phase difference of the gap functions. Since $\rho_F \sim k_F^{d-2}$ and $n \sim k_F^d$ for the Fermi wave number $k_F$, where $d$ is the space dimension, we write the action of this model in the form:

$$S_{SG} = \frac{\Lambda^{d-1}}{g}\int_0^\beta dx_0 \int d^dx\Big[\frac{1}{2}(\partial_\mu\varphi)^2 - \alpha\Lambda^2\cos\varphi\Big], \tag{8.115}$$

where we redefined $2\varphi$ to $\varphi$. We set $x_0 = Av_F\tau$ for a constant $A$ and $\beta = Av_F/(k_BT)$. $\Lambda$ is a cutoff and $g$ and $\alpha$ are coupling constants. $\alpha$ is proportional to the strength of the Josephson coupling $|\gamma_{12}|$. We adopt that $\alpha$ is positive; otherwise we consider $|\alpha|$. We define the dimensionless inverse temperature $u$ by $\beta = u/\Lambda$. The action $S_{SG}$ has a factor $\Lambda^{d-1}$, so that the coupling constant $g$ is dimensionless. Thus, the phase-difference mode is modeled by the (d+1)D sine-Gordon model. In the limit of small $\beta$ (high-temperature limit), the model is reduced to the $d$-dimensional sine-Gordon model:

$$S_{SG}^d = \frac{\Lambda^{d-2}}{t}\int d^dx\Big[\frac{1}{2}(\partial_\mu\varphi)^2 - \alpha\Lambda^2\cos\varphi\Big], \tag{8.116}$$

where we set $t \equiv g/u$.

### 8.10.2 Renormalization Group Equation

Let us investigate the renormalization group flow of the (d+1)D sine-Gordon model on the basis of the Wilson renormalization group method [116, 117] at finite temperature [118, 119]. In general, the Josephson coupling is small, and thus, the results will be relevant in the region where the Josephson coupling is still small and finite. We neglect the effect of the renormalization on $T_c$.

The renormalization group equations are

$$\Lambda\frac{\partial g}{\partial\Lambda} = (d-1)g + c(u)\alpha^2 g\coth\left(\frac{u}{2}\right), \tag{8.117}$$

$$\Lambda\frac{\partial}{\partial\Lambda}\frac{\alpha}{g} = -\left(d+1-g\frac{\Omega_d}{4(2\pi)^d}\coth\left(\frac{u}{2}\right)\right)\frac{\alpha}{g}, \tag{8.118}$$

$$\Lambda\frac{\partial u}{\partial\Lambda} = u. \tag{8.119}$$

Here, $c(u)$ is a constant for large $u$ and is proportional to $u$ for small $u$; $c(u) = c_1 u$. $\Omega_d$ is the solid angle in $d$ dimensions.

Let us consider fluctuations of the Leggett mode $\varphi$. A typical fluctuation mode is the kink (soliton) excitation where $\varphi$ changes from 0 to $2\pi$ or $2\pi$ to 0 in some regions in a superconductor. The one-dimensional kink, namely the domain wall, is expected to appear easily due to quantum fluctuation. The one-dimensional means that $\varphi(\tau, x_1, \ldots)$ depends on only one space variable $x_1$. The renormalization group equations for $d = 1$ read

$$\Lambda\frac{\partial g}{\partial\Lambda} = c(u)\alpha^2 g\coth\left(\frac{u}{2}\right), \tag{8.120}$$

$$\Lambda\frac{\partial}{\partial\Lambda}\frac{\alpha}{g} = -\left(2-\frac{g}{4\pi}\coth\left(\frac{u}{2}\right)\right)\frac{\alpha}{g}. \tag{8.121}$$

At low temperature where $u = \beta\Lambda \gg 1$, there is a fixed point at $g = 8\pi$ and $\alpha = 0$. We show the renormalization group flow when the cutoff $\Lambda$ decreases in Fig. 8.6. As the cutoff $\Lambda$ is reduced, $g(> 0)$ also decreases.

At high temperature, the equations reduce to

$$\Lambda\frac{\partial\Lambda}{\partial\Lambda} = (d-2)t + 2c_1\alpha^2 t, \tag{8.122}$$

$$\Lambda\frac{\partial\alpha}{\partial\Lambda} = -\alpha\left(2-\frac{\Omega_d}{2(2\pi)^d}t\right). \tag{8.123}$$

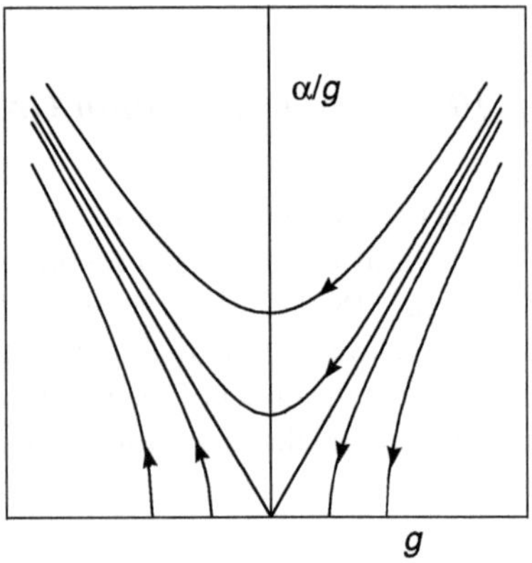


**Fig. 8.6** Renormalization group flow in the plane of $g$ and $\alpha/g$ ($d$=1). *Arrows* indicate the direction of flow when the cutoff decreases

There is a fixed point at $t = 8\pi$ and $\alpha = 0$ for $d = 2$. If this set of equations can be applied to a two-gap superconductor, there is a Kosterlitz-Thouless-like transition at $t = k_B Tg/\Lambda = 8\pi$.

## 8.11 Chiral Transition

### 8.11.1 N-variable Sine-Gordon Model

In this section, we present a field theoretic model that shows a chiral transition. This model is extracted from a model for multi-gap superconductors. The model should be regarded as a model in field theory, and we also discuss applicability to real superconductors. We adopted the London approximation to derive the model, where the fluctuation modes (Higgs modes) $\eta_j$ of the gap functions are neglected. A role of the fluctuation mode concerning the existence of the phase transition would be a problem for future discussion.

Let us consider an action for phase variable $\theta_j$:

$$S[\theta] = \frac{1}{k_B T}\int d^d x\left(\sum_j \frac{n_{sj}}{2m_j}(\nabla\theta_j)^2 + \sum_{i\neq j}\gamma_{ij}\bar{\Delta}_i\bar{\Delta}_j\cos(\theta_i - \theta_j)\right), \qquad (8.124)$$

where we neglect $\tau$ dependence of $\theta_j$. We simply assume that $K_j \equiv n_{sj}/(2m_j) = K$, $\bar{\Delta}_j = \bar{\Delta}$ and $\gamma_{ij} = \gamma_{ji} = \gamma$, namely, all the bands are equivalent. Then, the action for the phase variables $\theta_j$ is

$$S[\theta] = \frac{\Lambda^{d-2}}{t}\int d^d x\left(\sum_j (\nabla\theta_j)^2 + \alpha\Lambda^2\sum_{i<j}\cos(\theta_i - \theta_j)\right), \qquad (8.125)$$

where $t/\Lambda^{d-2} = k_B T/K$ and $\lambda\Lambda^2 = 2\gamma\Delta_0^2/K$. We have introduced the cutoff $\Lambda$ so that $t$ and $\alpha$ are dimensionless parameters. We assume that $\alpha > 0$ in this paper. We consider the case $N = 3$ and discuss the phase transition in this model. Apparently, this model has $S_3$ symmetry. If we neglect the kinetic term, the ground states are twofold degenerate. The two ground states are indexed by the chirality $\kappa$.

We perform a unitary transformation: $\theta_1 = -2\pi/3 - (1/\sqrt{2})\eta_1 + (1/\sqrt{6})\eta_2 + (1/\sqrt{3})\eta_3$, $\theta_2 = -(2/\sqrt{6})\eta_2 + (1/\sqrt{3})\eta_3$, and $\theta_3 = 2\pi/3 + (1/\sqrt{2})\eta_1 + (1/\sqrt{6})\eta_2 + (2/\sqrt{3})\eta_3$, where $\eta_i$ $(i = 1, 2, 3)$ indicate fluctuation fields. $\eta_3$ describes the total phase mode, $\eta_3 = (\theta_1 + \theta_2 + \theta_3)/\sqrt{3}$, and is not important because this mode turns out to be a plasma mode by coupling with the long-range Coulomb potential. The action $S[\eta] \equiv S[\theta]$ becomes

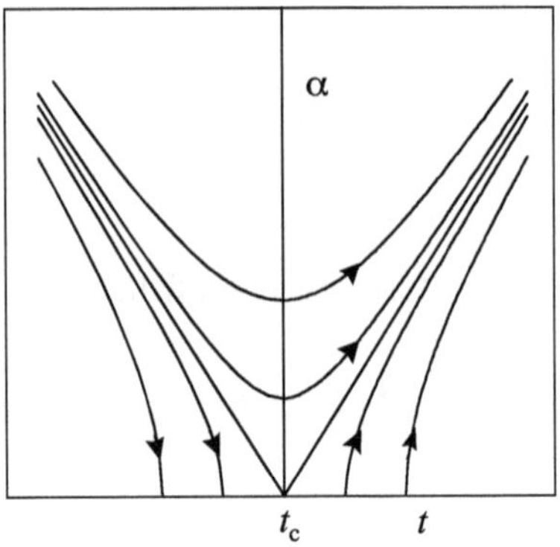


**Fig. 8.7** Renormalization group flow for the $N = 3$ generalized sine-Gordon model ($d = 2$). The flow is indicated as $\mu$ is increased ($\mu \to \infty$)

$$S[\eta] = \frac{\Lambda^{d-2}}{t} \int d^d x \Big[ \sum_j (\nabla \eta_j)^2 + \alpha \Lambda^2 \Big( \cos\left(\sqrt{2}\eta_1 + \frac{4\pi}{3}\right)$$
$$+ 2\cos\left(\frac{1}{\sqrt{2}}\eta_1 + \frac{2\pi}{3}\right) \cos\left(\sqrt{\frac{3}{2}}\eta_2\right)\Big)\Big]. \tag{8.126}$$

This model shows the chiral transition [55] as well as the Kosterlitz-Thouless transition [120]. The renormalization group method [121, 122] is applied to obtain the beta functions. They are given by

$$\mu \frac{\partial t}{\partial \mu} = (d-2)t + At\alpha^2 \tag{8.127}$$
$$\mu \frac{\partial \alpha}{\partial \mu} = -2\alpha + \frac{1}{4\pi}\alpha t, \tag{8.128}$$

for the mass parameter $\mu$. Here $A$ is a constant. The equation for $\alpha$ has a fixed point at $t = 8\pi$. In two dimension $d = 2$, the renormalization group flow is the same as that for the Kosterlitz-Thouless transition (see Fig. 8.7).

### 8.11.2 Chirality Transition

There is a chirality transition at a finite temperature where the states with chirality $\kappa = \pm 1$ disappear and simultaneously the chirality vanishes. This is shown by taking account of the fluctuation around the minimum of the potential. Using $\cos(\sqrt{3/2}\eta_2) = 1 - (4/3)\eta_2^2 + \cdots$, the action is written as

$$S = \frac{\Lambda^{d-2}}{t} \int d^d x \Big[ \sum_j (\nabla \eta_j)^2 + \alpha \Lambda^2 \Big( \cos\Big( \sqrt{2}\eta_1 + \frac{4\pi}{3} \Big) - 2 \Big| \cos\Big( \frac{1}{\sqrt{2}}\eta_1 + \frac{2\pi}{3} \Big) \Big| \Big) + \frac{3\alpha\Lambda^2}{2} \Big| \cos\Big( \frac{1}{\sqrt{2}}\eta_1 + \frac{2\pi}{3} \Big) \Big| \eta_2^2 \Big]. \quad (8.129)$$

We integrate out the field $\eta_2$ to obtain the effective action. The effective free-energy density in two dimensions is obtained as

$$\begin{aligned} \frac{f[\varphi]}{\Lambda^2} &= \frac{1}{2} K \Lambda^{-2} (\nabla\varphi)^2 + \epsilon_0 \Big( \cos\varphi - 2 \Big| \cos\Big( \frac{\varphi}{2} \Big) \Big| \Big) \\ &+ \frac{1}{2} k_B T \frac{c}{4\pi} \ln\Big( \frac{c\Lambda^d}{t} + \frac{3\alpha\Lambda^d}{2t} \Big| \cos\Big( \frac{\varphi}{2} \Big) \Big| \Big) \\ &+ k_B T \frac{3\alpha}{16\pi} \Big| \cos\Big( \frac{\varphi}{2} \Big) \Big| \ln\Big( 1 + \frac{2c}{3\alpha} \Big| \cos\Big( \frac{\varphi}{2} \Big) \Big|^{-1} \Big), \end{aligned} \quad (8.130)$$

for $\varphi \equiv 4\pi/3 + \sqrt{2}\eta_1$ where $\Lambda$ is a cutoff, $c$ is a constant, and $\epsilon_0 = k_B T\alpha/t = 2\gamma\Delta_0^2/\Lambda^2$. The critical temperature $T_{chiral}$ of the chirality transition is determined by the condition that we have a minimum at $\varphi = \pi$ (first-order transition). $T_{chiral}$ is shown as a function of $\alpha$ in Fig. 8.8. $T_{chiral} = (K/k_B)t_c$ is dependent on $\alpha$, where $\alpha$ is proportional to the Josephson coupling, while the temperature of the Kosterlitz-Thouless transition $T_{KT} = (K/k_B)8\pi$ is independent of $\alpha$. Thus, $T_{chiral}$ and $T_{KT}$ are different in general.

We have shown a model which shows a transition due to growing fluctuations. The disappearance of the chirality results in the emergency of a Nambu-Goldstone boson. This represents the phenomenon that the Nambu-Goldstone boson appears from a fluctuation effect. Please note that this does not say that a discreet symmetry can be broken by Nambu-Goldstone boson proliferation. A Nambu-Goldstone

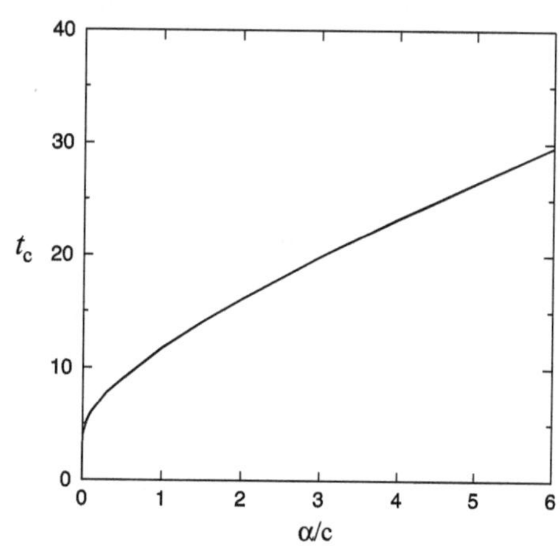


**Fig. 8.8** $t_c \equiv k_B T_{chiral}/K$ as a function of $\alpha/c$ with $c = 4\pi$

would emerge as a result of a discree symmetry breaking. At $T > T_{chiral}$, two spins in Fig. 8.3 are antiferromagnetically aligned and one spin vanishes. This means that the one spin is rotating freely accompanied with the existence of a massless boson. Our model shows that the $Z_2$-symmetry breaking induces a massless boson. If we neglect the kinetic term in the action, $T_{chiral}$ is determined uniquely as $T_{chiral} = \epsilon_0/2$. $\epsilon_0$ corresponds to $J$ in the two-dimensional XY model. This suggests that there is a chirality transition in the 2D XY model on a two-dimensional triangular lattice at near $T = J/2$, which has been confirmed by a numerical simulation [125]. The existence of the Kosterlitz-Thouless transition has also been shown at near $T = J/2$.

We discuss whether our model is applicable to real superconductors. We expect that our model applies to, for example, layered superconductors like cuprates with small Josephson couplings. This type of transition has been discussed for three-band superconductors with frustrated interband Josephson couplings [123]. Recent experiments have indicated that a first-order phase transition below the superconducting transition temperature occurs in multilayer cuprate superconductor $HgBa_2Ca_4Cu_5O_y$ [124]. We hope that this phase transition is related to the dynamics of multicomponent order parameters.

## 8.12 *SU*(*N*) Sine-Gordon Model

In this section, let us consider a generalized Josephson interaction where the Josephson term is given by a $G$-valued sine-Gordon potential for a compact Lie group $G$. This model includes multiple excitation modes and is a nonabelian generalization of the sine-Gordon model. The Lagrangian is written as

$$\mathcal{L} = \frac{1}{2t}\mathrm{Tr}\partial_\mu g\partial^\mu g^{-1} + \frac{\alpha}{2t}\mathrm{Tr}(g + g^{-1}), \tag{8.131}$$

for $g \in G$. When $g = e^{i\varphi} \in U(1)$, this Lagrangian is reduced to that of the conventional sine-Gordon model. This model can be regarded as the chiral model with the mass term. Here, we consider the $SU(N)$ or $O(N)$ model: $G = SU(N)$ or $O(N)$. In the limit $t \to \infty$ with keeping $\lambda \equiv \alpha/t$ constant, the $SU(N)$ sine-Gordon model is reduced to a unitary matrix model. It has been shown by Gross and Witten that, in the large $N$ limit with the coupling constant $\lambda = N\beta$, for the model $N\beta\mathrm{Tr}(g + g^\dagger)$, there is a third-order transition at some critical $t_c$ [126]. Brezin and Gross considered the model to generalize the coupling constant $\lambda$ to be a matrix and also found that there is a phase transition [127–129]. Recently, the vortex structure for a nonabelian sine-Gordon model was investigated numerically [130].

An element $g \in G$ is represented in the form:

$$g = g_0 \exp\left(i\lambda \sum_a T_a\pi_a\right), \tag{8.132}$$

where $\lambda$ is a real number $\lambda \in \mathbf{R}$ and $g_0 \in G$ is a some element in $G$. We put $g_0 = 1$ in this paper. $T_a$ $(a = 1, 2, \ldots, N_T)$ form a basis of the Lie algebra of $G$. $N_T = N^2 - 1$ for $SU(N)$ and $N_T = N(N-1)/2$ for $O(N)$. $\{T_a\}$ are normalized as

$$\mathrm{Tr} T_a T_b = c\delta_{ab}, \tag{8.133}$$

with a real constant $c$. The scalar fields $\pi_a$ indicate fluctuations around the classical solution, that is, the nonabelian perturbation to the state $g_0$. We expand $g$ by means of $\pi_a$ as

$$g = g_0\left[1 + i\lambda T_a\pi_a - \frac{1}{2}\lambda^2(T_a\pi_a)^2 + \cdots\right], \tag{8.134}$$

and evaluate the beta functions of renormalization group theory.

The renormalization group equations read [92]

$$\mu\frac{\partial t}{\partial\mu} = (d-2)t - \frac{C_2(G)}{2}t^2 + A_0 C(N) t\alpha^2, \tag{8.135}$$

$$\mu\frac{\partial\alpha}{\partial\mu} = -\alpha\,(2 - C(N)t)\,, \tag{8.136}$$

where $A_0 = A_0(N)$ is a constant (depending on $N$) and the volume element $\Omega_d/(2\pi)^d$ is included in the definition of $t$ for simplicity. $C(N)$ is the Casimir invariant in the fundamental representation given by

$$C(N) = c\frac{N^2-1}{N} \quad \text{for} \quad G = SU(N), \tag{8.137}$$

$$= c\frac{N-1}{2} \quad \text{for} \quad G = 0(N). \tag{8.138}$$

The coefficient of $t^2$ term in $\mu\partial t/\partial\mu$ is the Casimir invariant in the adjoint representation defined by $\sum_{ab} f_{abc} f_{abd} = C_2(G)\delta_{cd}$. $C_2(G)$ is given as

$$C_2(G) = 2Nc \quad \text{for} \quad G = SU(N), \tag{8.139}$$

$$= (N-2)c \quad \text{for} \quad G = O(N), \tag{8.140}$$

Thus, beta functions are determined by Casimir invariants.

There is a zero of beta functions in two dimensions ($d = 2$):

$$t_c = \frac{2}{C(N)}, \quad \alpha_c = \sqrt{\frac{C_2(G)}{A_0(N)}\frac{1}{C(N)}}. \tag{8.141}$$

This is a bifurcation point that divides the parameter space into two regions. One is the strong coupling region where $\alpha \to \infty$ as $\mu \to \infty$, and the other is the weak coupling region where $\alpha \to 0$ as $\mu \to \infty$. In the weak coupling region, we can use

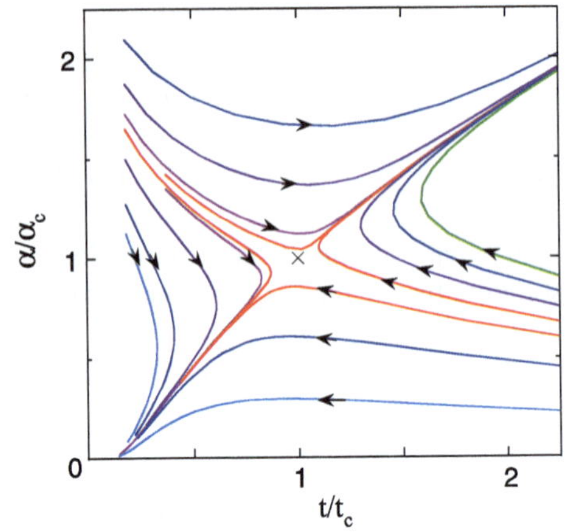


**Fig. 8.9** Renormalization flow as $\mu$ increases. The *cross* indicates the bifurcation point

a perturbation theory by expanding $g$ by means of the fluctuation fields $\pi_a$. This results in the existence of multiple frequency modes. We expect that these modes may be observed. There may be a possibility to classify excitation modes using a group theory. We show the renormalization flow in Fig. 8.9.

## 8.13 Summary

The Nambu-Goldstone-Leggett modes and the Higgs mode are typical fluctuation modes in multi-gap superconductors. We expect that they play an important role. We derived the effective action and showed the dispersion relations of these modes. The mass of the Higgs mode is proportional to the gap amplitude. The Nambu sum rule does not hold in general in a multi-gap superconductor. One mode among the Nambu-Goldstone modes becomes the massive mode in the presence of the Coulomb potential (Higgs mechanism). An $N$-gap superconductor has $N-1$ phase-difference variables, and the $U(1)^{N-1}$ phase invariance can be partially or totally broken spontaneously. The $N-1$ phase modes become in general massive due to the symmetry breaking by the Josephson interaction. When the Josephson couplings are frustrated, symmetry is partially broken and some of phase modes can be massless modes. A kink solution exists in the phase space of gap functions.

The kink solution provides a new excitation mode. A fractionally quantized flux vortex can exist at the edge of the kink in a magnetic field. The half-flux vortex can be regarded as a monopole with the Chern number in a superconductor. We discussed several versions of the sine-Gordon model and derived the renormalization group equations for these models. An effect of fluctuation is investigated, on the basis of a toy model, where the fluctuation restores the time-reversal symmetry in the ground state with time-reversal symmetry breaking.

**Acknowledgements** The author expresses his sincere thanks to J. Kondo, K. Yamaji, I. Hase and Y. Tanaka for helpful discussions.

## References

1. V.A. Moskalenko, Fiz. Metal Metallored **8**, 2518 (1959)
2. H. Suhl, B.T. Mattis, L.W. Walker, Phys. Rev. Lett. **3**, 552 (1959)
3. J. Peretti, Phys. Lett. **2**, 275 (1962)
4. J. Kondo, Prog. Theor. Phys. **29**, 1 (1963)
5. J. Bardeen, L. Cooper, R. Schrieffer, Phys. Rev. **106**, 162 (1957)
6. G. Binning, A. Baratoff, H.E. Hoenig, J.G. Bednorz, Phys. Rev. Lett. **45**, 1352 (1980)
7. L. P. Gorkov, arXiv:1508.00529
8. S.V. Shulga et al., Phys. Rev. Lett. **80**, 1730 (1998)
9. J. Nagamatsu, N. Nakagawa, T. Muranaka, Y. Zenitani, J. Akimitsu, Nature **410**, 63 (2001)
10. Y. Kamihara, T. Watanabe, M. Hirano, H. Hosono, J. Ame. Chem. Soc. **130**, 3296 (2008)
11. K.H. Bennemann, J.B. Ketterson (eds.), *The Physics of Superconductors* (Springer-Verlag, Berlin, 2004)
12. A. Bianconi, Nat. Phys. **9**, 536 (2013)
13. A. Bianconi, A.C. Castellano, M. De, Santis, P. Delogu, A. Garano and R. Giorgi. Solid State Commun. **63**, 1135 (1987)
14. The Physics of superconductors, vols. I and II, ed. by K.H. Bennemann, J.B. Ketterson (Springer, Berlin, 2003)
15. V.J. Emery, Phys. Rev. Lett. **58**, 2794 (1987)
16. J.E. Hirsch, D. Loh, D.J. Scalapino, S. Tang, Phys. Rev. B **39**, 243 (1989)
17. K. Yamaji et al., Physica C **304**, 225 (1998)
18. T. Yanagisawa, S. Koike, K. Yamaji, Phys. Rev. B **64**, 184509 (2001)
19. T. Yanagisawa, S. Koike, K. Yamaji, Phys. Rev. B **67**, 132408 (2003)
20. C. Weber, A. Lauchi, G.A. Sawatzky, Phys. Rev. Lett. **102**, 017005 (2009)
21. T. Yanagisawa, J. Phys. Soc. Jpn. **85**, 114707 (2016)
22. S. Koike et al., J. Phys. Soc. Jpn. **68**, 1657 (1999)
23. K. Yamaji et al., Physica C **235**, 2221 (1994)
24. T. Yanagisawa, Y. Shimoi, K. Yamaji, Phys. Rev. B **52**, R3860 (1995)
25. A. Bussmann-Holder, H. Keller, R. Khasanov, A. Simon, A. Bianconi, A.R. Bishop, New J. Phys. **13**, 093009 (2011)
26. H.Y. Choi et al., Phys. Rev. B **80**, 052505 (2009)
27. P.M. Shirage, K. Kihou, K. Miyazawa, C.-H. Lee, H. Kito, H. Eisaki, T. Yanagisawa, Y. Tanaka, A. Iyo. Phys. Rev. Lett. **103**, 257003 (2009)
28. T. Yanagisawa, K. Odagiri, I. Hase, K. Yamaji, P.M. Shirage, Y. Tanaka, A. Iyo, H. Eisaki, J. Phys. Soc. Jpn. **78**, 094718 (2009)
29. A.D. Hillier, J. Quintanilla, R. Cywinskii, Phys. Rev. Lett. **102**, 117007 (2009)
30. I. Hase, T. Yanagisawa, J. Phys. Soc. Jpn. **78**, 084724 (2009)
31. A.D. Hillier, J. Quintanilla, B. Mazidian, J.F. Annett, R. Cywinskii, Phys. Rev. Lett. **109**, 097001 (2012)
32. I. Hase, T. Yanagisawa, J. Phys. Soc. Jpn. **81**, 103704 (2012)
33. D.J. Singh, Phys. Rev. B **86**, 174507 (2012)
34. V.K. Pecharsky, L.L. Miller, K.A. Gschneider, Phys. Rev. B **58**, 497 (1998)
35. Y. Iwamoto, Y. Iwasaki, K. Ueda, T. Kohara, Phys. Lett. A **250**, 439 (1998)
36. H.H. Sung, S.Y. Chou, K.J. Syu, W.H. Lee, J. Phys. COndens. Matter **20**, 165207 (2008)
37. S. Katano et al., Phys. Rev. B **90**, 220508 (2014)
38. V. Stanev, Z. Tesanovic, Phys. Rev. B **81**, 134522 (2010)
39. Y. Tanaka, T. Yanagisawa, J. Phys. Soc. Jpn. **79**, 114706 (2010)

40. Y. Tanaka, T. Yanagisawa, Solid State Commun. **150**, 1980 (2010)
41. R.G. Dias, A.M. Marques, Supercond. Sci. Technol. **24**, 085009 (2011)
42. T. Yanagisawa, Y. Tanaka, I. Hase, K. Yamaji, J. Phys. Soc. Jpn. **81**, 024712 (2012)
43. X. Hu, Z. Wang, Phys. Rev. B **85**, 064516 (2012)
44. V. Stanev, Phys. Rev. B **85**, 174520 (2012)
45. C. Platt, R. Thomale, C. Homerkamp, S.C. Zhang, Phys. Rev. B **85**, 180502 (2012)
46. S. Maiti, A.V. Chubukov, Phys. Rev. B **87**, 144511 (2013)
47. B.J. Wilson, M.P. Das, J. Phys. Condens. Matter **25**, 425702 (2013)
48. Y. Takahashi, Z. Huang, X. Hu, J. Phys. Soc. Jpn. **83**, 034701 (2014)
49. R. Ganesh, G. Baskaran, J. van den Brink, D.V. Efremov, Phys. Rev. Lett. **113**, 177001 (2014)
50. Y. S. Yerin, A. N. Omelyanchouk and E. Il'chev, arXiv:1503.04653 (2015)
51. T. Yanagisawa, I. Hase, J. Phys. Soc. Jpn. **82**, 124704 (2013)
52. S.Z. Lin, X. Hu, New J. Phys. **14**, 063021 (2012)
53. K. Kobayashi, M. Machida, Y. Ota, F. Nori, Phys. Rev. B **88**, 224516 (2013)
54. T. Koyama, J. Phys. Soc. Jpn. **83**, 074715 (2014)
55. T. Yanagisawa, Y. Tanaka, New J. Phys. **16**, 123014 (2014)
56. Y. Tanaka et al., Physica C **516**, 10 (2015)
57. S.-Z. Lin, X. Hu, Phys. Rev. Lett. **108**, 177005 (2012)
58. YuA Izyumov, V.M. Laptev, Phase Transitions **20**, 95 (1990)
59. G.E. Volovik, *The Universe in a Helium Droplet* (Oxford University Press, Oxford, 2009)
60. Y. Tanaka, Phys. Rev. Lett. **88**, 017002 (2002)
61. E. Babaev, Phys. Rev. Lett. **89**, 067001 (2002)
62. S.V. Kuplevakhsky, A.N. Omelyanchouk, Y.S. Yerin, J. Low Temp. Phys. **37**, 667 (2011)
63. A.J. Leggett, Prog. Theor. Phys. **36**, 901 (1966)
64. Y. Tanaka et al., Physica C **471**, 747 (2011)
65. V. Moshchalkov et al., Phys. Rev. Lett. **102**, 117001 (2009)
66. M. Silaev, E. Babaev, Phys. Rev. B **84**, 094515 (2011)
67. Y. Nambu, Phys. Rev. **117**, 648 (1960)
68. Y. Nambu, in Physics in the 21st Century edited by K, ed. by H. Kunitomo, H. Otsubo (World Scientific, Singapore, Kikkawa, 1997)
69. T. Yanagisawa, Novel Supercon. Materials **1**, (2015)
70. B. T. Geilikman, R. O. Zaitsev and V. Z. Kresin, Sov. Phys.: Solid State **9**642, (1967)
71. D.R. Tilley, Proc. Phys. Soc. **84**, 573 (1964)
72. I.P. Ivanov, Phys. Rev. E **79**, 021116 (2009)
73. N.V. Orlova, A.A. Shanenko, M.V. Milosevic, F.M. Peeters, A. V. Vagov ad V. M. Axt. Phys. Rev. B **87**, 134510 (2013)
74. Z. Huang, X. Hu, Appl. Phys. Lett. **104**, 162602 (2014)
75. Y.S. Yerin, A.N. Omelyanchouk, Low Temp. Phys. **40**, 943 (2014)
76. Y. Tanaka, Super. Sci. Tech. **28**, 034002 (2015)
77. J.G. Storey, J.W. Loram, J.R. Cooper, Z. Bukowski, J. Karpinski, Phys. Rev. B **88**, 144502 (2013)
78. R.G. Cai, L. Li, L.F. Li, Y.Q. Wang, JHEP **9**, 074 (2013)
79. W.-Y. Wen, M.-S. Wu, S.-Y. Wu, Phys. Rev. D **89**, 066005 (2014)
80. M. Nishida, JHEP **08**, 136 (2015)
81. R. Rajaraman, *Solitons and Instantons* (North-Holland, Elsevier, Amsterdam, 1987)
82. Y. Tanaka, J. Phys. Soc. Jpn. **70**, 2844 (2001)
83. E. Babaev, L.D. Faddeev, A.J. Niemi, Phys. Rev. B **65**, 100512 (2002)
84. Y.M. Cho, P. Zhang, Phys. Rev. B **73**, 180506 (2006)
85. M. Eto, Y. Hirono, M. Nitta and S. Yasui, Prog. Theor. Exp. Phys. 012D01 (2014)
86. M.C.N. Fiolhais, J.L. Birman, Phys. Lett. A **378**, 2632 (2014)
87. E. Babaev, M. Speight, Phys. Rev. B **72**, 180502 (2005)
88. V.G. Kogan, J. Schmalian, Phys. Rev. B **83**, 054515 (2011)
89. E. Babaev, M. Silaev, Phys. Rev. B **86**, 016501 (2012)
90. V.G. Kogan, J. Schmalian, Phys. Rev. B **86**, 016502 (2012)

91. M. Marciani, L. Fanfarillo, C. Castellani, L. Benfatto, Phys. Rev. B **88**, 214508 (2013)
92. T. Yanagisawa, EPL **113**, 41001 (2016)
93. R. Kleiner, F. Steinmeyer, G. Kunkel, P. Muller, Phys. Rev. Lett. **68**, 2349 (1992)
94. G. Ohya, N. Aoyama, A. Irie, S. Kishida, H. Tokutaka, Jpn. J. Appl. Phys. **31**, L829 (1992)
95. K. Tamasaku, Y. Nakamura, S. Uchida, Phys. Rev. Lett. **69**, 1455 (1992)
96. Y. Matsuda, M.B. Gaifullin, K. Kumagai, K. Kadowaki, T. Mochiku, Phys. Rev. Lett. **75**, 4512 (1995)
97. T. Koyama, M. Tachiki, Phys. Rev. B **54**, 16183 (1996)
98. S.G. Sharapov, V.P. Gusynin, H. Beck, Eur. Phys. J. B **39**, 062001 (2002)
99. H.J. Choi, D. Roundy, H. Sun, M.L. Cohen, S.G. Louie, Nature **418**, 758 (2002)
100. G. Blumberg et al., Phys. Rev. Lett. **99**, 014507 (2007)
101. Y. Ota, M. Machida, T. Koyama, H. Aoki, Phys. Rev. B **83**, 060507 (2011)
102. R. Matsunaga, Y.I. Hamada, K. Makise, Y. Uzawa, H. Terai, Z. Wang, R. Shimano, Phys. Rev. Lett. **111**, 057002 (2013)
103. M.A. Measson, Y. Gallais, M. Cazayous, B. Clair, P. Rodiere, L. Cario, A. Sacuto, Phys. Rev. B **89**, 060503 (2014)
104. Y. Barlas, C.M. Varma, Phys. Rev. B **87**, 054503 (2013)
105. T. Koyama, Phys. Rev. B **90**, 064514 (2014)
106. N. Tsuji, H. Aoki, Phys. Rev. B **92**, 064508 (2015)
107. Y. Nambu, G. Jona-Lasinio, Phys. Rev. **122**, 345 (1961)
108. Y. Nambu, G. Jona-Lasinio, Phys. Rev. **124**, 246 (1961)
109. Y. Nambu, Physica D **15**, 147 (1985)
110. T. Koyama, J. Phys. Soc. Jpn. **88**, 064715 (2016)
111. U. Welp et al., Phys. Rev. B **78**, 140510 (R) (2008)
112. A. Gurevich, V. Vinokur, Phys. Rev. Lett. **90**, 047004 (2003)
113. H.-Y. Kee, Y.B. Kim, K. Maki, Phys. Rev. B **62**, R9275 (2000)
114. J. Jang et al., Science **331**, 186 (2011)
115. M. Nitta, M. Eto, T. Fujimori, T. Ohashi, J. Phys. Soc. Jpn. **81**, 084711 (2012)
116. K.G. Wilson, J.B. Kogut, Phys. Rep. **12C**, 75 (1974)
117. J.B. Kogut, Rev. Mod. Phys. **51**, 659 (1979)
118. C. Chakravarty, B.I. Halperin, D.R. Nelson, Phys. Rev. B **39**, 2344 (1989)
119. T. Yanagisawa, Phys. Rev. B **46**, 13896 (1992)
120. J.M. Kosterlitz, D. Thouless, J. Phys. C **6**, 1181 (1973)
121. J. Zinn-Justin, *Quantum Field Theory and Critical Phenomena* (Oxford University Press, Oxford, 1989)
122. D.J. Amit, Y.Y. Goldschmidt, S. Grinstein, J. Phys. A: Math. Gen. **13**, 585 (1980)
123. T.A. Bojesen, E. Babaev, A. Sudbo, Phys. Rev. B **88**, 220511 (2013)
124. Y. Tanaka et al., J. Phys. Soc. Jpn. **83**, 074705 (2014)
125. S. Miyashita, H. Shiba, J. Phys. Soc. Jpn. **53**, 1145 (1984)
126. D.J. Gross, E. Witten, Phys. Rev. D **21**, 446 (1980)
127. E. Brezin, D.J. Gross, Phys. Lett. B **97**, 120 (1980)
128. R.C. Brower, M. Nauenberg, Nucl. Phys. B **180**, 221 (1981)
129. E. Brezin, S. Hikami, JHEP **7**, 67 (2010)
130. M. Nitta, Nucl. Phys. B **895**, 288 (1995)

# Author Index





A. Crisan (ed.), *Vortices and Nanostructured Superconductors*, Springer Series in Materials Science 261, DOI 10.1007/978-3-319-59355-5

# Index





A. Crisan (ed.), *Vortices and Nanostructured Superconductors*, Springer Series in Materials Science 261, DOI 10.1007/978-3-319-59355-5

MIX
Papier aus verantwortungsvollen Quellen
Paper from responsible sources
FSC® C105338

If you have any concerns about our products,
you can contact us on
**ProductSafety@springernature.com**

In case Publisher is established outside the EU,
the EU authorized representative is:
**Springer Nature Customer Service Center GmbH**
**Europaplatz 3, 69115 Heidelberg, Germany**

Printed by Libri Plureos GmbH
in Hamburg, Germany